A Modern Approach to Physical Chemistry

A Modern Approach to Physical Chemistry

Editor: Jamie Langdon

NY RESEARCH
P R E S S

New York

Published by NY Research Press
118-35 Queens Blvd., Suite 400,
Forest Hills, NY 11375, USA
www.nyresearchpress.com

A Modern Approach to Physical Chemistry
Edited by Jamie Langdon

Cataloging-in-Publication Data

A modern approach to physical chemistry / edited by Jamie Langdon.
    p. cm.
Includes bibliographical references and index.
ISBN 978-1-63238-841-4
1. Chemistry, Physical and theoretical. 2. Chemistry. I. Langdon, Jamie.
QD453.3 .M63 2022
541--dc23

Contents

Preface...IX

Chapter 1 **Mechanical Property and Analysis of Asphalt Components based on Molecular Dynamics Simulation**..1
Rui Li, Qiqi Guo, Hui Du and Jianzhong Pei

Chapter 2 **Theoretical Analysis of the Effect Provoked by Bromine-Addition on the Thermolysis and Chemiexcitation of a Model Dioxetanone**........................10
Luís Pinto da Silva, Rui F. J. Pereira and Joaquim C. G. Esteves da Silva

Chapter 3 **Removal of Cd (II) from Aqueous Media by Adsorption onto Chemically and Thermally Treated Rice Husk**...18
María Camila Hoyos-Sánchez, Angie Carolina Córdoba-Pacheco, Luis Fernando Rodríguez-Herrera and Ramiro Uribe-Kaffure

Chapter 4 **An Assessment of Physicochemical Parameters of Selected Industrial Effluents**.........26
Abhinay Man Shrestha, Sanjila Neupane and Gunjan Bisht

Chapter 5 **Density Functional Theory Investigation into the B and Ga Doped Clean and Water Covered γ-Alumina Surfaces**..35
Lihong Cheng, Tianliang Xu, Wenkui Li, Zhiqin Chen, Jianping Ai, Zehua Zhou and Jianwen Liu

Chapter 6 **Influence of Surfactant Structure on the Stability of Water-in-Oil Emulsions under High-Temperature High-Salinity Conditions**.............................42
Abdelhalim I. A. Mohamed, Abdullah S. Sultan, Ibnelwaleed A. Hussein and Ghaithan A. Al-Muntasheri

Chapter 7 **Study of Phase Equilibrium of NaBr + KBr + H_2O and NaBr + $MgBr_2$ + H_2O at 313.15K**...53
Qing Chen, Jiping She and Yang Xiao

Chapter 8 **Decay Experiments of Effective N-Removing Microbial Communities in Sequencing Batch Reactors**...59
Chen Lv, Ming Li, Shuang Zhong, Jianlong Wang and Lei Wu

Chapter 9 **Amperometric Formaldehyde Sensor based on a Pd Nanocrystal Modified C/Co_2P Electrode**...64
Huan Wang, Yaodan Chi, Xiaohong Gao, Sa Lv, Xuefeng Chu, Chao Wang, Lu Zhou and Xiaotian Yang

Chapter 10 **Synthesis and Characterization of Ag-Modified V_2O_5 Photocatalytic Materials**...........73
Dora Alicia Solis-Casados, Luis Escobar-Alarcon, Antonia Infantes-Molina, Tatyana Klimova, Lizbeth Serrato-Garcia, Enrique Rodriguez-Castellon, Susana Hernandez-Lopez and Alejandro Dorazco-Gonzalez

Chapter 11   **Relevance of the Physicochemical Properties of Calcined Quail Eggshell (CaO) as a Catalyst for Biodiesel Production**...83
Leandro Marques Correia, Juan Antonio Cecilia, Enrique Rodríguez-Castellón, Célio Loureiro Cavalcante Jr. and Rodrigo Silveira Vieira

Chapter 12   **Assessment of a Physicochemical Indexing Method for Evaluation of Tropical River Water Quality**..95
Siong Fong Sim and Szewei Elaine Tai

Chapter 13   **Synthesis, Spectroscopic, and Thermal Investigations of Metal Complexes with Mefenamic Acid**...107
Karolina Kafarska, Michał Gacki and Wojciech M. Wolf

Chapter 14   **The Effect of Menisci on Kinetic Analysis of Evaporation for Molten Alkali Metal Salts ($CsNO_3$, CsCl, LiCl, and NaCl) in Small Cylindrical Containers**...........................114
In-Hwan Yang, Hee-Chul Yang and Hyung-Ju Kim

Chapter 15   **Study on Shale Adsorption Equation based on Monolayer Adsorption, Multilayer Adsorption, and Capillary Condensation**...124
Qing Chen, Yuanyuan Tian, Peng Li, Changhui Yan, Yu Pang, Li Zheng, Hucheng Deng, Wen Zhou and Xianghao Meng

Chapter 16   **Application of Spent Li-Ion Batteries Cathode in Methylene Blue Dye Discoloration**...........................135
Eric M. Garcia, Hosane A. Taroco, Ana Paula C. Madeira, Amauri G. Souza, Rafael R. A. Silva, Júlio O. F. Melo, Cristiane G. Taroco and Quele C. P. Teixeira

Chapter 17   **In Vitro Study of Adsorption Kinetics of Dextromethorphan Syrup onto Activated Charcoal in Simulated Gastric and Intestinal Fluids**.............................142
Shobha Regmi, Balmukunda Regmi, Sajan Lal Shyaula, Shiva Pathak, Bishnu Prasad Bhattarai and Saroj Kumar Sah

Chapter 18   **Effect of Bentonite on the Pelleting Properties of Iron Concentrate**..149
Hao Liu, Bing Xie and Yue-lin Qin

Chapter 19   **Thermal Stability and Luminescent Properties of Tri-cellulose Acetate Composites Containing $Dy(SSA)_3Phen$ Complex**....................................155
Xiangwei Sun, Feiyue Wu, Yan Luo, Mengjun Huang, Yuntao Li, Xiang Liu and Kunpeng Liu

Chapter 20   **Metal-Organic Framework-101 (MIL-101): Synthesis, Kinetics, Thermodynamics, and Equilibrium Isotherms of Remazol Deep Black RGB Adsorption**..........................162
Vo Thi Thanh Chau, Huynh Thi MinhThanh, Pham Dinh Du, Tran Thanh Tam Toan, Tran Ngoc Tuyen, Tran Xuan Mau and Dinh Quang Khieu

Chapter 21   **Viscosities and Conductivities of $[BMIM]Zn(Ac)_xCl_y$ ($x = 0, 1, 2, 3$; $y = 3, 2, 1, 0$) Ionic Liquids at Different Temperatures**.........................176
Hang Xu and Dandan Zhang

Chapter 22   **Enthalpies of Combustion and Formation of Histidine Stereoisomers** 181
A. Neacsu, D. Gheorghe, I. Contineanu, A. M. Sofronia, F. Teodorescu and S. Perişanu

Chapter 23   **Removal of Methylene Blue from Aqueous Solution using Agricultural Residue Walnut Shell: Equilibrium, Kinetic, and Thermodynamic Studies**..187
Ranxiao Tang, Chong Dai, Chao Li, Weihua Liu, Shutao Gao and Chun Wang

Permissions

List of Contributors

Index

Preface

The discipline that deals with the study of atomic, subatomic and macroscopic phenomena in chemical systems related to the concepts, practices and principles of physics is known as physical chemistry. A few of the major branches that fall under physical chemistry include chemical kinetics, thermochemistry, materials science, physical organic chemistry and biophysical chemistry. It is concerned with resolving the effects of intermolecular forces that act on the physical properties of materials. It also focuses on the effects of reaction kinetics on the rate of reaction, surface science and electrochemistry of the cell membranes. This book traces the progress of this field and highlights some of its key concepts and applications. Different approaches, evaluations, methodologies and advanced studies on physical chemistry have been included in this book. It is appropriate for students seeking detailed information in this area as well as for experts.

The researches compiled throughout the book are authentic and of high quality, combining several disciplines and from very diverse regions from around the world. Drawing on the contributions of many researchers from diverse countries, the book's objective is to provide the readers with the latest achievements in the area of research. This book will surely be a source of knowledge to all interested and researching the field.

In the end, I would like to express my deep sense of gratitude to all the authors for meeting the set deadlines in completing and submitting their research chapters. I would also like to thank the publisher for the support offered to us throughout the course of the book. Finally, I extend my sincere thanks to my family for being a constant source of inspiration and encouragement.

Editor

Mechanical Property and Analysis of Asphalt Components Based on Molecular Dynamics Simulation

Rui Li,[1,2] Qiqi Guo,[1] Hui Du,[3] and Jianzhong Pei[1]

[1]School of Highway, Chang'an University, Xi'an 710064, China
[2]School of Materials Science and Engineering, Nanyang Technological University, Singapore 639798
[3]School of Transportation Engineering, Southeast University, Nanjing 210096, China

Correspondence should be addressed to Jianzhong Pei; peijianzhong@126.com

Academic Editor: Tomokazu Yoshimura

The asphalt-aggregate interface interaction plays a significant role in the overall performances of asphalt mixture. In order to analyze the chemical constitution of asphalt effects on the asphalt-aggregate interaction, the average structure $C_{64}H_{52}S_2$ is selected to represent the asphalt, and the colloid, saturated phenol, and asphaltene are selected to represent the major constitutions in asphalt. The molecular models are established for the three compositions, respectively, and the Molecular Dynamics (MD) simulation was conducted for the three kinds of asphaltene-aggregate system at different presses. Comparing the E value of Young modulus of these three polymers, the maximum modulus value of asphaltene was 2.80 GPa, the modulus value of colloid was secondary, and the minimum modulus of saturated phenol was 0.52 GPa. This result corresponds to conventional understanding.

1. Introduction

At present, scholars at home and abroad have more and more research perspectives on polymer materials [1–3]. Besides, the research process and methods are also diverse. Since the materials prepared indoors and tested will be influenced by the factors of temperature and pressure, there are also some potential uncontrollable variables which will result in the fact that the depth of analytical investigation on experimental data will be limited [4, 5]. However, researchers have extended the material research field continuously thanks to the introduction of Molecular Dynamics method in recent years. With this simulation method, various parameters can be controlled, and even the environment that can not be realized in laboratory can be simulated [6]. Thus, it will be an important tendency of studying polymer materials to apply Molecular Dynamics method to predict and study the polymer materials' properties gradually [7–9].

In order to analyze the chemical constitution of asphalt on the AAI, the average structure $C_{65}H_{74}N_2S_2$ is selected in this paper to represent the asphalt and the colloid; saturated phenol and asphaltene are selected to represent the major constitution in asphalt. The molecular models are established for the three compositions, respectively, and the MD simulation was conducted for the three kinds of asphaltene-aggregate system at different presses.

2. Modeling

When the force acts on the system and if the whole system is balanced, the external force suffered in the system will have an absolute balance with the internal force generated inside the system. Normally, stress can be expressed as a second-order tensor containing 9 components as shown in

$$\begin{pmatrix} \sigma_{11} & \sigma_{12} & \sigma_{13} \\ \sigma_{21} & \sigma_{22} & \sigma_{23} \\ \sigma_{31} & \sigma_{32} & \sigma_{33} \end{pmatrix}. \tag{1}$$

In the process of molecular calculation, internal stress tensor can be expressed by the virial expression of

$$\sigma = -\frac{1}{V_0} \left[\left(\sum_{i=1}^{N} m_i \left(V_i V_i^T \right) \right) + \left(\sum_{i<j} r_{ij} f_{ij}^T \right) \right]. \tag{2}$$

In formula (2), i represents the sequence numbers (from 1 to N) of particles; m_i, V_i, and f_i represent the mass, speed, and force suffered of number i. V_0 refers to the volume of system without deformation.

Once stress acts on the molecular system, the position of internal particles of system will change relatively, where the change is expressed by the strain tensor of

$$\begin{pmatrix} \varepsilon_{11} & \varepsilon_{12} & \varepsilon_{13} \\ \varepsilon_{21} & \varepsilon_{22} & \varepsilon_{23} \\ \varepsilon_{31} & \varepsilon_{32} & \varepsilon_{33} \end{pmatrix}. \tag{3}$$

For the parallel hexahedron structured in this paper, strain tensor is only determined by the column vectors a_0, b_0, and c_0 and deformation state vectors a, b, and c under a certain condition, as shown in

$$\varepsilon = \frac{1}{2} \left[\left(h_0^T \right)^{-1} G h_0^{-1} - 1 \right], \tag{4}$$

where h_0 represents the matrix structured by vectors a_0, b_0, and c_0. h refers to the matrix structured by deformation state vectors a, b, and c. G represents $h^T h$.

Elastic stiffness constant is associated with the different compositions of system stress and strain. With regard to small deformation, the relationship between stress and strain meets Hooke's law, as shown in

$$\sigma_{im} = C_{imnk} \varepsilon_{nk}, \tag{5}$$

where C_{imnk} refers to the elastic stiffness constant.

Since stress tensor and strain tensor have some symmetry, stress formula (1) can be simplified as

$$\begin{pmatrix} \sigma_{11} & \sigma_{12} & \sigma_{13} \\ \sigma_{21} & \sigma_{22} & \sigma_{23} \\ \sigma_{31} & \sigma_{32} & \sigma_{33} \end{pmatrix} \longrightarrow \begin{pmatrix} \sigma_1 & \sigma_6 & \sigma_5 \\ \sigma_6 & \sigma_2 & \sigma_4 \\ \sigma_5 & \sigma_4 & \sigma_3 \end{pmatrix}. \tag{6}$$

Stain formula (3) can be simplified as

$$\begin{pmatrix} \varepsilon_{11} & \varepsilon_{12} & \varepsilon_{13} \\ \varepsilon_{21} & \varepsilon_{22} & \varepsilon_{23} \\ \varepsilon_{31} & \varepsilon_{32} & \varepsilon_{33} \end{pmatrix} \longrightarrow \begin{pmatrix} \varepsilon_1 & \dfrac{\varepsilon_6}{2} & \dfrac{\varepsilon_5}{2} \\ \dfrac{\varepsilon_6}{2} & \varepsilon_2 & \dfrac{\varepsilon_4}{2} \\ \dfrac{\varepsilon_5}{2} & \dfrac{\varepsilon_4}{2} & \varepsilon_3 \end{pmatrix}. \tag{7}$$

Suppose that the materials prepared are isotropic, and the stress and strain only depend on two independent coefficients. Stiffness matrix is shown as

$$\begin{pmatrix} \lambda + 2\mu & \lambda & \lambda & 0 & 0 & 0 \\ \lambda & \lambda + 2\mu & \lambda & 0 & 0 & 0 \\ \lambda & \lambda & \lambda + 2\mu & 0 & 0 & 0 \\ 0 & 0 & 0 & \mu & 0 & 0 \\ 0 & 0 & 0 & 0 & \mu & 0 \\ 0 & 0 & 0 & 0 & 0 & \mu \end{pmatrix}, \tag{8}$$

where λ and μ in formula (8) are called lame constants.

In addition, with regard to isotropic materials, the Yong modulus E, Poisson's ratio v, bulk modulus K, and shear modulus G can be obtained through lame constants λ and μ, as shown in

$$E = \frac{\mu (3\lambda + 2\mu)}{\lambda + \mu},$$

$$v = \frac{\lambda}{2 (\lambda + \mu)}, \tag{9}$$

$$K = \lambda + \frac{2}{3\mu},$$

$$G = \mu,$$

where Poisson's ratio can combine with these four parameters, and the relationship is shown in

$$E = 2G (1 + v) = 3K (1 - 2v). \tag{10}$$

3. Results and Discussion

3.1. Mechanical Property Simulation Analysis of Colloid

(1) Simulation Process and Result of Molecular Dynamics. The 1,7-dimethylnaphthalene was used as the repetitive unit to represent colloid in this paper. According to the suggestions of literature, macromolecular structure obtained at the polymerization degree in the range of 10~15 was employed [10, 11]. The character after calculation could meet the requirements such as research precision. The degree of polymerization selected for structuring the macromolecular chain of colloid was 15, and the degree of polymerization that was 15 to structure the long chain of colloid is shown in Figure 1.

Forcite module was used to analyze geometry optimization for the long chain of colloid. Figure 2 is the long chain of colloid after geometry optimization.

Afterwards, a cubic vacuum space in 60 Å × 60 Å × 60 Å was structured. The density was set at 0.1 g/cm³, and the initial model of colloid unit cell at the density of 0.1 obtained is shown in Figure 3.

The density of initial model is 0.1 g/cc, and the density was in the range of 0.99~1.1 g/cm³ at abnormal temperature; however, the low density system of Figure 3 is usually in a high energy state. Thus, structural optimization should be conducted to the system first to lower the potential energy, and then forcite can be used for optimization. When dynamic calculation was conducted, the temperature was 298 K (close to normal temperature), simulation time was 200 ps and pressure was 0.01 GPa, and the initial density of model at 0.1 would be compressed. In the simulation process, temperature, energy, system density, and the length of side of system were changing continually. The output images are shown from Figures 4–7, where the system energy of Figure 5 changed with simulation time. Figure 4(a) was the whole temperature changing process, and Figure 4(b) was the result after enlarging part of it.

It can be seen from Figure 4(a) that the temperature gradually fluctuated around 298 K. It can be seen from

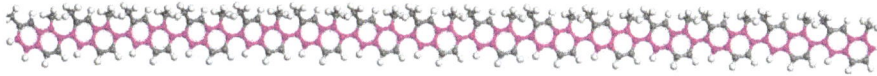

FIGURE 1: Long chain of colloid for the degree of polymerization at 15.

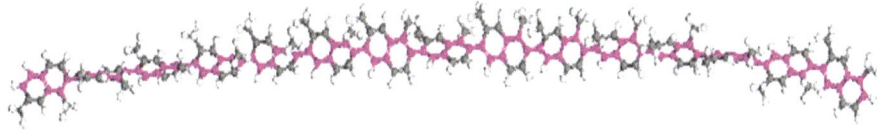

FIGURE 2: Long chain of colloid after being geometrically optimized.

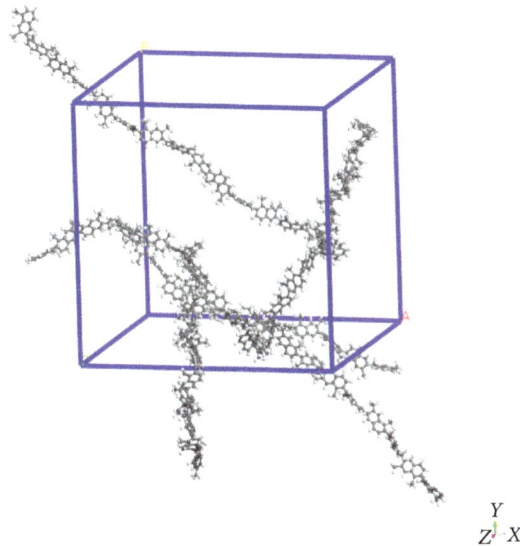

FIGURE 3: Initial model of colloid at the density of 0.1.

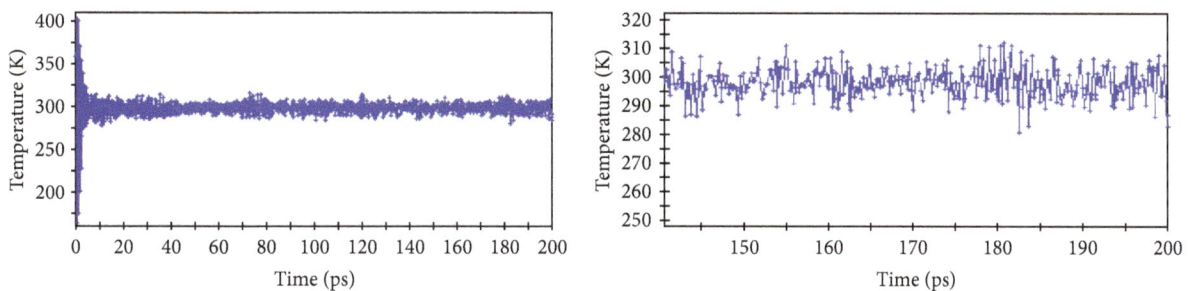

(a) System temperature fluctuated

(b) Partial enlarged detail of temperature change

FIGURE 4: Change graph of temperature with simulation time after NPT dynamics finished.

Figure 4(b) that the fluctuation range was smaller (±10°C). The energy tended to be convergent gradually, suggesting that the system reached a balance after dynamic simulation.

It was obtained from Figure 6 that the system was continuously pressured, the density of system increased, the length of side decreased, and system was compressed 100 ps before the process of simulation. After 100 ps, the system tended to be balanced, and the parameters tended to be stable.

When the pressure increased to 0.08 GPa, the simulation time was 200 ps, the system was compressed, the length of side decreased, and the density increased again. The curves of system density and the length of side changing with time are shown in Figure 7.

It can be seen from Figure 7 that the system was further compressed when the pressure of system increased to 0.08 GPa. It was found in Figure 7 that the system density

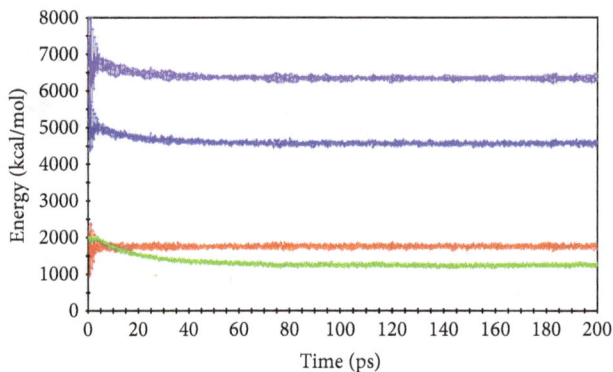

FIGURE 5: Change of system energy with simulation time.

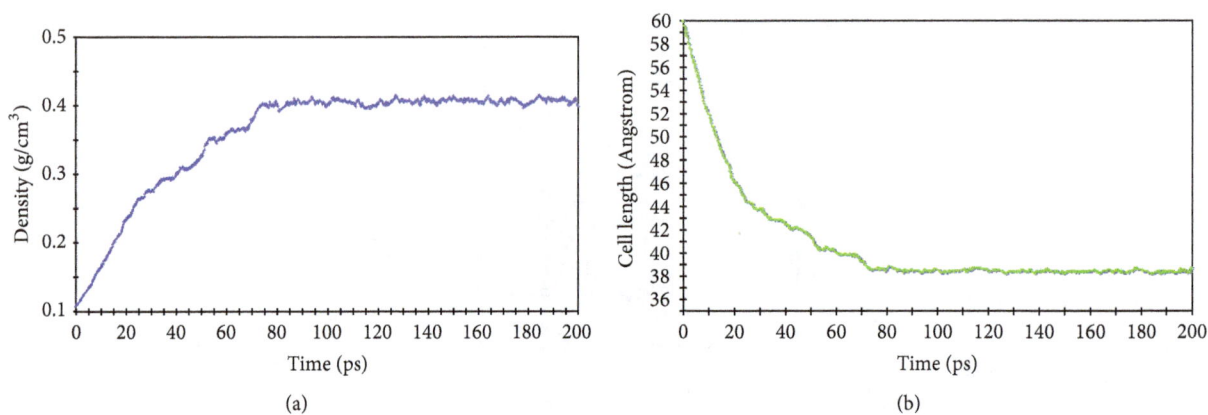

(a)

(b)

FIGURE 6: Change of (a) system density and (b) system length of side with time.

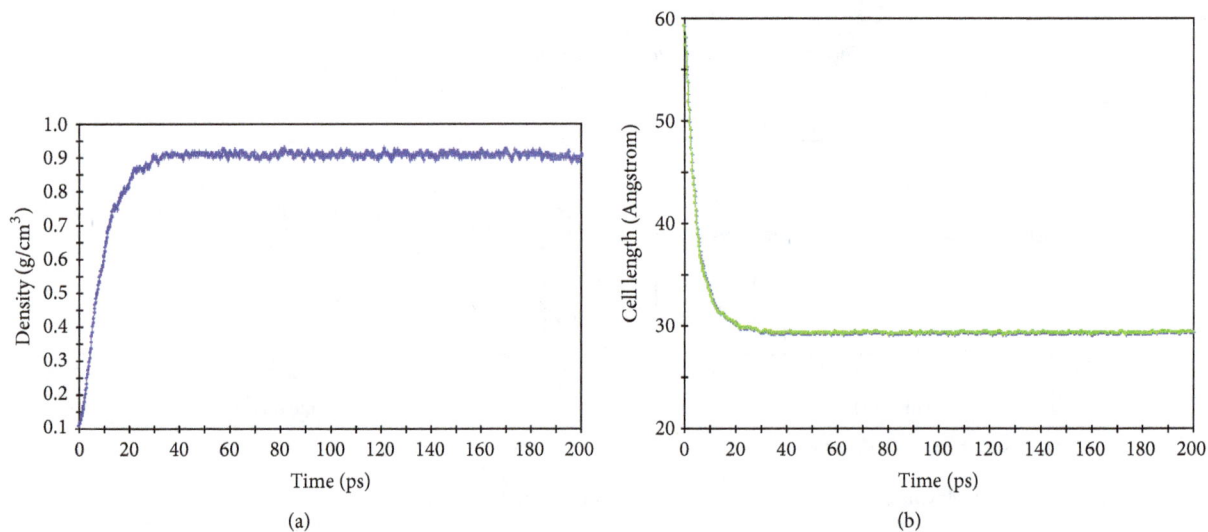

(a)

(b)

FIGURE 7: Change of (a) system density and (b) system length of side with time at 0.08 GPa.

was very close to the true density of colloid. At this moment, the system reached a balance again, and then there was a NPT balance lasting for 500 ps at 0.09 GPa. The increasing amplitude of density was limited. After verification, when the

pressure was up to 0.09 GPa and increasing constantly, the changes of system density and length of side were very small. Finally, the system density and length of side at 0.09 GPa were used as the ultimate model to test the mechanical property of

TABLE 1: Stiffness matrix obtained from colloid model (GPa).

C_{ij}	1	2	3	4	5	6
1	4.5089	2.8995	3.1170	0.1436	0.4446	−0.3247
2	2.8995	6.6093	3.5637	−0.0619	−0.4925	0.2330
3	3.1170	3.5637	4.5407	0.9075	0.6968	−0.1275
4	0.1436	−0.0619	0.9075	1.5682	0.5402	0.2406
5	0.4446	−0.4925	0.6968	0.5402	1.1374	−0.0765
6	−0.3247	0.2330	−0.1275	0.2406	−0.0765	0.0809

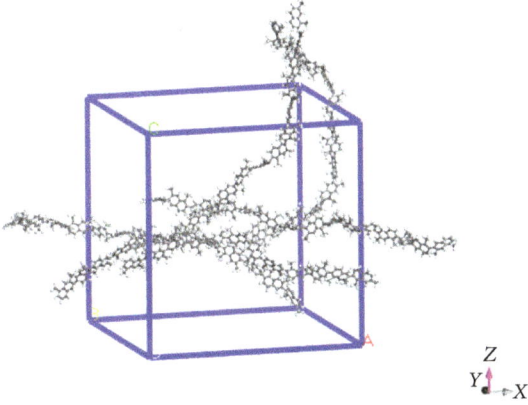

FIGURE 8: Ultimately obtained colloid system.

colloid. The system at the moment is shown in Figure 8. The density at this moment was 0.99 g/cm^3.

(2) Stiffness Matrix and Mechanical Parameters of Colloid. With regard to the subsequently conducted operation and calculation processes to mechanical properties, the program will impose strains in the modes of 100000, 010000, 001000, 000100, 000010, and 000001 on the system model. These strains act on the directions of $XX \cdot YY \cdot ZZ \cdot YZ \cdot ZX \cdot XY$ of the mode. In each mode, four strain values ±0.001 and ±0.003 will be imposed. The default value of software is the maximum value of 0.003. Under this strain loading, the corresponding stress will be obtained, and the corresponding stiffness constant C_{ij} will be obtained with this result to further obtain the elastic stiffness matrix of colloid mode, as shown in Table 1.

It can be seen from Table 1 that the values of C_{14}, C_{15}, C_{16}, C_{24}, C_{25}, C_{26}, C_{34}, C_{35}, C_{36}, C_{45}, C_{46}, and C_{56} are close to 0, but, for the groups of C_{11}, C_{22}, C_{33}; C_{12}, C_{13}, C_{23} and C_{44}, C_{55}, and C_{66}, the values in each group are very close, suggesting that the colloid model of Figure 10 is close to isotropy. Suppose the material is isotropic, and the calculation of lame constant is shown in

$$\lambda = \frac{1}{3}\left(C_{11} + C_{22} + C_{33}\right) - \frac{2}{3}\left(C_{44} + C_{55} + C_{66}\right)$$
$$= 3.3620 \, \text{GPa}, \qquad (11)$$

$$\mu = \frac{1}{3}\left(C_{44} + C_{55} + C_{66}\right) = 0.9288 \, \text{GPa}.$$

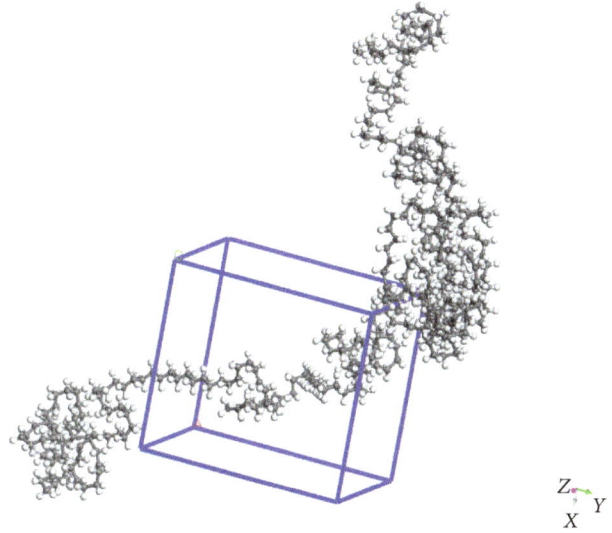

FIGURE 9: Initial model of saturated phenol at the density of 0.1.

After substituting λ = 3.3620 GPa and μ = 0.9288 GPa into formula (9), the mechanical parameters of colloid were obtained below:

$$E = 2.59 \, \text{GPa};$$
$$\nu = 0.39;$$
$$K = 4.08 \, \text{GPa}; \qquad (12)$$
$$G = 0.93 \, \text{GPa}.$$

3.2. Mechanical Property Simulation Analysis of Saturated Phenol

(1) Simulation Process and Result of Molecular Dynamics. The linear chain (n-$C_{22}H_{46}$) was selected to represent saturated phenol, and the degree of polymerization at 10 was selected to structure saturated phenol model. The model was optimized and placed into the cubic vacuum space where the length of three sides was 6 nm. The initial density was set at 0.1 g/cm^3, and the initial saturated phenol model was obtained as shown in Figure 9.

The true saturated density is 0.7944 g/cc, and thus the system should be compressed to increase the density and

TABLE 2: Stiffness matrix obtained for saturated phenol model.

C_{ij}	1	2	3	4	5	6
1	1.7170	1.3630	1.8868	−0.2839	0.4201	−0.4593
2	1.3630	2.0288	2.0845	−0.3727	0.2747	0.0954
3	1.8868	2.0845	1.7931	−0.1096	0.2204	−0.0308
4	−0.2829	−0.3727	−0.1096	0.1437	−0.0936	0.1666
5	0.4201	0.2747	0.2204	−0.0936	0.1250	−0.1628
6	−0.4593	0.0954	−0.0308	0.1666	−0.1628	0.2710

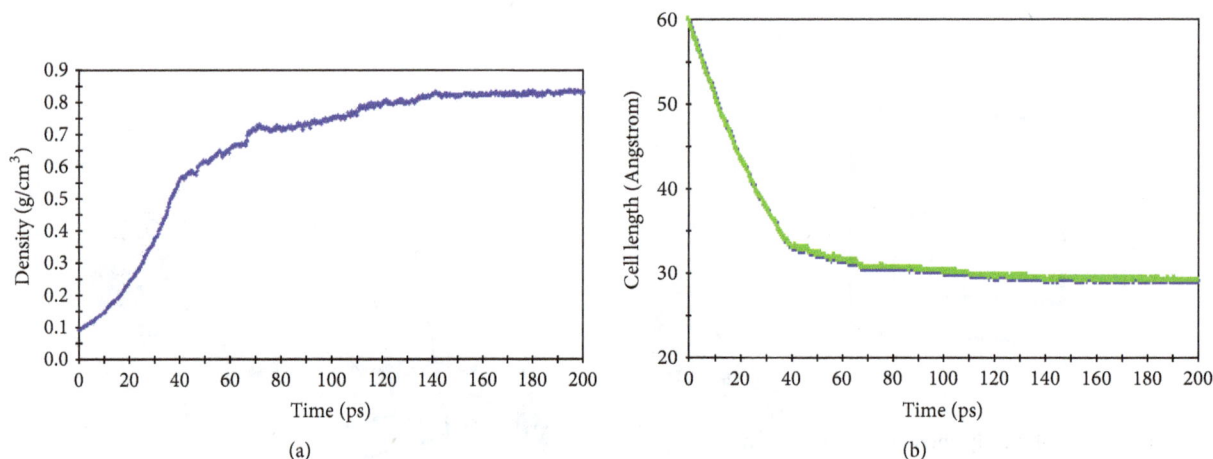

(a)

(b)

FIGURE 10: Change of (a) system density and (b) system length of side with time at 0.01 GPa.

decrease the length of side. After optimizing the structure, NVT ensemble was conducted, and subsequently NPT ensemble was conducted stable again under 0.01 GPa and 298 K. The output temperature and energy curves tended to be stable and balanced with the changes of time after dynamics. The changes of density and the length of side of system can be seen in Figure 10.

It can be seen from Figure 10 that the density was increasing continuously, but the trend density curve was not stable. After a NVT balance was conducted to the system to operate NPT ensemble, the pressure increased to 0.02 GPa, and simulation time increased to 400 ps, and the curve of system density changing with time is shown in Figure 11.

It can be seen from Figure 11 that the density tended to be balanced, which was very close to the actual density at 0.7944, and the compression was stopped immediately. At this moment, the size of model was 29.01 Å. The ultimately balanced saturated phenol model is shown in Figure 12.

(2) Stiffness Matrix and Mechanical Parameters of Saturated Phenol. The mechanical parameters of model in Figure 12 were calculated, and the stiffness matrix obtained for saturated phenol is shown in Table 2.

It can be seen from Table 2 that the values of C_{14}, C_{15}, C_{16}, C_{24}, C_{25}, C_{26}, C_{34}, C_{35}, C_{36}, C_{45}, C_{46}, and C_{56} are close to 0, but, for the three groups of C_{11}, C_{22}, and C_{33}; C_{12}, C_{13}, and C_{23}; and C_{44}, C_{55}, and C_{66}, the values in each group are very

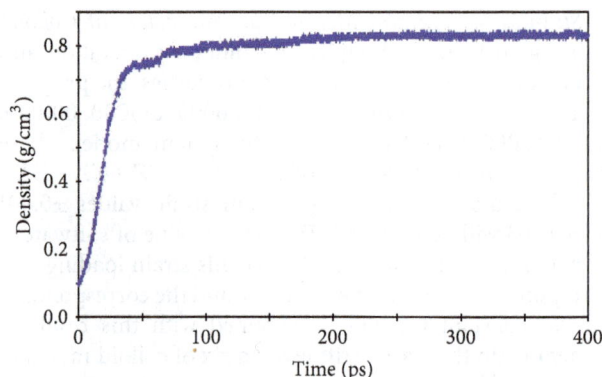

FIGURE 11: Change of system density with time at 0.02 GPa.

close, suggesting that the saturated phenol model of Figure 12 is close to isotropy.

Suppose the material is isotropic, and the calculation of lame constant can be seen in

$$\lambda = \frac{1}{3}\left(C_{11} + C_{22} + C_{33}\right) - \frac{2}{3}\left(C_{44} + C_{55} + C_{66}\right)$$

$$= 1.4864 \text{ GPa}, \tag{13}$$

$$\mu = \frac{1}{3}\left(C_{44} + C_{55} + C_{66}\right) = 0.1799 \text{ GPa}.$$

TABLE 3: Stiffness matrix obtained for asphaltene model.

C_{ij}	1	2	3	4	5	6
1	7.8405	1.5765	1.9309	−0.3414	1.3945	−0.9280
2	1.5765	1.8645	0.6143	−0.2868	0.4326	−1.0948
3	1.9309	0.6143	2.5698	−0.2413	0.1393	−0.0903
4	−0.3414	−0.2868	−0.2413	1.1054	0.0539	0.1551
5	1.3945	0.4326	0.1393	0.0539	1.2063	−0.1860
6	−0.9280	−1.0948	−0.0903	0.1551	−0.1860	0.8541

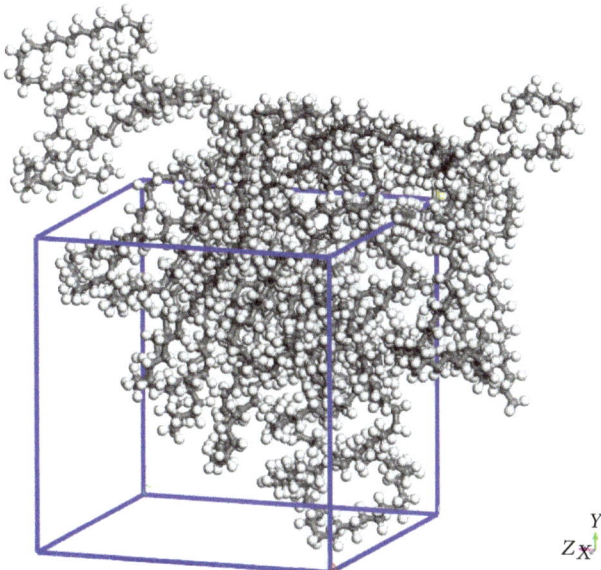

FIGURE 12: Ultimately saturated phenol model.

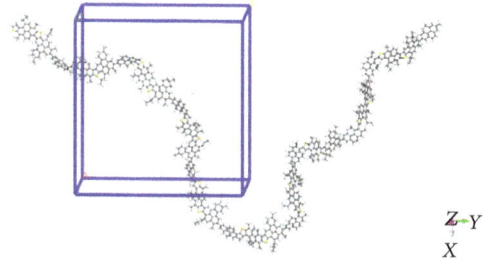

FIGURE 13: Initial model of asphaltene at the density of 0.1.

After substituting λ = 1.4865 GPa and μ = 0.1799 GPa into formula (9), the mechanical parameters of colloid were obtained below:

$$E = 0.52 \text{ GPa};$$
$$\nu = 0.45;$$
$$K = 5.19 \text{ GPa};$$
$$G = 0.18 \text{ GPa}. \tag{14}$$

3.3. Mechanical Property Simulation and Analysis of Asphaltene

(1) Simulation Process and Result of Molecular Dynamics. The $C_{64}H_{52}S_2$ was used to represent asphaltene, and the degree of polymerization at 10 was used to structure asphaltene model. First, the degree of polymerization of asphaltene structure at 10 was placed into the vacuum space where the length of three sides was 6 nm, and the initial density was set at 0.1 g/cm³. The initial model of asphaltene obtained is shown in Figure 13.

Under the temperature of 298 K, the true density of asphaltene was approximately 0.89 g/cm³ that the system must be compressed to increase the density and decrease the length of side. After optimizing the structure, NVT ensemble

was conducted. Subsequently, NPT was conducted to stable system under 0.06 GPa and 298 K. The changes of system density and length of side are shown in Figure 14.

It can be seen from Figure 14 that the density was increasing continuously, and it continued to increase. NVT and NPT were conducted to the stable system again, and the pressure increased to 0.08 GPa, and simulation time increased to 400 ps. The density curve was closed to actual density after the process. At this moment, the compression was stopped. The final output of the asphaltene model is shown in Figure 15. The system density was approximately 0.88 g/cm³, and the size of three sides of system was around 26.2 Å.

(2) Stiffness Matrix and Mechanical Parameters of Asphaltene. After the calculation of mechanical parameters was conducted to the model in Figure 15, the stiffness matrix of asphaltene obtained is shown in Table 3.

Suppose the material is isotropic, and the calculation of lame constant is shown in

$$\lambda = \frac{1}{3}\left(C_{11} + C_{22} + C_{33}\right) - \frac{2}{3}\left(C_{44} + C_{55} + C_{66}\right)$$
$$= 1.9811 \text{ GPa}, \tag{15}$$

$$\mu = \frac{1}{3}\left(C_{44} + C_{55} + C_{66}\right) = 1.0553 \text{ GPa}.$$

After substituting λ = 1.9811 GPa and μ = 1.0553 GPa into formula (9), the mechanical parameters of colloid were obtained below:

$$E = 2.80 \text{ GPa};$$
$$\nu = 0.33;$$
$$K = 2.61 \text{ GPa};$$
$$G = 1.05 \text{ GPa}. \tag{16}$$

(a) System density

(b) System length of side

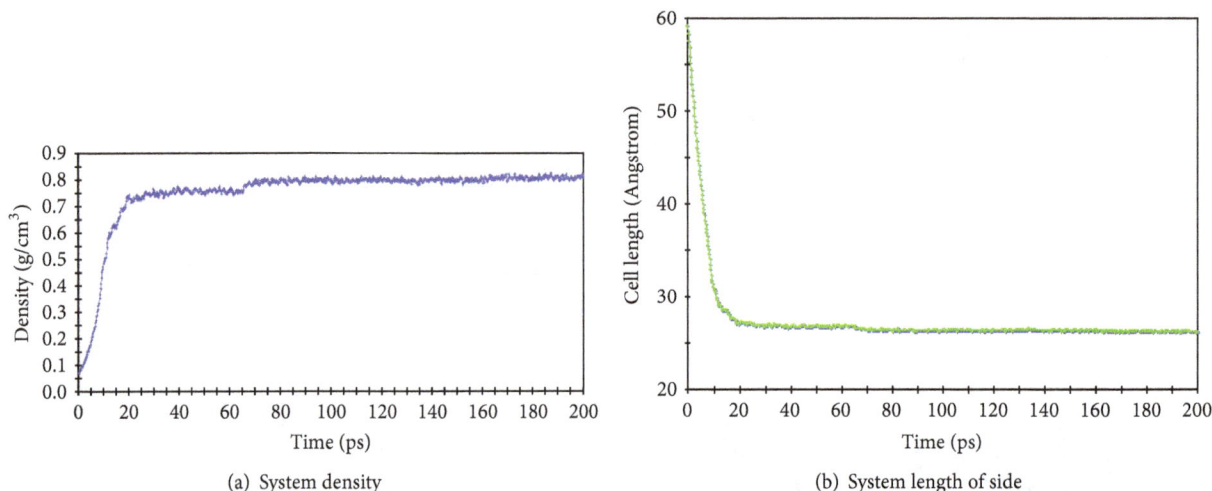

FIGURE 14: Change of (a) system density and (b) system length of side with time at 0.06 GPa.

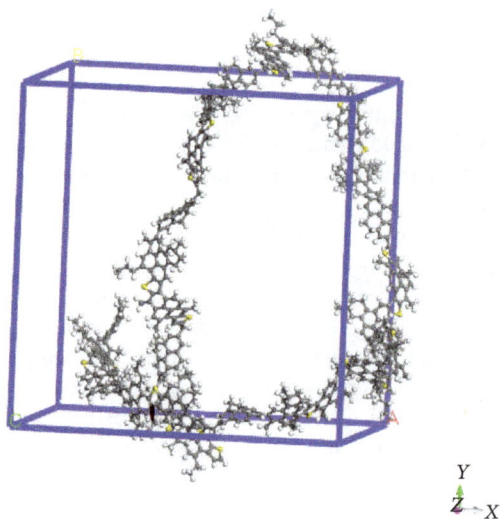

FIGURE 15: Ultimate model of asphaltene.

4. Conclusions

The calculation of mechanical properties was conducted to the groups of colloid, saturated phenol, and asphaltene, respectively, and the parameters of stiffness and modulus of all the materials were obtained. Comparing the E value of Young modulus of these three polymers, the maximum modulus value of asphaltene was 2.80 GPa, the modulus value of colloid was secondary, and the minimum modulus value of saturated phenol was 0.52 GPa. In asphalt, asphaltene is similar to solid, and Young's modulus is largest, followed by colloid. Saturated phenol is similar to liquid, of which the modulus is far less than asphaltene. In addition, comparing the values of Poisson's ratio, it can be found that the maximum Poisson's ratio of saturated phenol is 0.45, which is close to 0.5, further suggesting that saturated phenol is similar to liquid that it is incompressible. However, more molecular morphologies require further investigation.

Conflicts of Interest

The authors declare that there are no conflicts of interest regarding the publication of this paper.

Acknowledgments

The research was supported by the National Natural Science Foundation of China (Grant nos. 51378073 and 51408048), the Key Program for Science and Technology Projects of Shaanxi province (Grant no. 2017KCT-13), the science and technology projects of Shaanxi Provincial Transport Department (Grant no. 15-35T), and the Special Financial Grant from the China Postdoctoral Science Foundation (Grant no. 2016T90880).

References

[1] A. Bhasin and D. N. Little, "Application of microcalorimeter to characterize adhesion between asphalt binders and aggregates," *Journal of Materials in Civil Engineering*, vol. 21, no. 6, pp. 235–243, 2009.

[2] J. Yu, X. Zeng, S. Wu, L. Wang, and G. Liu, "Preparation and properties of montmorillonite modified asphalts," *Materials Science and Engineering: A Structural Materials: Properties, Microstructure and Processing*, vol. 447, no. 1-2, pp. 233–238, 2007.

[3] A. E. Alvarez, E. Ovalles, and A. Epps Martin, "Comparison of asphalt rubber-aggregate and polymer modified asphalt-aggregate systems in terms of surface free energy and energy indices," *Construction and Building Materials*, vol. 35, pp. 385–392, 2012.

[4] A. Bhasin, *Development of Method to Qualify Bitumen-Aggregate Adhesion and Loss of Adhesion due to Water*, Texas A&M University, College Station, Tex, USA, 2006.

[5] R. Li, H. Du, Z. P. Fan, and J. Z. Pei, "Molecular dynamics simulation to investigate the interaction of asphaltene and oxide in aggregate," *Advances in Materials Science & Engineering*, vol. 7, pp. 1–10, 2016.

[6] R. Li, H. F. Wang, P. Wang, H. Liu, and J. Pei, "Influence of PZT piezoelectric ceramics on the structure and electric properties

of piezoelectric lead zirconate titanate/poly(vinylidene fluoride) composites," *Materials Express*, vol. 6, no. 6, pp. 483–492, 2016.

[7] D. Cheng, D. Little N, and C. Holste J, "Use of surface free energy of asphalt-aggregate systems to predict moisture damage potential," *Journal of Association of Asphalt Paving Technologists*, vol. 71, pp. 59–84, 2002.

[8] R. Li, J. Li, Y. C. Xiao, and J. Z. Pei, "Reliability index study of asphalt pavement construction based on analytical hierarchy process," *Applied Mechanics and Materials*, vol. 470, pp. 884–888, 2014.

[9] R. Li, C. C. Wang, P. Z. Wang, and J. Z. Pei, "Preparation of a novel flow improver and its viscosity-reducing effect on bitumen," *Fuel*, vol. 181, pp. 935–941, 2016.

[10] X. Yu, C. Xiong, Y. You, L. Dong, and J. Yao, "Aromatic azo-polyamide electrolyte with liquid crystal structure and photoelectrical properties," *Synthetic Metals*, vol. 158, no. 8-9, pp. 375–378, 2008.

[11] S. Yang, J. Choi, and M. Cho, "Elastic stiffness and filler size effect of covalently grafted nanosilica polyimide composites: Molecular dynamics study," *ACS Applied Materials & Interfaces*, vol. 4, no. 9, pp. 4792–4799, 2012.

Theoretical Analysis of the Effect Provoked by Bromine-Addition on the Thermolysis and Chemiexcitation of a Model Dioxetanone

Luís Pinto da Silva,[1,2] Rui F. J. Pereira,[1] and Joaquim C. G. Esteves da Silva[1,2,3]

[1]*Chemistry Research Unit (CIQUP), Departamento de Química e Bioquímica, Faculdade de Ciências da Universidade do Porto, R. Campo Alegre 687, 4169-007 Porto, Portugal*

[2]*LACOMEPHI, Departamento de Geociências, Ambiente e Ordenamento do Território, Faculdade de Ciências da Universidade do Porto, R. Campo Alegre 687, 4169-007 Porto, Portugal*

[3]*Chemistry Research Unit (CIQUP), Departamento de Geociências, Ambiente e Ordenamento do Território, Faculdade de Ciências da Universidade do Porto, R. Campo Alegre 687, 4169-007 Porto, Portugal*

Correspondence should be addressed to Luís Pinto da Silva; luis.silva@fc.up.pt

Academic Editor: Teik-Cheng Lim

Chemi-/bioluminescence are phenomena in which chemical energy is converted into electronically excited singlet states, which decay with light emission. Given this feature, along with high quantum yields and other beneficial characteristics, these systems have gained numerous applications in bioanalysis, in biomedicine, and in the pharmaceutical field. Singlet chemiexcitation is made possible by the formation of cyclic peroxides (as dioxetanones) as thermolysis provides a route for a ground state reaction to produce singlet excited states. However, such thermolysis can also lead to the formation of triplet states. While triplet states are not desired in the typical applications of chemi-/bioluminescence, the efficient production of such states can open the door for the use of these systems as sensitizers in photocatalysis and triplet-triplet annihilation, among other fields. Thus, the goal of this study is to assess the effect of heavy atom addition on the thermolysis and triplet chemiexcitation of a model dioxetanone. Monobromination does not affect the thermolysis reaction but can improve the efficiency of intersystem crossing, depending on the position of monobromination. Addition of bromine atoms to the triplet state reaction product has little effect on its properties, except on its electron affinity, in which monobromination can increase between 3.1 and 8.8 kcal mol^{-1}.

1. Introduction

Bioluminescence is a widespread natural phenomenon in which living organisms convert chemical energy into light emission via biochemical reactions [1–5]. Bioluminescence can be found in organisms as different as bacteria, dinoflagellates, fungi, crustaceans, worms, insects, and fishes. Light emission from these systems is the result of enzyme-catalyzed reactions, which can be divided into two main classes: luciferase-luciferin reactions [2, 5–8] and photoprotein systems [2, 9]. The luciferase enzyme is responsible for catalyzing the oxidation of its substrate, luciferin, which generates an electronically excited singlet state product. This product, generally termed oxyluciferin, subsequently relaxes to the ground state by photon emission. It should be noted that luciferin, luciferase, and oxyluciferin are only generic terms, showing significant structural differences between bioluminescent species. Moreover, luciferase-luciferin reactions are the most prevalent bioluminescent systems [1–3, 5–8].

Photoprotein systems have been found to be exclusive to marine organisms and are characterized by the formation of a stable enzyme-substrate complex [2]. Such complex is formed between the apoprotein and an oxygenated marine luciferin (2-hydroxyperoxycoelenterazine). Binding of calcium ions to the photoprotein surface triggers the decomposition of the stable complex, which occurs with light emission [2].

Bioluminescence can be considered a subtype of chemiluminescence, in which light emission arises from a chemical

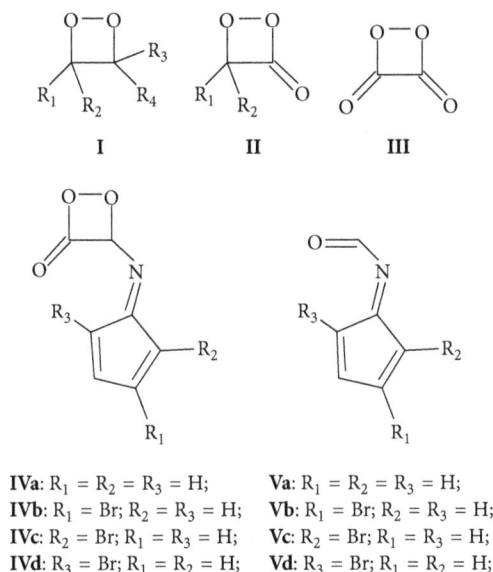

SCHEME 1: Schematic representation of dioxetanes (I), dioxetanones (II), dioxetanedione (III), and the model dioxetanone (IV) and its thermolysis product (V) studied here.

reaction [1, 10, 11]. The efficiency of light emission of both bioluminescent and chemiluminescent reactions is described in terms of quantum yield, which is controlled by three factors [1, 10, 11]: chemical yield of the ground state reaction, chemiexcitation yield of the singlet excited state product, and finally the fluorescent quantum yield of the emitter. Typically, bioluminescent systems present significantly higher quantum yields than chemiluminescent reactions, with some reactions reaching quantum yields of 45–61% [12]. Given this efficient production of electronically singlet excited state products, relative nontoxicity of luciferin compounds, and the relatively simple chemistry of these systems, among other beneficial characteristics (as sensitivity and sensibility), several chemi-/bioluminescent systems have gained numerous biomedical, pharmaceutical, and bioanalytical applications. More specifically, these systems are used in the analytical determination of ATP and other metabolites, in environmental monitoring, in bioimaging, and in biosensing, as a gene reporter, tested as alternative excitation sources in photodynamic therapy of cancer, and used in investigations of infectious diseases, among others [13–17].

The efficient formation of singlet excited state products, necessary for the use of these systems in the many applications referred above, is only made possible by the formation of cyclic peroxide intermediates during the different chemi- and bioluminescent reactions [18–28]. Within the large number of different chemi- and bioluminescent systems, these peroxide intermediates can take the form of dioxetanes (I), dioxetanones (II), or dioxetanedione (III) [18–28], which can be seen in Scheme 1. These cyclic peroxides are responsible for chemi- and bioluminescence as their thermolysis provides a route for a thermally activated ground state reaction to produce singlet excited state products [18–28]. This chemiexcitation process is thought to arise from crossing

points between the ground state and excited state potential energy surfaces (PES) on the reaction coordinate.

It should be noted that while chemi- and bioluminescent systems are better known for their production of singlet excited states, experimental studies have shown that more structurally simple dioxetanes and dioxetanones have the ability to produce triplet excited states [1, 10, 25–27]. While no experimental results are found for more complex chemi-/bioluminescent systems, different theoretical studies have found pathways for triplet chemiexcitation in the thermolysis of dioxetanone rings in several systems [1, 11, 18, 19, 24, 28, 29]. This production of triplet excited states can be problematic for the several practical applications based on the formation of light-emitting singlet states, as triplet states are very easily quenched, and their formation will not be detected with the luminescent and fluorescent approaches typically used to detect chemi-/bioluminescence. Moreover, triplet states are more reactive and are able to produce harmful reactive species (as singlet oxygen), which can lead to some problems when using these systems in biological samples.

While the formation of triplet state (instead of singlet ones) is not desired in the typical applications of chemi-/bioluminescence, such states can have important roles in other applications. One such example is upconversion by triplet-triplet annihilation, which typically proceeds as follows [30–32]: a sensitizer molecule is photoexcited and undergoes intersystem crossing to a triplet state. Subsequently, the sensitizer transfers its energy to an emitter molecule via fast triplet-triplet energy transfer, which stores the energy in the lowest triplet state of the emitter molecules. Then, two emitters interact and triplet-triplet annihilation occurs, which brings one emitter molecule to an excited singlet state while quenching another to its ground state. The emitter then emits light via fluorescence, at a higher energy than that of the photons initially absorbed by the sensitizer. Triplet-triplet annihilation has been already applied with success in several research fields, as in luminescence bioimaging [33], photovoltaics [34, 35], and photoinduced drug release [36].

Another useful application of triplet excited states is on the field of photocatalysis [37–39]. In this field, photosensitizers are used to mediate photochemical reactions by absorbing light and using that energy to activate ground state reactants toward some specific chemical reactions. One common method of photoactivation is via energy transfer from the longer-lived triplet state of the photosensitizer to the substrate [37–39]. Another photoactivation pathway involves an electron transfer from or to the photoexcited sensitizer [37, 38].

Given this, if one can shift the spin of the chemi-/bioluminescent products, from light-emitting singlet to triplet states, one can open the door for new types of applications for chemi- and bioluminescent systems, as in upconversion processes by triplet-triplet annihilation and photocatalysis. One way to facilitate intersystem crossing to triplet states is to introduce heavy atoms (as bromine and iodine) into the molecular structures, the so called "heavy atom effect" [40–42]. Thus, the objective of this work is to model theoretically the effect of bromine-substitution in a model dioxetanone (IV, Scheme 1) and, more specifically, in its thermolysis and triplet chemiexcitation steps. To our knowledge, this is the first theoretical study trying to understand the possible role of the "heavy atom effect" in the reactions of dioxetanone molecules and in what way it affects triplet chemiexcitation. To this end, a methodology combining density functional theory- (DFT-) based and multireference methods was used.

2. Theoretical Methodology

All calculations were made with the Gaussian 09 program package [43], with no solvent effects. DFT methods (particularly long-range-corrected hybrid exchange-correlation density functionals) have been gaining traction in the study of chemi-/bioluminescent reactions, given their ability to provide quite accurate qualitative pictures for these systems [18–21, 24, 28, 44–47]. In this study was used the CAM-B3LYP long-range-corrected density functional [44], which provides good estimates for $\pi \rightarrow \pi^*$ and $n \rightarrow \pi^*$ local excitations and charge transfer and Rydberg states [45]. Moreover, this functional was already used with success in the study of different dioxetanones [20, 21].

Geometry optimizations and frequency calculations were made with the CAM-B3LYP functional, with the 6-31G(d,p) basis set being used for all atoms except bromine, for which the LanL2DZ basis set was used. The combination of 6-31G(d,p) and LanL2DZ basis set was termed Basis Set-1 (BS-1). A restricted (R) approach was used for closed-shell species, while an unrestricted (U) species was used for open-shell structures. The U approach was used with a broken-symmetry technology, which mixes the HOMO and LUMO, making an initial guess for a biradical.

The thermolysis reaction of the model dioxetanones was studied by performing intrinsic reaction coordinates (IRC) [48], at the CAM-B3LYP/BS-1 level of theory, which assessed if the obtained transition states connected the desired reactants and products. The transition states were located by using the STQN method [49]. In this work was used the QST3 variant, which requires three molecular specifications: the reactants, the products, and an initial structure for the transition state. The Cartesian coordinates of these transition state structures, used in the IRC calculations, can be found in Tables S1 and S2 of Supplementary Material available online at https://doi.org/10.1155/2017/1903981.

The energies of the geometry optimizations, IRC, and QST3-obtained structures were reevaluated by single point calculations with the CAM-B3LYP density functional and the 6-31+G(d,p) basis set for all atoms, except for bromine. For this atom, the LanL2DZdp basis set was used, which includes polarization and diffuse functions. The combination of 6-31+G(d,p) and LanL2DZdp basis set was termed Basis Set-2 (BS-2). Thus, the energies of the singlet ground state (S_0) and first triplet state (T_1) were both calculated at the CAM-B3LYP/BS-2 level of theory.

The spin-orbit coupling between S_0 and T_1 states was calculated by using the CASSCF method [50]. The LanL2DZ basis set was used for all atoms. The active space consisted of two electrons on two orbitals. These were single point energy

calculations made on the DFT-computed IRC- or QST3-obtained structures.

3. Results and Discussion

We started this work by analyzing the thermolysis reaction of unsubstituted dioxetanone IVa and monobrominated species IVb, whose energetic profiles are presented in Figures 1(a) and 1(c), respectively. In Figures 1(b) and 1(d) are presented important geometric parameters: the bond lengths of O_1-O_4 and C_2-C_3 (Scheme 1). It should be noted that the imino-cyclopentadienyl moiety was based on the scaffold of aza-BODIPY [40–42], which are molecules capable of producing triplet states upon photoexcitation, when they are functionalized with heavy atoms.

Analysis of the geometric parameters shows that the thermolysis of both species occurs via a stepwise mechanism. The reaction begins by O_1-O_4 bond breaking, while the length of C_2-C_3 remains constant. Only after O_1-O_4, does the length of C_2-C_3 increases, subsequently leading to its cleavage. Analysis of the (S^2) value for the transition state of both IVb (~0.57) and IVa (~0.55) showed that these structures have a biradical character. Given this, we can ascribe a stepwise-biradical mechanism for the thermolysis of both dioxetanone species, which is in line with previous theoretical studies of such molecules [18, 20, 21, 24, 28, 29].

While a stepwise-biradical mechanism is usually found in the decomposition of these cyclic peroxides, it can be further subdivided: the biradical is formed due to an electron transfer from an electron-rich moiety to the dioxetanone, thereby forming a radical cation and a radical anion, respectively; the biradical is formed due to the homolytic cleavage of the O_1-O_4 bond. In this case, both molecules appear to undergo thermolysis via homolysis, as the electron spin density of the transition state resides only on the O_1 and O_4 heteroatoms (Figures 2(c) and 2(d)). This finding is in line with the limited charge transfer found between the dioxetanone and imino-cyclopentadiene moieties, as demonstrated in Figures 2(a) and 2(b). The atomic charges were measured within the Natural Population Analysis (NPA) scheme. The finding that these molecules undergo a homolysis-based stepwise-biradical thermolysis can be attributed to the absence of an ionizable group, as seen before in the theoretical analysis of such molecules [18, 20, 21, 24, 28, 29].

Both species have similar activation energies of 24.7 kcal mol^{-1} for IVa and 24.5 kcal mol^{-1} for IVb. These energies were calculated at the CAM-B3LYP/BS-2 level of theory, with thermal corrections calculated at the CAM-B3LYP/BS-1 level of theory. Considering that no solvents effects were considered and that these species are only model dioxetanones, the obtained activation parameters compare well with the experimentally obtained ones for several cyclic peroxides (within ~20.0 kcal mol^{-1}) [51–53].

So far, the main conclusion is that the addition of a bromine atom affects slightly the decomposition of the model dioxetanone, as there are only very minor differences between the thermolysis of species IVa and IVb. In fact, both species present the same characteristics as other dioxetanone without an ionized group [18–21, 24, 28, 29].

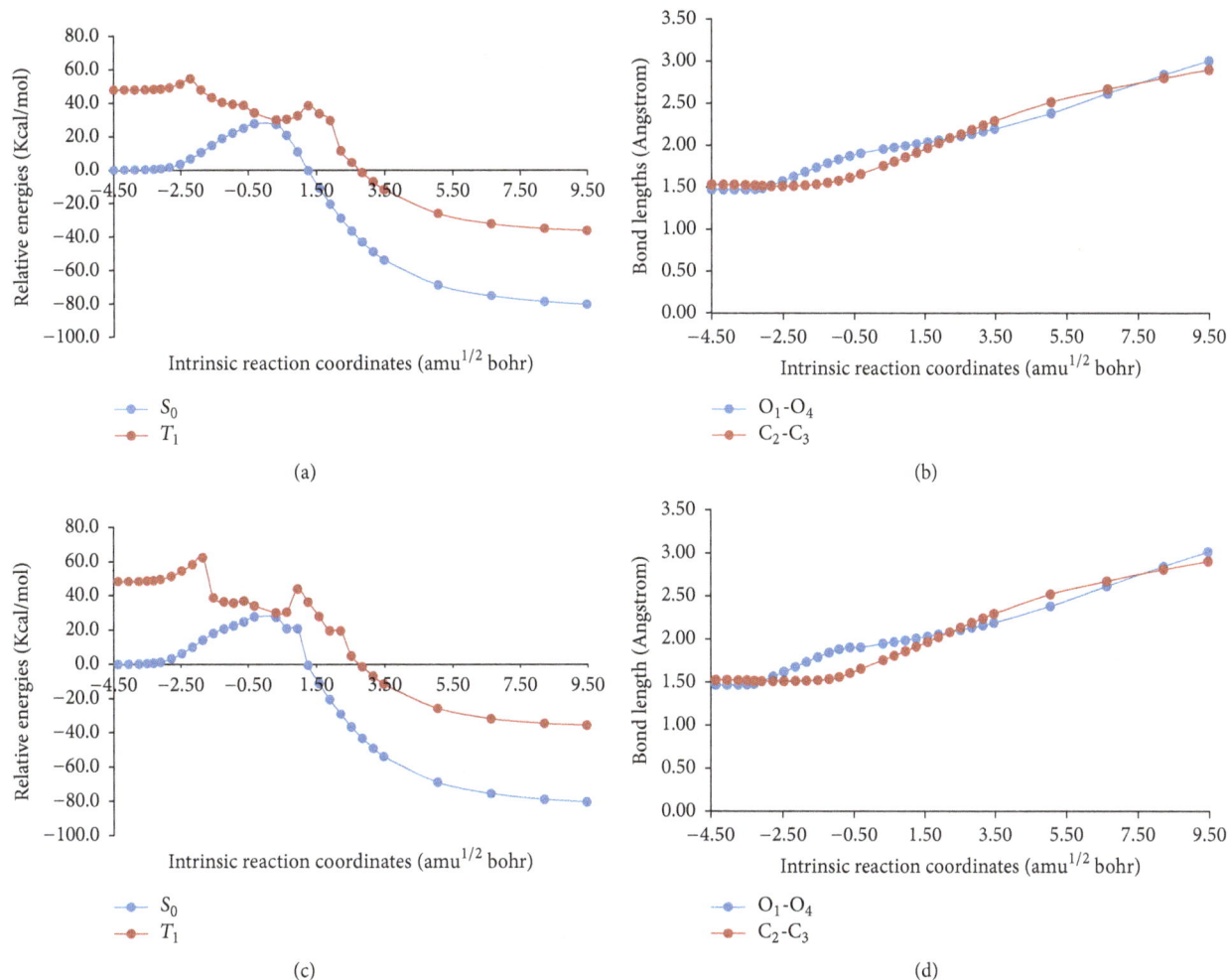

FIGURE 1: Energy profiles of the S_0 and T_1 states, determined in the S_0-computed IRC path, for species IVa (a) and IVb (c). O_1-O_4 and C_2-C_3 bond length variations recorded during the IRC calculations for IVa (b) and IVb (d).

In the same vein, we have found a pathway for triplet chemiexcitation for species IVa and IVb (Figures 1(a) and 1(c), resp.), in line with other theoretical works on different dioxetanones, dioxetanes, and dioxetanedione [11, 18, 19, 24, 28, 29]. Upon starting the reaction, the energetic difference between S_0 and T_1 states was large for both molecules (about 48 kcal mol^{-1}). However, from the reactant onward, the energy of the T_1 state decreased significantly to a point in both species where the S_0-T_1 energy gap is low enough to allow chemiexcitation: 2.5 kcal mol^{-1} (at 0.32 amu$^{1/2}$ bohr) for IVa and 2.6 kcal mol^{-1} (at 0.32 amu$^{1/2}$ bohr) for IVb. In conclusion, by analyzing the S_0 and T_1 energetic profiles, our results indicate that both IVa and IVb species are capable of triplet chemiexcitation. Moreover, the very similar S_0-T_1 energy gap suggests a triplet chemiexcitation of similar magnitude, thereby indicating a small effect provoked by the addition of a bromine heavy atom.

It should be noted, however, that intersystem crossing is a process formally forbidden in nonrelativistic quantum theory, and, so, inferring singlet-triplet transition probabilities from energy gaps is not sufficiently accurate [40, 54].

So, to assess the efficiency of intersystem crossing, we must take into account the spin-orbit coupling (SOC) between S_0 and T_1 [40, 54]. These were computed at the multireference CASSCF level of theory, and the SOC values for IVa and IVb are presented in Table 1. The CASSCF-computed S_0-T_1 energy gaps (3.1 kcal mol^{-1} for IVb and 2.9 kcal mol^{-1} for IVa) are in line with the DFT-computed ones gap (2.5 kcal mol^{-1} for IVb and 2.6 kcal mol^{-1} for IVa). Once again, it appears that the addition of a bromine heavy atom has little effect, as the SOC values for IVa (5.3 cm^{-1}) and IVb (6.1 cm^{-1}) are very similar, despite the SOC being higher for the monobrominated species.

So far, it does appear that the addition of a bromine atom has only a limited effect on the triplet chemiexcitation of model dioxetanone IV. However, it might be possible that this lack of effect is due to the position of bromine-substitution in the cyclopentadiene moiety and not due to a general absence of the "heavy atom effect" in this molecule. To test this hypothesis, we have calculated transition state structures (with the QST3 method, at the CAM-B3LYP/SD-1 level of theory) for species IVb, IVc, and IVd. The Cartesian

(a)

(b)

(c)

(d)

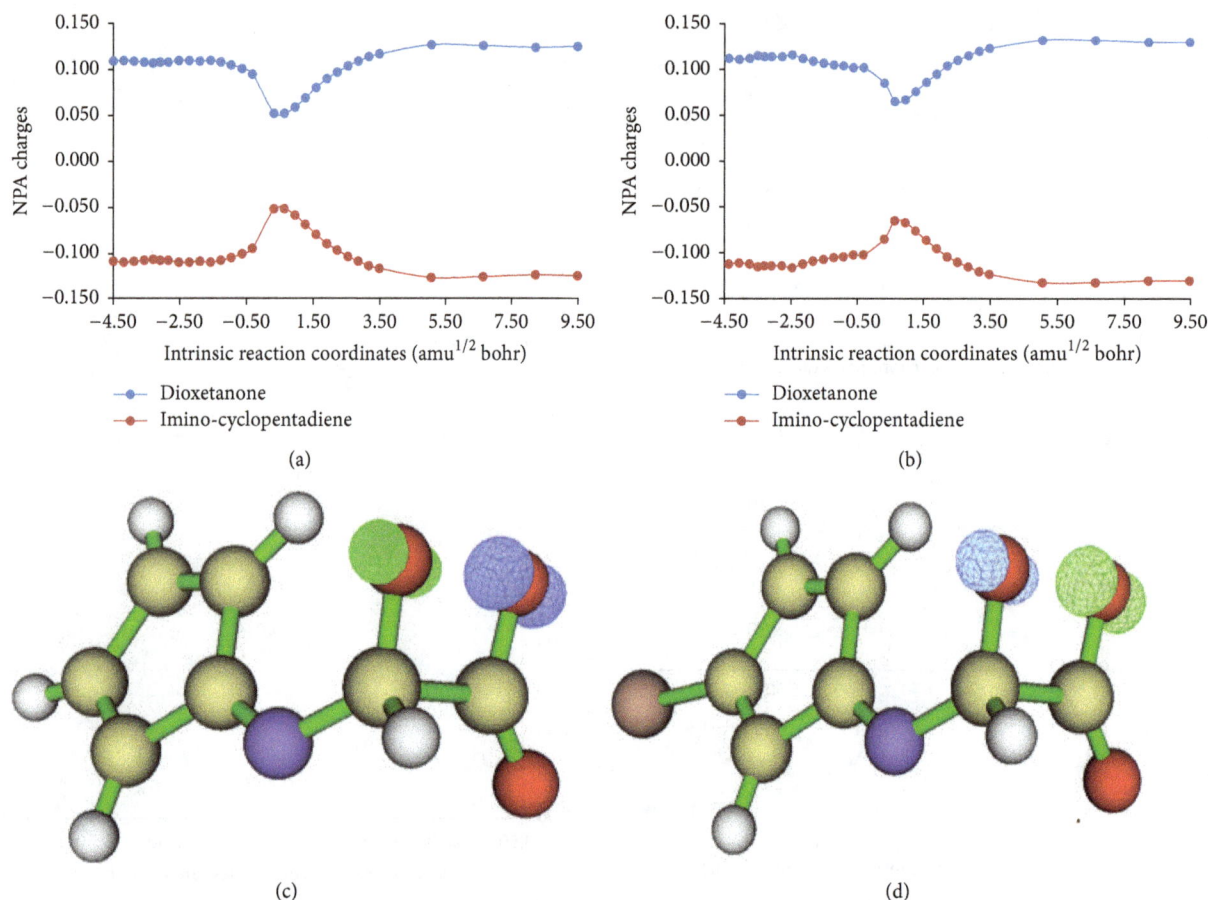

FIGURE 2: NPA charge density distribution between the dioxetanone and imino-cyclopentadiene moieties, for species IVa (a) and IVb (b). Electron spin density for the transition state structure of the thermolysis of IVa (c) and IVb (d).

TABLE 1: S_0-T_1 energy gaps ($\Delta E(S_0$-$T_1)$, in kcal mol^{-1}) and respective SOC values (SOC, in cm^{-1}), obtained with CASSCF(2,2)/LanL2DZ single point energy calculations on CAM-B3LYP/BS-1 structures.

	IVa[a]	IVb[a]	IVb[b]	IVc[b]	IVd[b]
$\Delta E(S_0$-$T_1)$	2.9	3.2	11.4	11.4	10.8
SOC	5.3	6.1	7.5	8.6	11.5

[a]Structures obtained during the IRC calculations, corresponding to points of lower S_0-T_1 energy gaps: 2.5 kcal mol^{-1} (at 0.32 amu$^{1/2}$ bohr) for IVa and 2.6 kcal mol^{-1} (at 0.32 amu$^{1/2}$ bohr) for IVb.
[b]Transition state structures obtained with the QST3 method.

TABLE 2: Adiabatic S_0-T_1 energy gaps ($\Delta E(S_0$-$T_1)$, in kcal mol^{-1}) and vertical ionization energies (IE, in kcal mol^{-1}) and electron affinity (EA, in kcal mol^{-1}) for the T_1 state of species Va–d, obtained at the CAM-B3LYP/BS-2//CAM-B3LYP/BS-1 level of theory.

	Va	Vb	Vc	Vd
$\Delta E(S_0$-$T_1)$	25.3	26.0	21.8	24.4
IE	190.5	191.1	188.9	187.2
EA	−59.1	−67.9	−63.9	−62.2

coordinates of these structures can be found in Tables S3–S5 of Supplementary Material. At those DFT-computed structures, CASSCF single point energy calculations were made to obtain the SOC values (Table 1). The CASSCF-computed S_0-T_1 energy gaps are similar to all species (between 10.8 and 11.4 kcal mol^{-1}), further indicating that the position of the bromine atom has little effect on the value of the S_0-T_1 energy gap. However, this is not true for the SOC values. These increased more significantly, from 7.5 to 11.5 cm^{-1}, with IVd presenting SOC values more than double the ones presented previously by IVa. Thus, these results indicate that

the addition of heavy atoms can indeed increase the efficiency of intersystem crossing and, so, the formation of triplet state products, but this effect is controlled by the position in which the heavy atom is inserted.

Having analyzed the effect of monobromination on the thermolysis and triplet chemiexcitation of model dioxetanone IV, we have studied some properties of the T_1 reaction product (species Va–d, Scheme 1). The Cartesian coordinates of these structures can be found on Tables S6–S9 of Supplementary Material. These properties are the electron affinity (EA), ionization energies (IE), and T_1-S_0 energy gaps (present in Table 2). The EA and IE were computed vertically, with single point energy calculations on the T_1 structure, with T_1 as the reference state. All V species present very high IEs, with a limited effect provoked by monobromination, which limits the use of these species as electron donors in photocatalysis. On the contrary, the EA values are more suitable for the use of these species as electron acceptors. Moreover, the addition of bromine atoms can significantly improve the EA of V up to $8.8 \, kcal \, mol^{-1}$, depending on the location of the substitution on the cyclopentadiene ring. As for T_1-S_0 energy gaps, these were computed adiabatically. Our calculations have indicated that monobromination has a limited effect on the T_1-S_0 energy gaps, except for species Vc, which decreased the gap by $3.5 \, kcal \, mol^{-1}$.

4. Conclusion

Chemi- and bioluminescence are phenomena in which chemical energy is converted into light emission, via chemical and biochemical reactions. Given this feature, coupled to high quantum yields, relative nontoxicity of the reaction substrates, and the relatively simple chemistry, among other beneficial characteristics (as sensitivity and sensibility), several chemi-/bioluminescent systems have gained numerous biomedical, pharmaceutical, and bioanalytical applications.

The efficient formation of light-emitting singlet excited states, needed for the applications referred above, is made possible by the formation of cyclic peroxides (as dioxetanes or dioxetanones). Their thermolysis provides a route for a thermally activated ground state reaction to produce singlet excited state products. However, both experimental and theoretical studies have demonstrated that the thermolysis of these species is also able to produce triplet states. While the formation of such states is not desired in the typical applications of chemi-/bioluminescence, efficient production of triplet states might open the door for the use of chemi-/bioluminescent systems as sensitizers in the fields of photocatalysis and upconversion processes by triplet-triplet annihilation, among others.

Given this, the objective of this theoretical study was to assess the effect induced by heavy atom substitution (in this case, bromine) on the thermolysis and triplet chemiexcitation of a model dioxetanone. Our calculations indicated that monobromination has little effect on the S_0 thermolysis reaction, with little effect on the energetics of the reaction and on the variation of other parameters (as selected bond lengths and electron spin and charge density). However, the addition of bromine atoms can increase the spin-orbit coupling of the S_0 and T_1, thereby increasing the efficiency of intersystem crossing. Nevertheless, this effect is dependent on the position of the bromine-substitution. Study of the T_1 reaction product showed a general limited effect provoked by monobromination on the ionization energies and S_0-T_1 gaps. On the contrary, monobromination improves the electron affinity of the T_1 product, with the degree of improvement being controlled by the position of monobromination.

Conflicts of Interest

The authors declare that they have no conflicts of interest.

Acknowledgments

This work was made in the framework of Project PTDC/QEQ-QFI/0289/2014, which is funded with national funds by FCT/MEC (PIDDAC). The project is also cofunded by "Fundo Europeu de Desenvolvimento Regional" (FEDER), through "COMPETE-Programa Operacional Fatores de Competitividade" (COMPETE-POFC). This work was also made in the framework of the Project Sustainable Advanced Materials (NORTE-01-00145-FEDER-000028), funded by "Fundo Europeu de Desenvolvimento Regional (FEDER)," through "Programa Operacional do Norte" (NORTE2020). Acknowledgment to Project POCI-01-0145-FEDER-006980, funded by FEDER through COMPETE2020, is also made. The Laboratory for Computational Modeling of Environmental Pollutants-Human Interactions (LACOMEPHI) is acknowledged. Luís Pinto da Silva also acknowledges a postdoctoral grant funded by Project Sustainable Advanced Materials (NORTE-01-00145-FEDER-000028).

References

[1] L. Pinto Da Silva and J. C. G. Esteves Da Silva, "Firefly chemiluminescence and bioluminescence: efficient generation of excited states," *ChemPhysChem*, vol. 13, no. 9, pp. 2257–2262, 2012.

[2] A. S. Tsarkova, Z. M. Kaskova, and I. V. Yampolsky, "A tale of two luciferins: fungal and earthworm new bioluminescent systems," *Accounts of Chemical Research*, vol. 49, no. 11, pp. 2372–2380, 2016.

[3] J. Vieira, L. P. Da Silva, and J. C. G. E. Da Silva, "Advances in the knowledge of light emission by firefly luciferin and oxyluciferin," *Journal of Photochemistry and Photobiology B: Biology*, vol. 117, pp. 33–39, 2012.

[4] E. P. Coutant and Y. L. Janin, "Synthetic routes to coelenterazine and other imidazo[1,2-a]pyrazin-3-one luciferins: essential tools for bioluminescence-based investigations," *Chemistry*, vol. 21, pp. 17158–17171, 2015.

[5] S. V. Markova and E. S. Vysotski, "Coelenterazine-dependent luciferases," *Biochemistry (Moscow)*, vol. 80, no. 6, pp. 714–732, 2015.

[6] S. M. Marques and J. C. G. Esteves Da Silva, "Firefly bioluminescence: a mechanistic approach of luciferase catalyzed reactions," *IUBMB Life*, vol. 61, no. 1, pp. 6–17, 2009.

[7] S. Tu and H. I. X. Mager, "Biochemistry of bacterial bioluminescence," *Photochemistry and Photobiology*, vol. 62, no. 4, pp. 615–624, 1995.

[8] D. E. Desjardin, A. G. Oliveira, and C. V. Stevani, "Fungi bioluminescence revisited," *Photochemical & Photobiological Sciences*, vol. 7, pp. 170–182, 2008.

[9] O. Shimomura and F. H. Johnson, "Peroxidized coelenterazine, the active group in the photoprotein aequorin," *Proceedings of the National Academy of Sciences of the United States of America*, vol. 75, pp. 2611–2615, 1978.

[10] M. Matsumoto, "Advanced chemistry of dioxetane-based chemiluminescent substrates originating from bioluminescence," *Journal of Photochemistry and Photobiology C: Photochemistry Reviews*, vol. 5, no. 1, pp. 27–53, 2004.

[11] I. Navizet, Y. J. Liu, N. Ferré, D. Roca-Sanjuán, and R. Lindh, "The chemistry of bioluminescence: an analysis of chemical functionalities," *ChemPhysChem*, vol. 12, pp. 3064–3076, 2011.

[12] K. Niwa, Y. Ichino, S. Kumata et al., "Quantum yields and kinetics of the firefly bioluminescence reaction of beetle luciferases," *Photochemistry and Photobiology*, vol. 86, no. 5, pp. 1046–1049, 2010.

[13] A. Roda, M. Guardigli, E. Michelini, and M. Mirasoli, "Bioluminescence in analytical chemistry and in vivo imaging," *TrAC - Trends in Analytical Chemistry*, vol. 28, no. 3, pp. 307–322, 2009.

[14] M. B. Gu, R. J. Mitchell, and B. C. Kim, "Whole-cell-based biosensors for environmental biomonitoring and application," *Advances in Biochemical Engineering/Biotechnology*, vol. 87, pp. 269–305, 2004.

[15] C. M. Magalhães, J. C. G. Esteves da Silva, and L. Pinto da Silva, "Chemiluminescence and Bioluminescence as an excitation source in the photodynamic therapy of cancer: a critical review," *ChemPhysChem*, vol. 17, pp. 2286–2294, 2016.

[16] K. E. Luker and G. D. Luker, "Bioluminescence imaging of reporter mice for studies of infection and inflammation," *Antiviral Research*, vol. 86, no. 1, pp. 93–100, 2010.

[17] L. Cevenini, M. M. Calabretta, A. Lopreside et al., "Exploiting nanoluc luciferase for smartphone-based bioluminescence cell biosensor for (anti)-inflammatory activity and toxicity," *Analytical and Bioanalytical Chemistry*, vol. 408, pp. 8859–8868, 2016.

[18] L. Pinto da, C. M. Silva, and J. C. G. Esteves da Silva, "Interstate crossing-induced chemiexcitation mechanism as the basis for imidazopyrazinone bioluminescence," *ChemistrySelect*, vol. 1, pp. 3343–3356, 2016.

[19] L. P. Da Silva and J. C. G. E. Da Silva, "Interstate crossing-induced chemiexcitation as the reason for the chemiluminescence of dioxetanones," *ChemPhysChem*, vol. 14, no. 5, pp. 1071–1079, 2013.

[20] B.-W. Ding, P. Naumov, and Y.-J. Liu, "Mechanistic insight into marine bioluminescence: photochemistry of the chemiexcited cypridina (sea firefly) lumophore," *Journal of Chemical Theory and Computation*, vol. 11, no. 2, pp. 591–599, 2015.

[21] B. W. Ding and Y. J. Liu, "Bioluminescence of firefly squid via mechanism of single electron-transfer oxygenation and charge-transfer-induced luminescence," *Journal of the American Chemical Society*, vol. 139, pp. 1106–1119, 2017.

[22] L. F. M. L. Ciscato, F. H. Bartoloni, A. S. Colavite, D. Weiss, R. Beckert, and S. Schramm, "Evidence supporting a 1,2-dioxetanone as an intermediate in the benzofuran-2(3H)-one chemiluminescence," *Photochemical and Photobiological Sciences*, vol. 13, no. 1, pp. 32–37, 2014.

[23] S. Schramm, I. Navizet, P. Naumov et al., "The light emitter of the 2-coumar-anone chemiluminescence: theoretical and experimental elucidation of a possible model for bioluminescent systems," *European Journal of Organic Chemistry*, vol. 4, 2016.

[24] L. Pinto Da Silva and J. C. G. Esteves Da Silva, "Mechanistic study of the unimolecular decomposition of 1,2-dioxetanedione," *Journal of Physical Organic Chemistry*, vol. 26, no. 8, pp. 659–663, 2013.

[25] W. Adam and W. J. Baader, "Effects of methylation on the thermal stability and chemiluminescence properties of 1,2-dioxetanes," *Journal of the American Chemical Society*, vol. 107, no. 2, pp. 410–416, 1985.

[26] S. P. Schmidt and G. B. Schuster, "Kinetics of unimolecular dioxetanone chemiluminescence. Competitive parallel reaction paths," *Journal of the American Chemical Society*, vol. 100, no. 17, pp. 5559–5561, 1978.

[27] W. Adam, "Thermal generation of electronic excitation with hyperenergetic molecules," *Pure and Applied Chemistry*, vol. 52, no. 12, pp. 2591–2608, 1980.

[28] H. Isobe, S. Yamanaka, S. Kuramitsu, and K. Yamaguchi, "Regulation mechanism of spin-orbit coupling in charge-transfer-induced luminescence of imidazopyrazinone derivatives," *Journal of the American Chemical Society*, vol. 130, no. 1, pp. 132–149, 2008.

[29] L. Yue, Y.-J. Liu, and W.-H. Fang, "Mechanistic insight into the chemiluminescent decomposition of firefly dioxetanone," *Journal of the American Chemical Society*, vol. 134, no. 28, pp. 11632–11639, 2012.

[30] T. F. Schulze and T. W. Schmidt, "Photochemical upconversion: present status and prospects for its application to solar energy conversion," *Energy & Environmental Science*, vol. 8, pp. 103–125, 2015.

[31] X. Cui, J. Zhao, Y. Zhou, J. Ma, and Y. Zhao, "Reversible photoswitching of triplet-triplet annihilation upconversion using dithienylethene photochromic switches," *Journal of the American Chemical Society*, vol. 136, no. 26, pp. 9256–9259, 2014.

[32] K. Xu, J. Zhao, D. Escudero, Z. Mahmood, and D. Jacquemin, "Controlling triplet-triplet annihilation upconversion by tuning the PET in aminomethyleneanthracene derivatives," *Journal of Physical Chemistry C*, vol. 119, no. 42, pp. 23801–23812, 2015.

[33] Q. Liu, T. Yang, W. Feng, and F. Li, "Blue-emissive upconversion nanoparticles for low-power-excited bioimaging in vivo," *Journal of the American Chemical Society*, vol. 134, no. 11, pp. 5390–5397, 2012.

[34] J. S. Lissau, J. M. Gardner, and A. Morandeira, "Photon upconversion on dye-sensitized nanostructured ZrO$_2$ films," *Journal of Physical Chemistry C*, vol. 115, no. 46, pp. 23226–23232, 2011.

[35] A. Monguzzi, D. Braga, M. Gandini et al., "Broadband upconversion at subsolar irradiance: triplet-triplet annihilation boosted by fluorescent semiconductor nanocrystals," *Nano Letters*, vol. 14, no. 11, pp. 6644–6650, 2014.

[36] S. H. C. Askes, A. Bahreman, and S. Bonnet, "Activation of a photodissociative ruthenium complex by triplet-triplet annihilation upconversion in liposomes," *Angewandte Chemie—International Edition*, vol. 53, pp. 1029–1033, 2014.

[37] M. H. Shaw, J. Twilton, and D. W. C. MacMillan, "Photoredox catalysis in organic chemistry," *Journal of Organic Chemistry*, vol. 81, no. 16, pp. 6898–6926, 2016.

[38] M. Kozlowski and T. Yoon, "Editorial for the Special Issue on Photocatalysis," *Journal of Organic Chemistry*, vol. 81, no. 16, pp. 6895–6897, 2016.

[39] A. Iyer, S. Jockusch, and J. Sivaguru, "A photo-auxiliary approach—enabling excited state classical phototransformations with metal free visible light irradiation," *Chemical Communications*, vol. 53, no. 10, 2017.

[40] B. C. De Simone, G. Mazzone, J. Pirillo, N. Russo, and E. Sicilia, "Halogen atom effect on the photophysical properties of substituted aza-BODIPY derivatives," *Physical Chemistry Chemical Physics*, vol. 19, pp. 2530–2536, 2017.

[41] N. Adarsh, M. Shanmugasundaram, R. R. Avirah, and D. Ramaiah, "Aza-BODIPY derivatives: enhanced quantum yields of triplet excited states and the generation of singlet oxygen and their role as facile sustainable photooxygenation catalysts," *Chemistry*, vol. 18, no. 40, pp. 12655–12662, 2012.

[42] B. Kuçukoz, G. Sevinç, E. Yildiz et al., "Enhancement of two photon absorption properties and intersystem crossing by charge transfer in pentaarylboron-dipyrromethene (BODIPY)

derivatives," *Physical Chemistry Chemical Physics*, vol. 18, pp. 13546–13553, 2016.

[43] M. J. Frisch, G. W. Trucks, H. B. Schlegel, G. E. Scuseria, M. A. Robb, J. R. Cheeseman et al., *Gaussian 09, Revision A.02*, Gaussian, Wallingford, Connecticut, USA, 2009.

[44] T. Yanai, D. P. Tew, and N. C. Handy, "A new hybrid exchange-correlation functional using the Coulomb-attenuating method (CAM-B3LYP)," *Chemical Physics Letters*, vol. 393, no. 1–3, pp. 51–57, 2004.

[45] C. Adamo and D. Jacquemin, "The calculations of excited-state properties with time-dependent density functional theory," *Chemical Society Reviews*, vol. 42, no. 3, pp. 845–856, 2013.

[46] C. G. Min, L. Pinto da Silva, X. K. Esteves da Silva et al., "A computational investigation of the equilibrium constants for the fluorescent and chemiluminescent states of coelenteramide," *ChemPhysChem*, vol. 1, pp. 117–123, 2017.

[47] L. Pinto da Silva, C. M. Magalhães, D. M. A. Crista, and Esteves da Silva J. C. G., "Theoretical modulation of singlet/triplet chemiexcitation of chemiluminescent imidazopyrazinone dioxetanone via C8-substitution," *Photochemical & Photobiological Sciencies*, 2017.

[48] K. Fukui, "The path of chemical reactions—the IRC approach," *Accounts of Chemical Research*, vol. 14, no. 12, pp. 363–368, 1981.

[49] C. Peng, P. Y. Ayala, H. B. Schlegel, and M. J. Frisch, "Using redundant internal coordinates to optimize equilibrium geometries and transition states," *Journal of Computational Chemistry*, vol. 17, no. 1, pp. 49–56, 1996.

[50] M. Klene, M. A. Robb, M. J. Frisch, and P. Celani, "Parallel implementation of the CI-vector evaluation in full CI/CAS-SCF," *Journal of Chemical Physics*, vol. 113, no. 14, pp. 5653–5665, 2000.

[51] F. A. Augusto, G. A. De Souza, S. P. De Souza Jr., M. Khalid, and W. J. Baader, "Efficiency of electron transfer initiated chemiluminescence," *Photochemistry and Photobiology*, vol. 89, no. 6, pp. 1299–1317, 2013.

[52] M. Matsumoto, Y. Ito, M. Murakami, and N. Watanabe, "Synthesis of 5-tert-butyl-1-(3-tert-butyldimethylsiloxy)phenyl-4,4-dimethyl-2,6,7- trioxabicyclo[3.2.0]heptanes and their fluoride-induced chemiluminescent decomposition: Effect of a phenolic electron donor on the CIEEL decay rate in aprotic polar solvent," *Luminescence*, vol. 17, no. 5, pp. 305–312, 2002.

[53] M. Tanimura, N. Watanabe, H. K. Ijuin, and M. Matsumoto, "Intramolecular Charge-transfer-induced decomposition promoted by an aprotic polar solvent for bicyclic dioxetanes bearing a 4-(benzothiazol-2-yl)-3- hydroxyphenyl moiety," *Journal of Organic Chemistry*, vol. 76, no. 3, pp. 902–908, 2011.

[54] C. M. Marian, "Spin-orbit coupling and intersystem crossing in molecules," *Wiley Interdisciplinary Reviews: Computational Molecular Science*, vol. 2, no. 2, pp. 187–203, 2012.

Removal of Cd (II) from Aqueous Media by Adsorption onto Chemically and Thermally Treated Rice Husk

María Camila Hoyos-Sánchez,[1] **Angie Carolina Córdoba-Pacheco,**[1]
Luis Fernando Rodríguez-Herrera,[2] **and Ramiro Uribe-Kaffure**[3]

[1]*Department of Biology, University of Tolima, Altos de Santa Helena, C. P. 730006 Ibagué, Colombia*
[2]*Department of Chemistry, University of Tolima, Altos de Santa Helena, C. P. 730006 Ibagué, Colombia*
[3]*Department of Physics, University of Tolima, Altos de Santa Helena, C. P. 730006 Ibagué, Colombia*

Correspondence should be addressed to Ramiro Uribe-Kaffure; rauribe@ut.edu.co

Academic Editor: Wenshan Guo

Chemically and thermally treated rice husks were evaluated as a potential decontaminant of toxic Cd (II) in aqueous media. Rice husk (RH), a by-product from rice milling, was chemically treated with HCl and NaOH. Then, thermal treatments to 300, 500, and 700°C were applied. The chemical composition and morphological characteristics of RH were evaluated by different techniques. The specific surface area analysis of RH samples by BET nitrogen adsorption method provided specific surface areas ranging from 6 to 14 m^2/g. SEM, FTIR, and EDX analyses of RH were carried out to determine the surface morphology, functional groups involved in metal binding mechanism, and C/O and C/Si ratios, respectively. The maximum Cd (II) adsorption capacity was 28.27 mg/g at an optimum pH, 6.0. The kinetic studies revealed that adsorption process followed the pseudo-second-order kinetic model.

1. Introduction

Although many different definitions have been proposed to the term "heavy metal," some based on density, some on atomic number or atomic weight, and some on chemical properties [1], the term is often used to denote a group of metals and semimetals (metalloids) that have been associated with contamination and potential toxicity or ecotoxicity.

While some are essential for growth, reproduction, and survival of living organisms, the contamination by heavy metals has become a significant environmental problem [2, 3]. In order to minimize the hazardous effects of such elements on the environment and human health, there is a particular need for the development of efficient techniques for removing heavy metals from water sources.

Among the heavy metals, cadmium, which exists commonly in Cd (II) form in aqueous media, poses severe risks to human health. It has been reported that cadmium intoxication can lead to kidney [4], bone [5], and pulmonary [6] damage among others. Therefore, the increase in environmental cadmium concentrations, mainly for its use in industrial processes, added to the fact that it has not shown any physiological function within the human body [7], validates the study of techniques for cadmium removal from aqueous media.

In recent years, a wide variety of techniques have been applied for removal of heavy metal from aqueous media: ion exchange [8], bioremediation by microorganism [9], biometallurgy [10], and bioelectrochemical metal recovery [11], among others. In this context, adsorption has been referenced as an efficient and easy-to-use alternative. By this method, environmental pollutants are removed and concentrated on a specific area for better handling and disposal. Materials such as activated carbon and clays, among others, have shown a potential contaminants adsorbent capacity. However, in the need to find inexpensive and high availability materials, there are a growing number of studies on agroindustrial waste as bagasse from sugar [12], stem of papaya [13], shell bean [14], banana peel [15], coffee residues [16], orange peel [17], among others, to be used as adsorbent materials.

Previous reports have shown the potential of raw rice husk and rice husk ash as adsorbent materials for metals in aqueous media [18–20]; however, chemical and physical changes on the material are required to optimize their adsorbent capacity [21]. This study was aimed to evaluate the Cd (II) adsorption capacity onto different adsorbent materials obtained from rice husk chemically and thermally treated (not necessarily as ash). In addition, it tries to determine the experimental pH and equilibrium time conditions required to achieve the highest percentage of Cd (II) adsorption onto rice husk.

2. Materials and Methods

2.1. Samples Preparation. Rice (*Oryzika*-I) husks were provided by a local industry (Molino Caribe), located in Espinal city, Colombia. To prepare the samples, the rice husks were manually cleaned and thoroughly washed with distilled water to remove all dirt. Then rice husks were crushed and passed through a steel sieve; particle sizes ≤ 100 μm were obtained. At this point, chemical composition analysis was performed. Without any other physical or chemical treatment on rice husk, we select the first sample, RHw (rice husk without treatment).

Chemical and Thermal Treatment on Rice Husks. According to previous reports, the treatment of rice husks with HCl helps in removing impurities such as inorganic salts [22]. Besides, it also helps to remove traces of different oxides such as potassium oxide (K_2O) from the use of fertilizers in rice cultivation [23]. On the other hand, washing with NaOH generates an increase in the adsorption capacity of rice husk, because it can partially degrade some components of the husk, exposing reactive functional groups, as OH, which could retain the contaminating molecules [24, 25]. Also, this washing removes surface impurities that can interfere with the adsorption [21].

According to the above, cleaned and sieved rice husk was soaked in 0.5 M HCl solution (1 : 20) at room temperature for 4 h in permanent agitation on a shaker to 200 RPM. Rice husk was then washed, filtered, and soaked in 0.5 M NaOH solution (1 : 20) at the same above-mentioned conditions. Finally, rice husk was repeatedly washed with distillated water and dried at 40°C for 48 h. This treated rice husk was designated as RHc (rice husk chemically treated).

In order to choose the adequate temperatures for the rice husk's thermal treatment, at this time a thermogravimetric analysis was carried out in oxygen flow, with temperature rise of 10°C per minute until 700°C, in a STA 7200 TGA analyzer. From thermogravimetric results, the working temperatures were chosen, coinciding with temperatures for which the greatest mass losses occur. Samples were then heated by increasing the temperature for 3 hours to the desired value and maintained at that value for 3 hours.

2.2. Samples Characterization. Specific surface area determination was conducted to assess the structural changes of the rice husk induced by chemical and thermal treatment. Specific surface area was measured according to the Brunauer-Emmett-Teller (BET) method, based on the nitrogen adsorption by sample surface. The assays were conducted at 77 K in a surface area analyzer (Autosorb-iQ 07165, Quantachrome Instrument).

The morphological characteristics and elemental analysis of rice husks were evaluated using a JEOL, 6490-LV scanning electron microscope equipped with an EDX (Inca Energy 250 EDS System LK-IE250, Oxford). The rice husk samples were covered with a thin layer of gold and an electron acceleration voltage of 20 kV was applied.

The functional groups on rice husks were characterized by a Fourier Transform Infrared Spectrometer (Thermo Nicolet NEXUS 670 FTIR). The spectral range was varied from 4000 to 400 cm^{-1}.

2.3. Adsorption Experiments. Working solutions of Cd (20, 40, 60, 80, and 100 mg/L) were prepared by diluting the stock solution (1000 mg/L) in deionized water. Each adsorbent material (at a concentration of 3 g of sorbent/L of solution) was soaked in a Cd (II) solution. Solutions were under constant stirring at room temperature (22°C) for 5 days; pH was maintained at 6 during the whole process. Finally, the solutions were filtered and quantification of residual Cd (II) in solutions was carried out using atomic absorption spectrometry (Perkin Elmer 3110).

The adsorption of Cd (II) ions onto rice husk materials was calculated using the following equation:

$$q_e = \frac{(C_o - C_e) V}{m}, \tag{1}$$

where q_e is the amount of metal ions adsorbed onto rice husk (mg/g), C_o and C_e (mg/L) are the initial and equilibrium concentrations of metal ions, V (L) is the volume of solution, and m (g) is the adsorbent mass. Similarly, the percentage of adsorption for each adsorbent material was determined by the following equation:

$$\% \text{ Ads} = \frac{C_o - C_e}{C_o} \times 100. \tag{2}$$

2.4. pH and Kinetics of Adsorption Studies for the Best Sample. Once the material with the best adsorption capacity was found, the effect of pH on Cd (II) adsorption onto this material was evaluated. The best adsorbent material was soaked in different Cd solutions (20 mg/L) at pH 4, 5, and 6. It was under constant stirring at room temperature (22°C) for 6 days. From capacity adsorption results, the best working pH value was determined.

In order to evaluate the reaction time required in the adsorption process, the best adsorbent material, at the best pH value, was placed in contact with eleven independent Cd (II) solutions (100 mg/L). It was under constant stirring at room temperature (22°C). Pseudo-first-order and pseudo-second-order kinetic models were applied to results of adsorption capacity in order to predict the nature of the kinetics.

All the experiments were performed in triplicate; ANOVA and Fisher's exact tests were employed for statistical analyses of data.

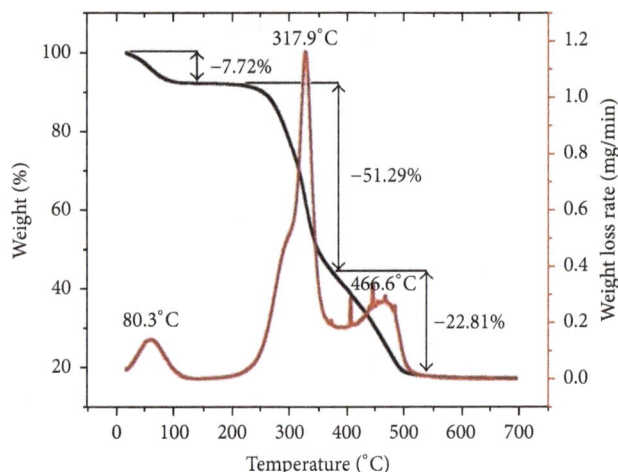

FIGURE 1: Raw rice husk's thermogravimetric analyses.

TABLE 1: Raw rice husk chemical composition.

Compound	Percentage (%)
Cellulose	30.6
Hemicellulose	18.3
Lignin	29.9
Ash	19.5
Moisture	4.9

3. Results and Discussion

3.1. Sample Preparation. Table 1 shows the RHw composition determined by chemical analysis. These results are in the same range of previously reported values [26] and are significant since superficial OH groups present in compounds such as cellulose, hemicellulose, and lignin could interact with the contaminants in the adsorption process [27]. In addition, chemical analysis shows that rice husk contains 19% ash, which, according to previous reports, is composed of 95% silica (SiO_2), an important aspect for its use as adsorbent material [28].

Thermogravimetric analyses (Figure 1) showed that the overall mass loss of rice husk can be divided into steps related to loss of moisture, hemicellulose, cellulose, and lignin. Thus, a mass loss of 7.72% at 100°C was confirmed due to the elimination of moisture retained in this material. Afterwards, a second step was obtained between 250 and 350°C, corresponding to the highest mass loss (51.29%). This step is related to hemicellulose and cellulose decomposition, as well as loss of the remaining adsorbed water [29]. Lignin decomposition occurs in the 360 to 520°C range. Finally, no significant mass loss was observed at higher temperatures indicating the presence of oxides.

From TGA results, the calcination temperatures for thermal treatment on rice husk were chosen: 300°C near to the maximum mass loss peak and 500°C after the loss of compounds that could influence the material adsorption capacity. Additionally, the 700°C value was selected, because no more mass losses are presented and only rice husk mineral

TABLE 2: Features of all prepared samples.

Material	Rice husk features
RHw	Without any treatment
RHc	With chemical treatment
RHc300	With chemical and thermal treatment to 300°C
RHc500	With chemical and thermal treatment to 500°C
RHc700	With chemical and thermal treatment to 700°C

residue remains. After this thermal treatment, calcined samples at 300, 500, and 700°C were obtained (namely, RHc300, RHc500, and RHc700, resp.).

In summary, after chemical and thermal treatments on rice husks, five materials were prepared and physiochemically characterized. Onto these materials, Cd (II) adsorption capacity was evaluated. Table 2 shows the five materials prepared for Cd (II) adsorption.

3.2. Samples Characterization. The specific surface area of rice husk samples was determined by nitrogen adsorption isotherm at 77 K using an Autosorb-iQ 07165, Quantachrome Instrument. Values of specific surface area of rice husk samples were calculated by Brunauer-Emmett-Teller (BET) method and are consigned in the last column of Table 4. There, an important decrease in the specific surface area value of rice husk chemically treated ($4 \, m^2/g$) in relation to the sample without treatment ($13 \, m^2/g$) can be seen. This change can be attributed to the effects of NaOH on the material either by partial degradation of some components of the rice husk, generating merging of smaller pores (micropores) into large pores (mesopores) [30], or because of the fixing of NaOH on the surface of the material which may cause blockage in the pores.

With the thermal treatment, RHc300 sample showed the greatest specific surface area of all studied materials, $14 \, m^2/g$, probably due to changes in pore volume. However, the higher calcination temperature, the smaller specific surface area, as can be seen for samples RHc500 ($7 \, m^2/g$) and RHc700 ($6 \, m^2/g$). In this regard, Della et al. [31] reported that higher increases in rice husk calcining temperature cause an agglomeration effect, diminishing porosity. But also our own previous research results (without publishing) indicate that, in rice husk treated with NaOH and calcined at 700°C, silicon dioxides were organized into a crystalline structure known as cristobalite. Due to the fact that the rice husk specific surface area increases when silica in the ash is amorphous [31, 32], the presence of cristobalite could explain the decrease in specific surface area in RHc700 sample.

Figure 2 shows SEM microphotographs of RHw, RHc, and RHc300 samples. Image of RHw sample (Figure 2(a)) shows a section of the rice husk where well-defined and organized cavities are observed. Although in image of RHc sample (Figure 2(b)) the structure is preserved, some cracking of cavities can be observed. This cracking, which generates deformation of the pores and loss of regularity, could be explained by the intracellular breakdown that NaOH causes [33]. Furthermore, some agglomerations observed on

(a)

(b)

(c)

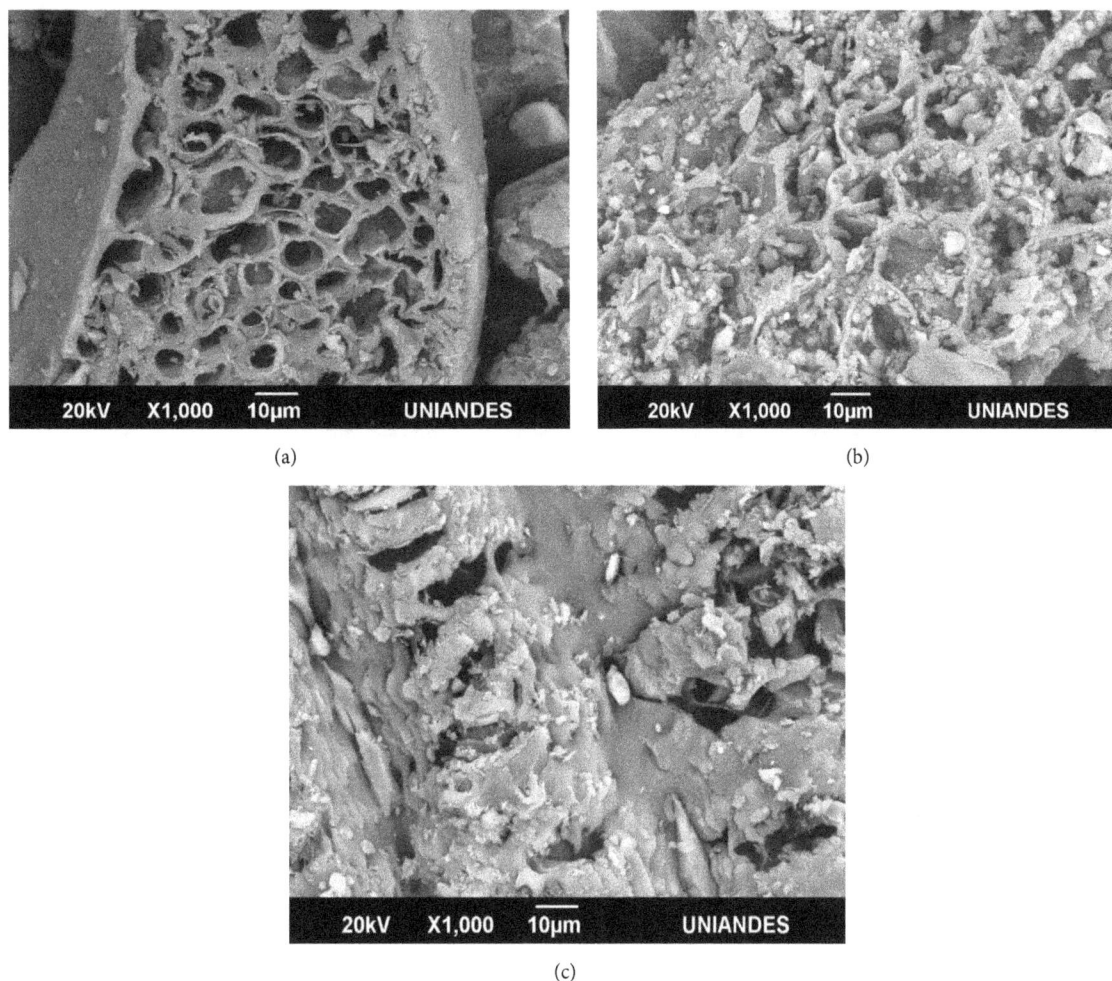

FIGURE 2: Microphotographs of RHw (a), RHc (b), and RHc300 (c) materials.

the material may possibly be due to NaOH fixed on the surface.

RHc300 sample microphotograph (Figure 2(c)) shows that the collapse of rice husk structure continues not only by intracellular breakdown caused by NaOH but also by the combustion that generates degradation in the material structure and loss of part of carbon phase.

Figure 3 shows SEM microphotographs of RHc500 and RHc700 samples. Both samples show a surface with less roughness than RHc300 (Figure 2(c)) sample, but RHc700 (Figure 3(b)) has the cleanest surface with well-defined pores on it.

Figures 4 and 5 show the IR spectrum for thermally untreated and treated samples, respectively. In all spectra, above $1,200 \, \mathrm{cm}^{-1}$ the organic part of rice husk can be distinguished, and below $1,200 \, \mathrm{cm}^{-1}$ the presence of vibrations corresponding to different forms of silicon appears [34].

Figure 4 shows the IR spectrum for RHw and RHc samples. RHw sample spectrum (dashed line) has some highlighted bands, as the characteristic of hydroxyl groups (-OH) at $3,418 \, \mathrm{cm}^{-1}$, the vibration of C-H at $2,918 \, \mathrm{cm}^{-1}$,

and the one at $1,650 \, \mathrm{cm}^{-1}$ corresponding to C=C bond vibration. These bands, also reported by other authors [35, 36], correspond to the main organic compounds of the rice husk: cellulose, hemicellulose, and lignin. Also, a significant band at $1,100 \, \mathrm{cm}^{-1}$, evidence of Si-O bonds present in cyclic siloxanes, is observed. The presence of cyclic siloxanes could be confirmed by peaks occurring between 470 and $800 \, \mathrm{cm}^{-1}$. Additionally, the peaks at 1,101 and $3,418 \, \mathrm{cm}^{-1}$, distinctive of -Si-O and -O-H, could be evidence of silanol groups [34, 36].

There are few differences in spectra shown in Figure 4; however, for RHc sample (solid line), widening of the band corresponding to the vibrations of the OH and a decrease in the intensity of the band associated with cyclic siloxanes at $1,100 \, \mathrm{cm}^{-1}$ can be seen. These changes are only due to the chemical treatment.

On the other hand, there are significant differences between thermally treated and untreated rice husks. Figure 5 shows the spectrum of all thermally treated samples, RHc300, RHc500, and RHc700.

It can be seen that as the heat treatment temperature increases, the characteristic bands of main rice husk organic

(a)   (b)

FIGURE 3: Microphotographs of RHc500 (a) and RHc700 (b) materials.

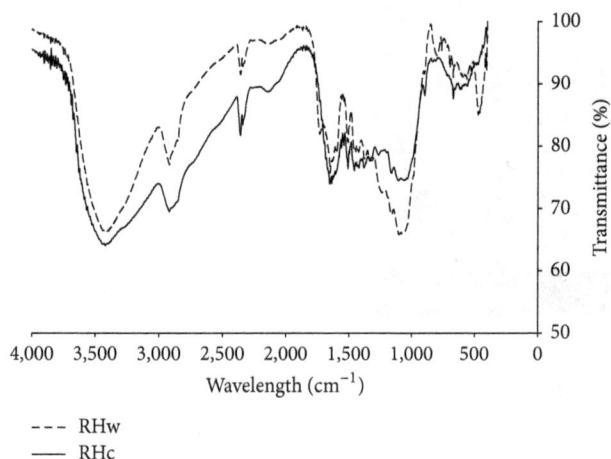

--- RHw
— RHc

FIGURE 4: IR spectrum of RHw (dashed line) and RHc (solid line) samples.

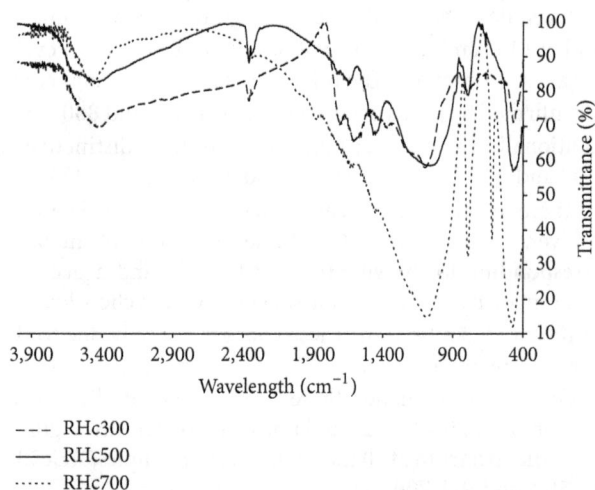

--- RHc300
— RHc500
······ RHc700

FIGURE 5: IR spectrum of RHc300 (dashed line), RHc500 (solid line), and RHc700 (dotted line) samples.

TABLE 3: C/O and C/Si ratios obtained by EDX analysis for all adsorbent materials.

Material	Ratio	
	C/O	C/Si
RHw	1.8	4.5
RHc	1.5	18.9
RHc300	1.1	1.2
RHc500	0.5	0.6
RHc700	0.3	0.4

compounds are decreasing. On the contrary, the band corresponding to the -Si-O vibration ($1,100 \, cm^{-1}$) is gaining intensity, probably due to the formation of silica-rich ashes.

The C/O and C/Si relationships for each sample were obtained by energy-dispersive X-ray spectroscopy (EDX) analysis and are contained in Table 3. According to the results, the RHw material has the highest C/O ratio, because the rice husk is composed mostly of carbonaceous rings of lignin, cellulose, and hemicellulose. The C/O relationship for chemically treated husk rice (RHc) is lower than that in the raw material (RHw); this agrees with IR results that showed an increase in OH groups for RHc sample. Besides, the C/Si ratio for the RHc material increases relative to the one for raw material (see Table 3), indicating a decrease in silicon (as reported in the IR analysis). The Si decrease may be due to effect of NaOH washing on the material, since it has been reported that NaOH is capable of reacting with the silicon present in the husk forming sodium silicates, which are soluble and could be removed with water washes carried out after chemical treatment of the husk [25, 37].

Table 3 also shows that for heat treated materials a progressive decrease in C/O and C/Si ratios is observed. This is due to the combustion process that generates progressive loss in carbon phase, while the silicon remains constant (relative to the RHc sample).

TABLE 4: Specific surface area, adsorption capacity, and percentage of Cd (II) adsorption onto adsorbent materials.

Material	Adsorption Capacity (mg/g)	Adsorption Percentage (%)	Specific surface area BET (m²/g)
RHw	7.44	31	13
RHc	17.45	75	4
RHc300	28.27	95	14
RHc500	2.03	5	7
RHc700	2.03	8	6

3.3. Adsorption Experiments. Experimental data of Cd (II) adsorption on all studied materials were fitted to the Langmuir and Freundlich isotherms models. According to the parameters obtained in the setting of these models, the RHw and RHc materials adapt to Freundlich isotherm model, while the remaining materials, RHc300, RHc500, and RHc700, fit to Langmuir isotherm model.

Results of Cd (II) adsorption analysis for all samples are shown in Table 4. Although the major factors influencing adsorption are the textural characteristics of adsorbents, the results obtained and given in Table 4 do not show a direct relationship between the adsorption capacity and specific surface area of the samples. This could indicate that adsorbate-adsorbent interaction occurs mainly thanks to the active sites and functional groups present in each material. These results agree with previously reported studies [38, 39].

From data in Table 4, it can be seen that adsorption capacity for RHc sample was increased by more than 2 times the value of RHw sample. This indicates that chemical treatment with NaOH effectively generates changes in the original structure of rice husk, probably causing surface groups that act as active sites and promote the affinity between Cd (II) and the sample. In this regard, it has been reported that modification of lignocellulosic residues by NaOH treatment produces changes in the fibers structure of biomass, which improve its accessibility properties, in addition to possibly adding active groups on the surface of material or leaving exposed surface groups because degradation caused on structure [24].

The RHc300 sample showed the highest Cd (II) adsorption capacity. This is probably due to chemical characteristics: its structure still has a carbon-oxygen phase from the organic composition of raw rice husk; it has phase silicon groups such as SiOH available on the surface, resulting from the thermal treatment; and it possibly has active OH- groups incorporated by treating with NaOH. The presence of all these active groups on the surface of CAq300 favors interaction between sample and Cd (II), generating greater efficiency in this material compared to others [40–43].

The similarities in the physicochemical characteristics of RHc500 and RHc700 samples could explain the coincidence in their Cd (II) adsorption capacities.

Finally, from adsorption analysis, RHc300 was determined as the best sample in adsorbing Cd˙ (II); therefore this material was selected to determine the effect of pH and kinetics of equilibrium on Cd (II) adsorption.

3.4. pH and Kinetics of Adsorption Studies for RHc300 Sample. The effect of solution pH on Cd (II) adsorption for RHc300 sample (selected as the best adsorbent) was studied. The

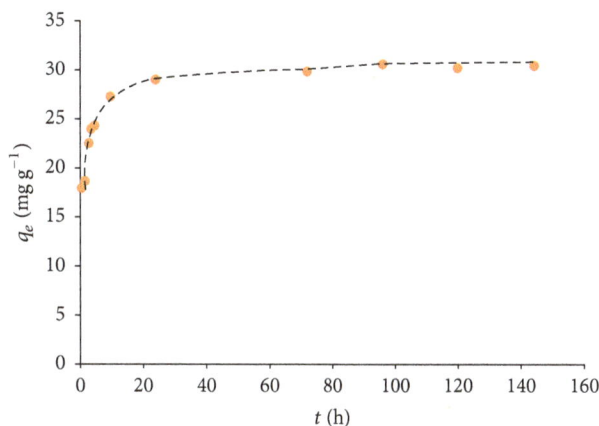

FIGURE 6: Equilibrium time for Cd (II) adsorption onto RHc300 sample at pH = 6. Dashed line only shows data trend.

results showed that at pH 5 and 6 the adsorption percentages are similar (98.83 and 98.89%, resp.), whereas the adsorption at pH 4 was smaller, 98.36%. Statistically, it was determined that the means of adsorption are different.

Figure 6 shows the Cd (II) adsorption capacity (q_e) on RHc300 sample as a function of time. From the graph, it was determined that equilibrium time for adsorption is 72 hours. After this time, changes in the adsorption capacity are less than 1 mg/g.

From experimental data of equilibrium time for Cd (II) adsorption onto RHc300 sample, the highest correlation coefficient ($R^2 = 0.999$) was obtained for the pseudo-second-order kinetic model. This implies that chemisorption is the rate-limiting step in the adsorption process [44]. As chemisorption is characterized by formation of monolayer, this result is consistent with our assumption of Langmuir isotherm model, which states that the monolayer adsorption occurs on specific sites of the homogeneous surface. Additionally, these results are also consistent with those obtained by Kumar et al. [45] who found that adsorption of Cd (II) on rice husk ash fits the pseudo-second-order kinetic model.

4. Conclusions

The chemical (HCl and NaOH) and thermal treatments (300°C for 3 h) on rice husk increase its Cd (II) removal efficiency almost four times. Although chemical and thermal treatments produced changes in the specific surface area of studied samples, the adsorption capacity of samples shows no direct relationship with their specific surface area. Thus, it could be considered that, in this system, the binding adsorbate-adsorbent is mainly due to the presence of active groups on the surface of adsorbent materials.

The maximum Cd (II) adsorption capacity was 28.27 mg/g at an optimum pH 6.0 for RHc300 sample. On the other hand, the kinetic studies revealed that adsorption process followed the pseudo-second-order kinetic model. This implies that chemisorption is the rate-limiting step in the adsorption process.

Although other adsorbents have shown the same or even greater adsorption efficiencies, as Duolite GT-73 [46], rice husk has high availability and low cost. This indicates that the RHc300 material, which requires less energy investment than rice husk ash, could be an excellent alternative for removing Cd (II) from aqueous media for pH values ranging from 4 to 6.

Conflicts of Interest

The authors declare that there are no conflicts of interest regarding the publication of this paper.

Acknowledgments

The authors acknowledge University of Tolima for the financial support (Project code 860113). The authors also thank Group on Porous Solids and Calorimetry, Department of Chemistry, Faculty of Science, University of the Andes, for experimental support.

References

[1] J. H. Duffus, ""Heavy metals"—a meaningless term?" *Pure and Applied Chemistry*, vol. 75, no. 9, article 1357, 2003.

[2] Z. Rengel, *Heavy Metals as Essential Nutrients*, Springer, Berlin, Heidelberg, Germany, 2004.

[3] U. Forstner and G. T. W. Wittmann, *Metal Pollution in the Aquactic Environment*, Springer, 2nd edition, 1979.

[4] O. Barbier, G. Jacquillet, M. Tauc, M. Cougnon, and P. Poujeol, "Effect of heavy metals on, and handling by, the kidney," *Nephron - Physiology*, vol. 99, no. 4, pp. 105–110, 2005.

[5] G. Kazantzis, "Cadmium, osteoporosis and calcium metabolism," *BioMetals*, vol. 17, no. 5, pp. 493–498, 2004.

[6] J. Y. Barbee Jr. and T. S. Prince, "Acute respiratory distress syndrome in a welder exposed to metal fumes," *Southern Medical Journal*, vol. 92, no. 5, pp. 510–512, 1999.

[7] J. Godt, F. Scheidig, C. Grosse-Siestrup et al., "The toxicity of cadmium and resulting hazards for human health," *Journal of Occupational Medicine and Toxicology*, vol. 1, article 22, 2006.

[8] A. Bożęcka, M. Orlof-Naturalna, and S. Sanak-Rydlewska, "Removal of lead, cadmium and copper ions from aqueous solutions by using ion exchange resin C 160," *Gospodarka Surowcami Mineralnymi*, vol. 32, no. 4, pp. 129–140, 2016.

[9] S. He, B. Ruan, Y. Zheng, X. Zhou, and X. Xu, "Immobilization of chlorine dioxide modified cells for uranium absorption," *Journal of Environmental Radioactivity*, vol. 137, pp. 46–51, 2014.

[10] W.-Q. Zhuang, J. P. Fitts, C. M. Ajo-Franklin, S. Maes, L. Alvarez-Cohen, and T. Hennebel, "Recovery of critical metals using biometallurgy," *Current Opinion in Biotechnology*, vol. 33, pp. 327–335, 2015.

[11] H. Wang and Z. J. Ren, "Bioelectrochemical metal recovery from wastewater: A review," *Water Research*, vol. 66, pp. 219–232, 2014.

[12] V. K. Gupta and I. Ali, "Utilisation of bagasse fly ash (a sugar industry waste) for the removal of copper and zinc from wastewater," *Separation and Purification Technology*, vol. 18, no. 2, pp. 131–140, 2000.

[13] S. Basha, Z. V. P. Murthy, and B. Jha, "Sorption of Hg(II) onto carica papaya: experimental studies and design of batch sorber," *Chemical Engineering Journal*, vol. 147, no. 2-3, pp. 226–234, 2009.

[14] A. Saeed, M. Iqbal, and W. H. Höll, "Kinetics, equilibrium and mechanism of Cd2+ removal from aqueous solution by mungbean husk," *Journal of Hazardous Materials*, vol. 168, no. 2-3, pp. 1467–1475, 2009.

[15] J. Anwar, U. Shafique, Waheed-uz-Zaman, M. Salman, A. Dar, and S. Anwar, "Removal of Pb(II) and Cd(II) from water by adsorption on peels of banana," *Bioresource Technology*, vol. 101, no. 6, pp. 1752–1755, 2010.

[16] N. Azouaou, Z. Sadaoui, A. Djaafri, and H. Mokaddem, "Adsorption of cadmium from aqueous solution onto untreated coffee grounds: equilibrium, kinetics and thermodynamics," *Journal of Hazardous Materials*, vol. 184, no. 1-3, pp. 126–134, 2010.

[17] V. Lugo-Lugo, C. Barrera-Díaz, F. Ureña-Núñez, B. Bilyeu, and I. Linares-Hernández, "Biosorption of Cr(III) and Fe(III) in single and binary systems onto pretreated orange peel," *Journal of Environmental Management*, vol. 112, pp. 120–127, 2012.

[18] C. R. T. Tarley and M. A. Z. Arruda, "Biosorption of heavy metals using rice milling by-products. Characterisation and application for removal of metals from aqueous effluents," *Chemosphere*, vol. 54, no. 7, pp. 987–995, 2004.

[19] X. Liu, X. Chen, L. Yang, H. Chen, Y. Tian, and Z. Wang, "A review on recent advances in the comprehensive application of rice husk ash," *Research on Chemical Intermediates*, vol. 42, no. 2, pp. 893–913, 2016.

[20] Renu, M. Agarwal, and K. Singh, "Heavy metal removal from wastewater using various adsorbents: a review," *Journal of Water Reuse and Desalination*, 2016, article jwrd2016104.

[21] U. Kumar and M. Bandyopadhyay, "Sorption of cadmium from aqueous solution using pretreated rice husk," *Bioresource Technology*, vol. 97, no. 1, pp. 104–109, 2006.

[22] A. Chakraverty, P. Mishra, and H. D. Banerjee, "Investigation of combustion of raw and acid-leached rice husk for production of pure amorphous white silica," *Journal of Materials Science*, vol. 23, no. 1, pp. 21–24, 1988.

[23] S. Chandrasekhar, K. G. Satyanarayana, P. N. Pramada, P. Raghavan, and T. N. Gupta, "Review processing, properties and applications of reactive silica from rice husk—an overview," *Journal of Materials Science*, vol. 38, no. 15, pp. 3159–3168, 2003.

[24] B. S. Ndazi, S. Karlsson, J. V. Tesha, and C. W. Nyahumwa, "Chemical and physical modifications of rice husks for use as composite panels," *Composites Part A: Applied Science and Manufacturing*, vol. 38, no. 3, pp. 925–935, 2007.

[25] B. S. Ndazi, C. Nyahumwa, and J. Tesha, "Chemical and thermal stability of rice husks against alkali treatment," *BioResources*, vol. 3, no. 4, pp. 1267–1277, 2008.

[26] P. T. Williams and N. Nugranad, "Comparison of products from the pyrolysis and catalytic pyrolysis of rice husks," *Energy*, vol. 25, no. 6, pp. 493–513, 2000.

[27] D. Sud, G. Mahajan, and M. P. Kaur, "Agricultural waste material as potential adsorbent for sequestering heavy metal ions from aqueous solutions—a review," *Bioresource Technology*, vol. 99, no. 14, pp. 6017–6027, 2008.

[28] N. Soltani, A. Bahrami, M. I. Pech-Canul, and L. A. González, "Review on the physicochemical treatments of rice husk for production of advanced materials," *Chemical Engineering Journal*, vol. 264, pp. 899–935, 2015.

[29] H. Yang, R. Yan, H. Chen, D. H. Lee, and C. Zheng, "Characteristics of hemicellulose, cellulose and lignin pyrolysis," *Fuel*, vol. 86, no. 12-13, pp. 1781–1788, 2007.

[30] T. N. Ang, G. C. Ngoh, and A. S. M. Chua, "Comparative study of various pretreatment reagents on rice husk and structural changes assessment of the optimized pretreated rice husk," *Bioresource Technology*, vol. 135, pp. 116–119, 2013.

[31] V. P. Della, I. Kühn, and D. Hotza, "Rice husk ash as an alternate source for active silica production," *Materials Letters*, vol. 57, no. 4, pp. 818–821, 2002.

[32] R.-S. Bie, X.-F. Song, Q.-Q. Liu, X.-Y. Ji, and P. Chen, "Studies on effects of burning conditions and rice husk ash (RHA) blending amount on the mechanical behavior of cement," *Cement and Concrete Composites*, vol. 55, pp. 162–168, 2015.

[33] S. Chakraborty, S. Chowdhury, and P. Das Saha, "Adsorption of Crystal Violet from aqueous solution onto NaOH-modified rice husk," *Carbohydrate Polymers*, vol. 86, no. 4, pp. 1533–1541, 2011.

[34] M. Rozainee, S. P. Ngo, A. A. Salema, and K. G. Tan, "Fluidized bed combustion of rice husk to produce amorphous siliceous ash," *Energy for Sustainable Development*, vol. 12, no. 1, pp. 33–42, 2008.

[35] M. Akhtar, S. Iqbal, A. Kausar, M. I. Bhanger, and M. A. Shaheen, "An economically viable method for the removal of selected divalent metal ions from aqueous solutions using activated rice husk," *Colloids and Surfaces B: Biointerfaces*, vol. 75, no. 1, pp. 149–155, 2010.

[36] X. Luo, Z. Deng, X. Lin, and C. Zhang, "Fixed-bed column study for Cu2+ removal from solution using expanding rice husk," *Journal of Hazardous Materials*, vol. 187, no. 1-3, pp. 182–189, 2011.

[37] T.-H. Liou and S.-J. Wu, "Characteristics of microporous/ mesoporous carbons prepared from rice husk under base- and acid-treated conditions," *Journal of Hazardous Materials*, vol. 171, no. 1-3, pp. 693–703, 2009.

[38] A. Erto, L. Giraldo, A. Lancia, and J. C. Moreno-Piraján, "A comparison between a low-cost sorbent and an activated carbon for the adsorption of heavy metals from water," *Water, Air, and Soil Pollution*, vol. 224, no. 4, article no. 1531, pp. 1–10, 2013.

[39] N. A. Medellin-Castillo, R. Leyva-Ramos, R. Ocampo-Perez et al., "Adsorption of fluoride from water solution on bone char," *Industrial and Engineering Chemistry Research*, vol. 46, no. 26, pp. 9205–9212, 2007.

[40] T. H. Baig, A. E. Garcia, K. J. Tiemann, and J. L. Gardea-Torresdey, "Adsorption of heavy metal ions by the biomass of Solanum claeagnifolium (Silverleaf night-shade)," in *Proceedings of the Hazardous Waste Research*, pp. 131–142, 1999.

[41] K. K. Krishnani, X. Meng, C. Christodoulatos, and V. M. Boddu, "Biosorption mechanism of nine different heavy metals onto biomatrix from rice husk," *Journal of Hazardous Materials*, vol. 153, no. 3, pp. 1222–1234, 2008.

[42] V. C. Srivastava, I. D. Mall, and I. M. Mishra, "Characterization of mesoporous rice husk ash (RHA) and adsorption kinetics of metal ions from aqueous solution onto RHA," *Journal of Hazardous Materials*, vol. 134, no. 1-3, pp. 257–267, 2006.

[43] V. C. Srivastava, I. D. Mall, and I. M. Mishra, "Adsorption thermodynamics and isosteric heat of adsorption of toxic metal ions onto bagasse fly ash (BFA) and rice husk ash (RHA)," *Chemical Engineering Journal*, vol. 132, no. 1-3, pp. 267–278, 2007.

[44] Y. S. Ho, "Review of second-order models for adsorption systems," *Journal of Hazardous Materials*, vol. 136, no. 3, pp. 681–689, 2006.

[45] P. S. Kumar, K. Ramakrishnan, S. D. Kirupha, and S. Sivanesan, "Thermodynamic and kinetic studies of cadmium adsorption from aqueous solution onto rice husk," *Brazilian Journal of Chemical Engineering*, vol. 27, no. 2, pp. 347–355, 2010.

[46] T. Vaughan, C. W. Seo, and W. E. Marshall, "Removal of selected metal ions from aqueous solution using modified corncobs," *Bioresource Technology*, vol. 78, no. 2, pp. 133–139, 2001.

4

An Assessment of Physicochemical Parameters of Selected Industrial Effluents in Nepal

Abhinay Man Shrestha,[1] Sanjila Neupane,[1] and Gunjan Bisht[2]

[1]Department of Environment Science and Engineering, School of Science, Kathmandu University, Dhulikhel, Nepal
[2]Department of Chemical Science and Engineering, School of Engineering, Kathmandu University, Dhulikhel, Nepal

Correspondence should be addressed to Gunjan Bisht; gunjanbisht31@gmail.com

Academic Editor: Wenshan Guo

It is a well-known fact that the effluents released from the industries and environmental degradation go hand in hand. With the ongoing global industrialization this problem has been further aggravated. As such, Nepal is no exception. Hundreds of industries are being registered in the country annually which inevitably brings the issues regarding environmental pollution. This study has been conducted with samples of wastewater from 5 different industrial sites in 4 districts of Nepal, namely, Makwanpur, Sunsari, Morang, and Kathmandu, among which two were Waste Water Treatment Plants which treated the combined effluents collected from various sources. The other three sites were from wires and cables industry, paint manufacturing industry, and plastic cutting industry. The physicochemical parameters analysed were pH, temperature, conductivity, turbidity, and Cu, Cr, SO_4^{2-}, and PO_4^{3-} levels. Possible onsite measurements were recorded using portable, handheld devices whereas other parameters were assessed in the laboratory. The observed parameter levels in the collected samples were compared against the available Nepal national standards for industrial effluents and in the absence of standards for industrial effluents, with other relevant standard levels. Most of the parameters analysed were within the permissible limits with the exception of pH and Cr levels in some sites.

1. Introduction

Effluents released from the industries into water bodies can cause serious environmental degradation and deterioration; especially with the ongoing increase in the industrialization around the world, water pollution too is becoming rampant. It is reported that around 70% of the industrial wastes in the developing nations are disposed of untreated into waters thereby contaminating the existing water supplies [1]. In the state of Colorado, US, around 23,000 abandoned mines have been accredited for the pollution of as much as 2,300 km of streams [2].

Many studies conducted have highlighted this problem. Kaplay and Patode (2004) [3] observed that groundwater from the region of New Nanded, Maharashtra, India, demonstrated higher content of total dissolved solids (TDS), Cl, Total Hydrocarbons (TH), Ca, Mg, and SO_4, with the source of pollution being reported as effluents from the nearby industries. Similar results have been observed in the

groundwater pollution due to the discharge of industrial effluents in Venkatapuram area, Andhra Pradesh, India [4]. It has been reported that, over the past few decades, the surface and groundwater of China have been considerably polluted owing to the discharge of industrial and municipal wastewaters, household wastes, and agricultural activities [5]. As per a national water quality survey in the country, water from about only 58.8% of major lakes, 64.2% of the river sections, and only 23.2% of the groundwater wells could meet the standard quality criteria of drinking water sources [6].

Meanwhile, Nepal has been witnessing some rapid urbanization and industrialization over the past few years. It is reported that Nepal is one of the fastest urbanizing countries in the world and for the period 2014–2050, it is expected that the country will continue to remain among the top ten fastest urbanizing countries in the world, with an expected urbanization rate (annual) of 1.9 percent [7]. As per the latest industrial statistics of the country, new industry registrations are increasing by the year with manufacturing industries

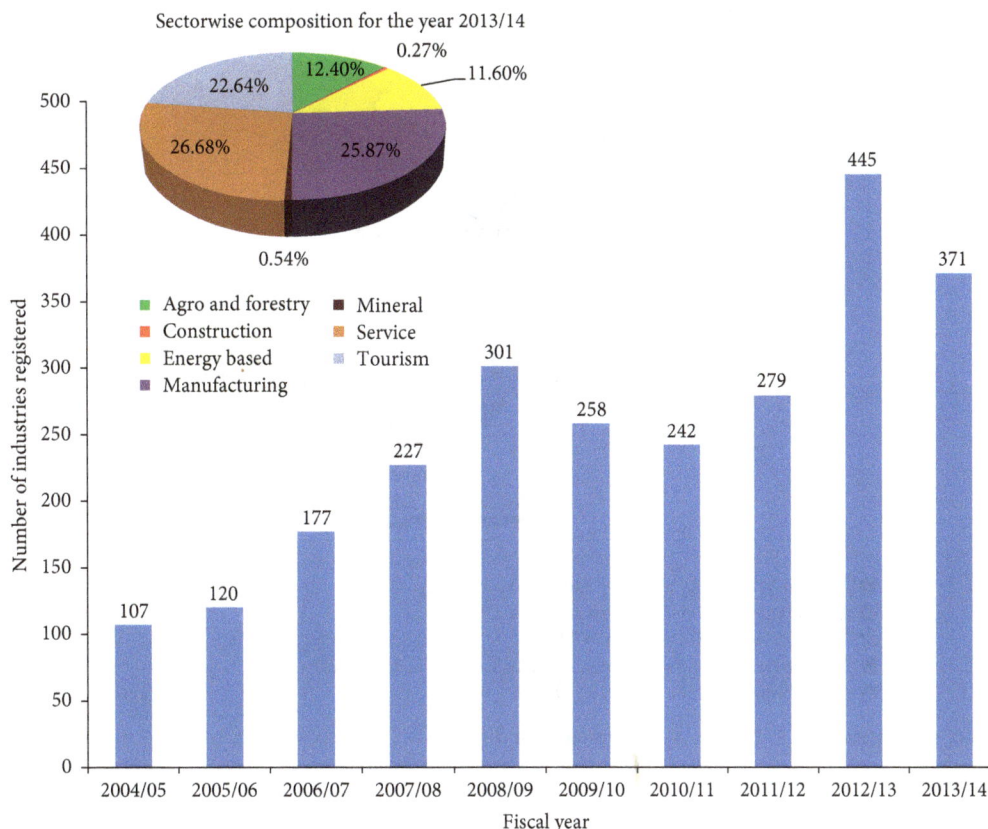

FIGURE 1: The increasing trend of industry registration in Nepal from the fiscal year 2004/05 to 2013/14, along with the sectorwise composition for the fiscal year 2013/14 (source: Department of Industry, GoN [8]).

TABLE 1: The categorization of the four industrial zones of Nepal.

S number	Zone	Designation/specification
(1)	Zone 1	Municipality areas within the Kathmandu Valley
(2)	Zone 2	Municipality areas outside the Kathmandu Valley
(3)	Zone 3	Municipality areas of Biratnagar, Dharan, Pokhara, Birgunj, Janakpur, Siddharthanagar, Butwal, and Nepalganj
(4)	Zone 4	Municipality areas other than those mentioned in Zone 3

Source: Ministry of Industry and Ministry of Commerce and Supplies, GoN 2009 [9].

constituting a major portion (25.87%) of the categorical composition of industries as highlighted by the data shown in Figure 1 [8].

Furthermore, majority of the industries and industrial regions are scattered in the hot Terai flatlands and Hilly mountainous belts of the country. Urban centres of the country's capitol Kathmandu Valley and other major cities as Biratnagar, Janakpur, Bhairahawa, Rajbiraj, Butwal, Nepalganj, and Hetauda harbour most of the industries. At present, the Government of Nepal (GoN) has established 11 Industrial estates, namely, Balaju, Patan, Bhaktapur, Hetauda, Pokhara, Dharan, Butwal, Nepalganj, Birendranagar, Rajbiraj, and Dhankuta [9], along with 4 industrial zones as specified in Table 1.

However, genuine voices are being raised regarding the fact that there is no effective or adequate study or strategy that

addresses the positive and negative consequences that occur with such rapid urbanization and the water pollution caused by such industries and industrial activities in the country. Sah et al. (2002) [10] have studied the detrimental effects on fish and other aquatic organisms in the Narayani River due to the receiving effluents from several industries including pulp and paper mills and they found that considerable concentrations of heavy metals such as Zn, Cu, and Cr were found in the tissues of the fishes inhabiting in the river. Furthermore, according to an article published by a leading national daily of the country, The Himalayan Times (dated November 14, 2007) [11], a report made public by the Narayani Environment Conservation and Local Resource Management Centre highlighted that chemical wastes disposed into the Narayani River by the nearby located brewery and paper mills industries were leading to fluctuations in the temperatures of the river water,

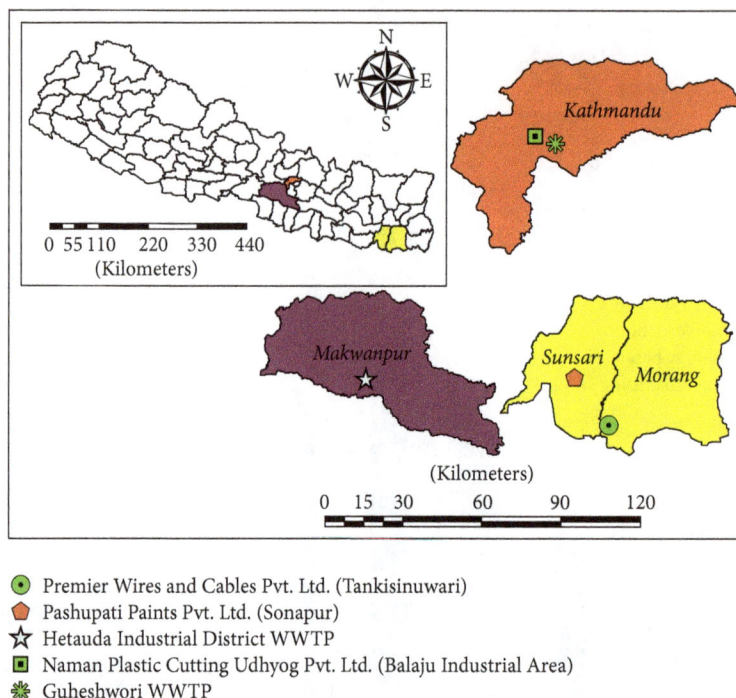

- Premier Wires and Cables Pvt. Ltd. (Tankisinuwari)
- Pashupati Paints Pvt. Ltd. (Sonapur)
- Hetauda Industrial District WWTP
- Naman Plastic Cutting Udhyog Pvt. Ltd. (Balaju Industrial Area)
- Guheshwori WWTP

FIGURE 2: Five sites of this study (Sites 1, 2, 3, 4, and 5) present in 4 districts of Nepal (inset: map of Nepal showing the four districts as highlighted).

rise in the acidity, and reduction in the oxygen levels along with a corresponding rise in the nitrogen and phosphorus levels.

Neupane in 2003 [12] has reported that the Lumbini Sugar Mills released their effluents directly into the nearby Somnath stream. They analysed physicochemical parameters like temperature, pH, acidity, alkalinity, hardness, free CO_2, total solid matter, DO, N, P, K, BOD, and COD in this contaminated stream water which was used for irrigation purposes and concluded that the crop productivity in those lands was relatively lower than on land irrigated with uncontaminated water.

Likewise, an assessment of the geochemical parameters of the water quality of Karra river, located near the Hetauda Industrial District, has highlighted that the pollution of the river was severe near the industrial stretches due to the direct discharge of the treated and untreated effluents from the industries. Certain samples have showcased a high pH of >9.0 which is reported to be detrimental and even lethal for the growth of amphibian larvae in the river. High conductivity levels (up to 2023 μS/cm), high bicarbonate levels (up to 327 mg/L), and silica levels (14.8 mg/L) were also reported to be found in the river [13].

Not many reports or studies have been published regarding the quality of industrial effluents of Nepal. The aforementioned studies are some of the few published literatures available regarding it. Hence, in the light of these issues, this study has been conducted in selected sites, among the different industrial regions of the country, to provide a further reach into assessing the various industries of Nepal and their effluent quality parameters.

2. Materials and Methods

2.1. Site Selection.
Samples of wastewater were collected from 5 different sites in 4 districts of Nepal, namely, Makwanpur, Sunsari, Morang, and Kathmandu. Site 1 was an industrial Waste Water Treatment Plant (WWTP) located in Makwanpur which treated the combined effluents and wastewater from an industrial estate in the district. Site 2 was a paint manufacturing and processing plant located in Sunsari. Site 3 included wires and cables manufacturing industry located in Morang. The other two sites, Site 4 and Site 5, were from within the Kathmandu Valley. Site 4 was another WWTP that treated industrial along with municipal wastes within the valley whereas Site 5 included a plastic cutting industry located in the Balaju Industrial Estate. These sites have been showcased in Figure 2.

2.2. Sample Collection.
A total of 6 samples were collected from Sites 1, 4, and 5 each, whereas 5 samples were collected from Sites 2 and 3 each. Sampling bottles and the preservative to be used (HNO_3) were provided by Environment and Public Health Organisation (ENPHO), Baneshwor, Nepal. For the purpose of onsite measurements, portable pH meter (Hanna S358236) and portable conductivity meter (HM Digital Aquapro Water Tester) were provided by the Department of Environmental Science and Engineering (DESE), Kathmandu University.

2.3. Parameter Analysis.
The samples collected from the 5 sites were analysed for parameters differentiated into 3 categories, namely, physical parameters which included pH,

temperature, conductivity, and turbidity; heavy metal analysis of Cu, Cr, Ni, and As; and analysis of anions SO_4^{2-} and PO_4^{3-}.

2.3.1. Determination of Physical Parameters. As stated earlier, pH, conductivity, and temperature readings were taken directly in the field using the aforementioned portable, handheld devices. The turbidity analyses of the samples were conducted using turbidity meter Hanna HI98703-01 in the DESE Lab itself.

2.3.2. Determination of Heavy Metals. For the detection of heavy metals, the samples were sent to ENPHO Lab, to be detected using Atomic Absorption Spectroscopy (APHA, AWWA, WEF 2012, 3111 B) [14]. Detection for four heavy metals, namely, Cu, Cr, Ni, and As, was carried out where Ni and As levels in the samples were found to be below the detection limit of the instrument. Hence further analysis was done with the Cu and Cr concentrations in the samples.

2.3.3. Determination of Anions. The analysis for anions was conducted in the Quantitative Analysis Lab at the Department of Biotechnology, Kathmandu University. For the detection of sulphate, conditional reagent was prepared by mixing appropriate amounts of glycerol with a solution containing conc. HCl, distilled water, ethanol (95%), and sodium chloride. A standard solution of sulphate was prepared by dissolving anhydrous sodium sulphate in distilled water. A calibration curve was plotted taking various concentrations of the standard sulphate solution with specified amount of conditional reagent and analytical grade barium chloride crystals. The same procedure was conducted using the samples for the detection of sulphate in them. The absorbance was measured at 420 nm using GENESYS 10S UV-Vis Spectrophotometer [15].

For the detection of phosphates, conditional reagents were prepared by dissolving appropriate amounts of ammonium molybdate in distilled water; sodium sulphide; and diluted (0.25 N) sulphuric acid. Stock solution of phosphate was prepared by dissolving disodium hydrogen phosphate in distilled water (whose working solution was prepared by diluting this solution to an appropriate concentration with water). A calibration curve was plotted taking various concentrations of the standard phosphate solution with specified amount of conditional reagents and the procedure was also followed for the detection of phosphates in the samples by measuring the absorbance at 715 nm using GENESYS 10S UV-Vis Spectrophotometer [16].

3. Results and Discussions

3.1. Assessment of the Physical Parameters of the Industrial Effluents. The pH and turbidity levels observed in the samples of 5 sites of this study have been shown in Figure 3. As per the Environmental Conservation Regulations (ECR), 1997, formulated by the Government of Nepal, the tolerable

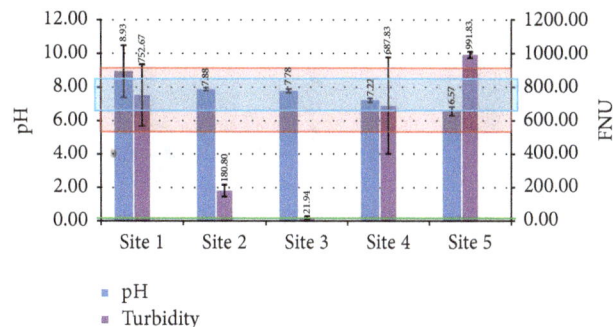

FIGURE 3: Graphs showing the pH and turbidity levels with their respective error bars on standard deviation, observed in the samples of 5 sites of this study, along with the ECR tolerable pH range (5.5–9) as shown by red rectangle, NDWQS pH range (6.5–8.5) as shown by blue rectangle, and NDWQS turbidity standard (10 FNU) as shown by green line.

limits for pH range from 5.5 to 9 for the industrial effluents [17], whereas the tolerable range as per the National Drinking Water Quality Standards (NDWQS) for rural surface water supply system is 6.5–8.5 [18]. All the samples were found to have mean pH values within these limits except for samples of Site 1. The pH levels were found to be maximum in the samples from Site 1 with a mean value of 8.93, slightly higher than the NDWQS limit of 8.5, and the lowest in the samples from plastic cutting industry (Site 5) having a mean of 6.57. Regarding the turbidity levels, no standards for industrial levels as such were found; hence they were compared against the National Drinking Water Quality Standards which had set a turbidity standard of 10 FNU for rural surface water supply systems [18]. Turbidity levels were found highest in the samples from Sites 5, 4, and 1 with mean values of 991.83 FNU, 752.67 FNU, and 687.83 FNU, respectively. Turbidity levels were found to be the lowest in Site 3 (cables and wire industries) having a mean of 21.94 FNU, as was evident from direct observation of the clear samples in the field itself. Turbidity levels are positively related to the Total Suspended Solids (TSS) [19]; hence it can be established that the samples from Sites 5, 4, and 1 had higher content of TSS in them. Among the samples, Sites 1 and 4 showed higher deviations from the mean. This could be attributed to the fact that these sites were WWTPs and therefore the influent that they received was not uniform in composition, given the large number of discharge sources and the change in the discharge composition by the minute, which would enter these WWTPs.

Figure 4 shows the conductivity and temperature levels observed in the samples of five sites of this study. As in the case of turbidity, there were no specific standards for the conductivity levels in industrial effluents. Hence, they were compared against the National Drinking Water Quality Standards which had set a conductivity standard of 1500 μS/cm for rural surface water supply systems [18]. The mean conductivity levels of all samples were below the tolerance limits. The conductivity levels too were found to be the highest in samples from Sites 1, 4, and 5 with

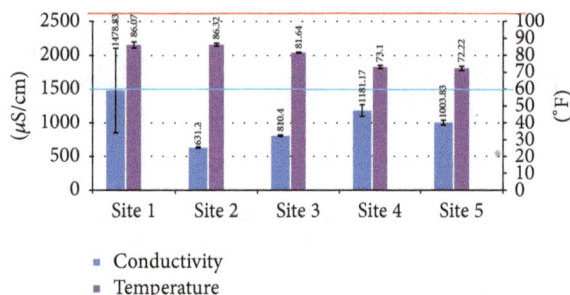

FIGURE 4: Graphs showing the conductivity and temperature levels with their respective error bars on standard deviation, observed in the samples of 5 sites of this study, along with the ECR tolerable limit for temperature (104°F) as shown by red line and NDWQS tolerable limit for conductivity (1500 μS/cm) as shown by blue line.

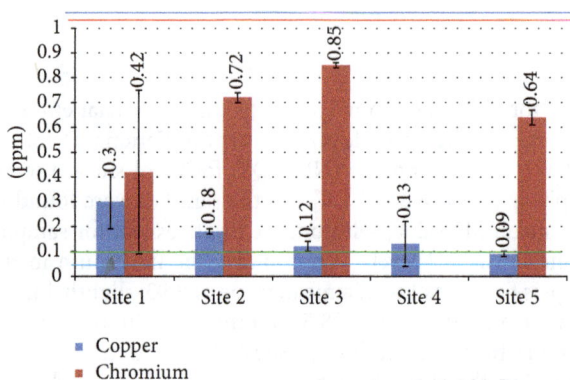

FIGURE 5: Graphs showing the copper and chromium levels with their respective error bars on standard deviation, observed in the samples of 5 sites of this study, along with the ECR tolerable limit for copper (3 ppm) as shown by red line, the ECR tolerable limit for chromium (0.1 ppm) as shown by green line, and the NDWQS tolerable limit for chromium (0.05 ppm) as shown by blue line.

means of 1478.83 μS/cm, 1181.17 μS/cm, and 1003.83 μS/cm, respectively. The conductivity levels are strongly related to the TDS levels [20]; hence it can be established that the samples from these sites contained high TDS contents given that these were effluents from a mixture of industries (as in the case of Site 1), wastewater sources (as in the case of Site 4), and plastic waste products (as in the case of Site 5). Furthermore, here too, significant deviations from the mean values can be seen in the samples from Sites 1 and 4, the reason being as mentioned previously. The temperature levels in all the samples were found to be significantly under the tolerance limit for industrial effluent standard (40°C, i.e., 104°F) as specified by Environment Conservation Regulations (ECR) 1997 [17].

3.2. Assessment of the Heavy Metals in the Industrial Effluents. Copper is found in chalcophile deposits along with other metals like Pd, Cd, Zn, and so on, which are used in various industries like alloys, ceramics, and pesticides [21], and is also released from plumbing systems [22]. Figure 5 shows the copper and chromium levels observed in the samples. All the

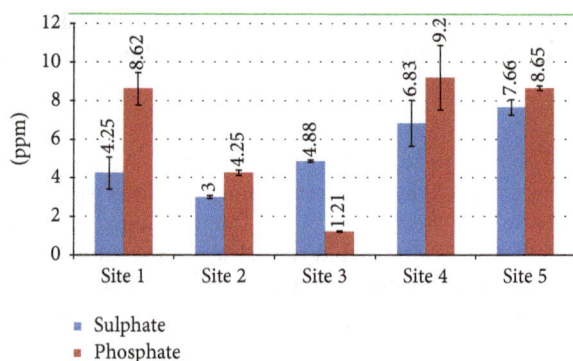

FIGURE 6: Graphs showing the sulphate and phosphate levels with their respective error bars on standard deviation, observed in the samples of 5 sites of this study, along with Nepal Gazette tolerance limit for sulphur (500 ppm) as shown by green line.

samples showed copper levels below the standard specified by Environment Conservation Regulations (ECR) 1997 [17]. The highest concentration (mean 0.3 ppm) was found in samples from Site 1 which can be attributed to the presence of the above-mentioned industries within the industrial estate. Copper can also be released into the environment from other industries as well such as chloralkali, electroplating, paints and dyes, pesticides, and fertilizers along with the disposal of municipal and industrial wastes [21]. Hence significant levels are found in samples from paint industry at Site 2 (mean 0.18 ppm) and the WWTP at Site 4 (mean 0.13 ppm). Regarding the presence of chromium in the samples, Sites 1, 2, 3, and 4 showed levels higher than the standard specified by the ECR 1997 [17], as well as the NDWQS standard for rural surface water supply [18]. From the samples, the highest concentration was found in wires and cables industry at Site 3 (mean 0.85 ppm) and paint industry at Site 2 (mean 0.72 ppm) which is supported by the fact that industrial activities like electroplating, metal cleaning and processing, and leather tanning are considered the major sectors that release chromium into the environment [23] and, furthermore, chromium is also used in paints and dying processes [24].

3.3. Assessment of Sulphate and Phosphate Levels in the Industrial Effluents. Figure 6 shows the sulphate and phosphate levels in the collected samples. In the absence of other industrial effluent standard for sulphate, the levels were compared against the Nepal Gazette tolerance limit for discharge of sulphate into public sewerage which is stated as 500 ppm [25]. All the samples showed sulphate levels very well below this standard. The highest level of sulphate (mean 7.66 ppm) was found in the samples from plastic cutting industry (Site 5). This could be due to the abundance of ink cartridge wastes in the plastic waste samples as in such ink cartridges sulphate can be used in the form of barium sulphate as a filler material [26]. Furthermore, it is also reported that barium sulphate can also be used as additive in various plastic products [27]. Phosphate content is seen as the highest one in samples from Site 4 (mean 9.2 ppm), Site 5 (8.65 ppm), and

Site 1 (8.62 ppm). This can be attributed to the increased use of phosphates in the form of fertilizers for enhanced plant growth [28] and hence its high content in the municipal wastewater (Site 4) or its use in meat processing and poultry productions industries [29] and hence its high content in the industrial effluents (Site 1). Since there were no Nepal standards for phosphate levels in drinking or effluent waters, for comparison purposes the Indian Standard Institution levels for organic phosphates were used, which stated a limit of 100 μg/L (i.e., 0.1 ppm) in drinking water [30]. The samples of this study have shown levels exceeding this standard. Furthermore, a total phosphate-phosphorus value exceeding 0.1 mg/L (i.e., 0.1 ppm) is associated with eutrophication and natural water degradation [31]; hence such problems could arise if these effluents were to be released into the water bodies.

3.4. Correlational Analysis for the Various Physicochemical Parameters of the 5 Sites. The correlations for the various physicochemical parameters of the 5 sites have been tabulated in the form of a correlation matrix shown in Table 2. From the correlation matrix it can be seen that not much correlation was found among the various physicochemical parameters analysed. However, it can be observed that the temperature and conductivity readings were found to be positively correlated (as marked with asterisks) in all the samples. This could be explained as in an aqueous solution; with increase in the temperature, the components of the solution could undergo dissociation into their respective free ions which could then conduct electricity. Hence a positive correlation is obtained. Barron and Ashton in 2007 have observed similar results whilst studying various aqueous solutions [32]. Mandal in 2014 has also analysed the effect of temperature on the electrical conductivity of industrial effluents and has found similar trend, the reason being attributed to the decrease in viscosity associated with an increase in temperature [33].

There is also a positive correlation observed with the electrical conductivity and the presence of ions (copper, chromium, sulphate, and phosphate) in all the samples (as highlighted by green). The relation of electrical conductivity and the presence of ions have been highlighted in many studies [34, 35] which further supports the observed trend.

4. Conclusion

Samples of industrial effluents from 5 industrial regions in the country were collected and their physicochemical analysis was done. The levels were compared against the standards for industrial effluents as specified by the Environmental Conservation Regulations (ECR 1997) [17] and in the absence of the standards for the concerned parameter, with other relevant standards.

The mean pH levels of all the samples were found to be well within the tolerable range as specified by the Environmental Conservation Regulations (ECR) 1997 [17] and met the National Drinking Water Quality Standards (NDWQS) for rural surface water supply systems [18], with the exception of Site 1 which had mean pH values slightly

higher than the drinking water quality standards as per the NDWQS. However, the turbidity levels were found to be immensely higher than the NDWQS [18]. Turbidity levels were found highest in samples from Sites 5, 4, and 1 and since turbidity levels are positively related to the Total Suspended Solids (TSS) [19], it can be established that the samples from Sites 5, 4, and 1 had higher content of TSS in them.

The mean conductivity levels in all samples were below the NDWQS for rural surface water supply systems [18]. The conductivity levels too were found to be the highest in samples from Sites 1, 4, and 5 and since conductivity levels are strongly related to the TDS (total dissolved solids) levels [20], it can be established that the samples from these sites contained high TDS content. The temperature levels in all the samples were found to be significantly under the tolerance limit for industrial effluent standard (40°C, i.e., 104°F) as specified by Environment Conservation Regulations (ECR) 1997 [17].

Copper levels were highest in the samples from Site 1 due to the presence of copper releasing industries within the industrial estate. However, overall, copper levels were found to be well below the ECR standard (Bhandari 2014) [17] in all 5 samples. Chromium levels were found to be exceeding the ECR standard [17] in 4 sites, namely, 1, 2, 3, and 4, as well as the NDWQS for rural water supply [18]. Chromium was found to be the highest in samples from the wires and cables industry at Site 3 and the paint manufacturing industry at Site 2. Significant chromium concentrations were not observed in samples from Site 4. Furthermore, sulphate levels in all 5 sites' samples were well under the Nepal Gazette tolerance limit for discharge of sulphate into public sewerage which is stated as 500 ppm [25]. Sulphate levels were highest in samples from the plastic cutting industry at Site 5, the reason being attributed to the use of barium sulphate as additive in various plastic products [26, 27]. Phosphate contents in all the samples were found very well exceeding the ISI limits for organic phosphate in drinking water [30] with the highest one in samples from Sites 4, 5, and 1, the reason being attributed to the increased use of phosphates in the form of fertilizers [28] and in meat processing and poultry production industries [29].

It should be noted here that the levels of parameters showed higher deviations from the mean in samples from Sites 1 and 4 which could be attributed to the fact that these sites were WWTPs and therefore the influent that they received was not uniform in composition; hence the deviations observed in the parameter levels are as such. Also, the samples showcased positive correlation between temperature and conductivity levels as was supported by other studies as well [32, 33] along with positive correlation between the electrical conductivity and the presence of ions as was supported by other studies [34, 35].

This study was conducted to give a brief outlook into the various industries and their effluents in Nepal. The variability of physicochemical parameters analysed was limited by the budgetary concerns and time constraints. However, this study can provide a base and a direction towards a broad and perhaps a more detailed inspection of the various industries

TABLE 2: The correlation matrix for the physicochemical parameters of the 5 sites.

	Temp	Conductivity	Turbidity	pH	Copper	Chromium	Sulphate	Phosphate
Temp	1							
Conductivity	Site 1 = 0.25** Site 2 = 0.76** Site 3 = 0.59** Site 4 = 0.55** Site 5 = 0.12**	1						
Turbidity	Site 1 = −0.69 Site 2 = −0.26 Site 3 = −0.73 Site 4 = 0.53 Site 5 = 0.01	Site 1 = −0.1 Site 2 = −0.28 Site 3 = −0.96 Site 4 = 0.05 Site 5 = 0.69	1					
pH	Site 1 = 0.49 Site 2 = −0.33 Site 3 = 0.31 Site 4 = 0.77 Site 5 = −0.47	Site 1 = −0.49 Site 2 = −0.52 Site 3 = 0.67 Site 4 = 0.15 Site 5 = −0.88	Site 1 = −0.72 Site 2 = 0.81 Site 3 = −0.52 Site 4 = 0.91 Site 5 = −0.63	1				
Copper	Site 1 = 0.49 Site 2 = −0.12 Site 3 = 0.26 Site 4 = 0.19 Site 5 = 0.45	Site 1 = 0.83** Site 2 = 0.15** Site 3 = 0.04** Site 4 = 0.75** Site 5 = 0.9**	Site 1 = −0.05 Site 2 = 0.84 Site 3 = 0.02 Site 4 = −0.53 Site 5 = 0.72	Site 1 = −0.29 Site 2 = 0.6 Site 3 = 0.03 Site 4 = −0.39 Site 5 = −0.88	1			
Chromium	Site 1 = 0.32 Site 2 = −0.01 Site 3 = 0.45 Site 4 = — Site 5 = 0.91	Site 1 = 0.87** Site 2 = 0.61** Site 3 = 0.89** Site 4 = — Site 5 = 0.21**	Site 1 = −0.44 Site 2 = −0.3 Site 3 = −0.87 Site 4 = — Site 5 = −0.11	Site 1 = −0.06 Site 2 = −0.37 Site 3 = 0.32 Site 4 = — Site 5 = −0.57	Site 1 = 0.69 Site 2 = 0.24 Site 3 = 0.19 Site 4 = — Site 5 = 0.39	1		
Sulphate	Site 1 = −0.28 Site 2 = −0.19 Site 3 = 0.48 Site 4 = 0.68 Site 5 = 0.27	Site 1 = 0.85** Site 2 = 0.33** Site 3 = 0.07** Site 4 = 0.92** Site 5 = 0.17**	Site 1 = 0.19 Site 2 = 0.13 Site 3 = −0.22 Site 4 = 0.26 Site 5 = −0.51	Site 1 = −0.73 Site 2 = −0.41 Site 3 = 0.23 Site 4 = 0.33 Site 5 = −0.01	Site 1 = 0.51 Site 2 = 0.36 Site 3 = −0.56 Site 4 = 0.52 Site 5 = 0.12	Site 1 = 0.69 Site 2 = 0.46 Site 3 = −0.27 Site 4 = — Site 5 = 0.59	1	
Phosphate	Site 1 = 0.85 Site 2 = −0.36 Site 3 = −0.1 Site 4 = 0.63 Site 5 = 0.46	Site 1 = 0.43** Site 2 = 0.49** Site 3 = 0.23** Site 4 = 0.93** Site 5 = 0.75**	Site 1 = −0.35 Site 2 = −0.46 Site 3 = −0.27 Site 4 = −0.1 Site 5 = 0.76	Site 1 = 0.15 Site 2 = −0.88 Site 3 = −0.47 Site 4 = 0.08 Site 5 = −0.84	Site 1 = 0.79 Site 2 = −0.37 Site 3 = 0.08 Site 4 = 0.8 Site 5 = 0.87	Site 1 = 0.36 Site 2 = 0.13 Site 3 = 0.63 Site 4 = — Site 5 = 0.45	Site 1 = −0.05 Site 2 = 0.69 Site 3 = −0.6 Site 4 = 0.9 Site 5 = −0.06	1

** Values showing positive correlation.

and its effluents in the country in the near future if the limitations as aforementioned were to be curtailed.

Conflicts of Interest

The authors declare that they have no conflicts of interest.

Acknowledgments

The authors acknowledge University Grant Commission of Nepal for funding this research.

References

[1] UN-Water, "World Water Day brochure," 2009, http://www.unwater.org.

[2] Pacific Institute, "World Water Quality Facts and Statistics," 2010, http://www.pacinst.org.

[3] R. D. Kaplay and H. S. Patode, "Groundwater pollution due to industrial effluent at Tuppa, New Nanded, Maharashtra, India," *Environmental Geology*, vol. 46, no. 6-7, pp. 871–882, 2004.

[4] N. Subba Rao, V. V. S. Gurunadha Rao, and C. P. Gupta, "Groundwater pollution due to discharge of industrial effluents in Venkatapuram area, Visakhapatnam, Andhra Pradesh, India," *Environmental Geology*, vol. 33, no. 4, pp. 289–294, 1998.

[5] J. Liu and J. Diamond, "China's environment in a globalizing world," *Nature*, vol. 435, no. 7046, pp. 1179–1186, 2005.

[6] Y. Hu and H. Cheng, "Water pollution during China's industrial transition," *Environmental Development*, vol. 8, no. 1, pp. 57–73, 2013.

[7] S. Bakrania, "Urbanisation and Urban Growth in Nepal," *GSDRC Helpdesk Research Report*, p. 1294, 2015.

[8] Department of Industry, Ministry of Industry, GoN. Industrial Statistics Fiscal Year 2070/071(2013/2014).

[9] Ministry of Industry (MoI) and Ministry of Commerce and Supplies (MoCS), *NEPAL: Foreign Investment Opportunities*, Enhancing Nepal's Trade-Related Capacity (ENTRec) project UNDP-Nepal, Nepal, South Asia, 2009.

[10] S. K. Sah, P. Acharya, and V. A. Lance, "Effect of Industrial Pollution on Fish in the Narayani River, Central Nepal," *Nepal Journal of Science and Technology*, vol. 4, pp. 5–14, 2002.

[11] The Himalayan Times, (dated 14th November, 2007), Industrial chemical wastes polluting Narayani river - by Chintamani Poudel, https://thehimalayantimes.com/nepal/industrial-chemical-wastes-polluting-narayani-river/.

[12] R. P. Neupane, *Effect of Industrial Effluents on Agricultural Crops and Soil [Dissertation, thesis]*, A Dissertation submitted for the partial fulfilment of the requirements for the Master of Science in Botany, Kathmandu, Nepal, South Asia, September 2003.

[13] S. P. Kayastha, "Geochemical Parameters of Water Quality of Karra River, Hetauda Industrial Area, Central Nepal," *J. of Inst. of Sci. and Tech*, vol. 20, no. 2, pp. 31–36, 2015.

[14] APHA, AWWA, and WEF, *Standard Methods for Examination of Water and Wastewater*, American Public Health Association, Washington, DC, USA, 22nd edition, 2012.

[15] S. A. B. Mussa, H. S. Elferjani, F. A. Haroun, and F. F. Abdelnabi, "Determination of Available Nitrate, Phosphate and Sulfate in Soil Samples," *Int. J. of PharmTech Research, IJPRIF*, vol. 1, no. 3, pp. 598–604, 2009.

[16] M. M. S. Yogendra Kumar, M. S. Abdul Galil, M. S. Suresha, M. A. Sathish, and G. Nagendrappa, "A simple spectrophotometric determination of phosphate in sugarcane juices, water and detergent samples," *E-Journal of Chemistry*, vol. 4, no. 4, pp. 467–473, 2007.

[17] G. Bhandari, "A Review of Urban Water Reuse - Limits, Benefits and Risks in Nepal," *International Journal of Geology, Agriculture and Environmental Sciences*, vol. 2, no. 1, 2014.

[18] Ministry of Physical Planning and Works, "National Drinking Water Quality Standards, 2005 Implementation Directives for National Drinking Water Quality Standards," GoN, 2005, http://mowss.gov.np/assets/uploads/files/NDWQS_2005_Nepal.pdf.

[19] L. H. X. Daphne, H. D. Utomo, and L. Z. H. Kenneth, "Correlation between Turbidity and Total Suspended Solids in Singapore Rivers," *J. of Wat. Sust*, vol. 1, no. 3, pp. 55–64, 2011.

[20] N. S. Ali, K. Mo, and M. Kim, "A case study on the relationship between conductivity and dissolved solids to evaluate the potential for reuse of reclaimed industrial wastewater," *KSCE Journal of Civil Engineering*, vol. 16, no. 5, pp. 708–713, 2012.

[21] A. K. Shrivastava, "A Review on Copper Pollution and its Removal from Water Bodies by Pollution Control Technologies," *Indian Journal of Environmental Protection*, vol. 29, no. 6, pp. 552–560, 2009.

[22] G. R. Calle, I. T. Vargas, M. A. Alsina, P. A. Pasten, and G. E. Pizarro, "Enhanced copper release from pipes by alternating stagnation and flow events," *Environmental Science & Technology*, vol. 41, no. 21, pp. 7430–7436, 2007.

[23] A. A. Belay, "Impacts of chromium from tannery effluent and evaluation of alternative treatment options," *Journal of Environmental Protection*, vol. 1, no. 1, pp. 53–58, 2010.

[24] J. Guertin, C. Avakian, and J. Jacobs, "Overview of Chromium (VI) in the Environment: Background and History," in *Chromium (VI) Handbook*, pp. 1–21, CRC Press, Florida, Fla, USA, 2004.

[25] Central Bureau of Statistics (2014), *Environmental Statistics of Nepal 2013*, CNN Printing Press, Kathmandu, Nepal, South Asia, 2014, GoN.

[26] M. Ellila, *The Use of Barium Sulphate in Printing Inks as Filler Material, Bachelor of Engineering Thesis*, Helsinki Metropolia University of Applied Sciences, Helsinki, Finland, 2011.

[27] COWI, "Hazardous substances in plastic materials," Prepared in cooperation with Danish Technological Institute, 2013, http://www.miljodirektoratet.no/old/klif/publikasjoner/3017/ta3017.pdf.

[28] S. H. Chien, L. I. Prochnow, and H. Cantarella, "Recent developments of fertilizer production and use to improve nutrient efficiency and minimize environmental impacts," *Advances in Agronomy*, vol. 102, pp. 267–322, 2009.

[29] N. H. Bach Son Long, R. Gál, and F. Buňka, "Use of phosphates in meat products," *African Journal of Biotechnology*, vol. 10, no. 86, pp. 19874–19882, 2011.

[30] M. Kumar and A. Puri, "A review of permissible limits of drinking water," *Indian Journal of Occupational and Environmental Medicine*, vol. 16, no. 1, pp. 40–44, 2012.

[31] Water Research Center, "Phosphates in the Environment," http://www.water-research.net/index.php/phosphates.

[32] J. J. Barron and C. Ashton, "The effect of temperature on conductivity measurement," *TSP*, vol. 7, no. 3, 2007, https://www.reagecon.com/pdf/technicalpapers/Effect_of_Temperature_TSP-07_Issue3.pdf.

[33] H. K. Mandal, "Effect of temperature on electrical conductivity in industrial effluents," *Recent Research in Sci and Tech*, vol. 6, no. 1, pp. 171–175, 2014.

[34] H. Golnabi, M. R. Matloob, M. Bahar, and M. Sharifian, "Investigation of electrical conductivity of different water liquids and electrolyte solutions," *Iranian Physical Journal*, vol. 3, no. 2, pp. 24–28, 2009.

[35] C. P. Helito, M. K. Demange, and M. B. Bonadio, "Electrical Conductivity of Potassium salt- dimethylsulfoxide-water systems at different temperatures," *Proc. of the Yerevan State Univ. Chemistry and Biology*, vol. 1, pp. 3–6, 2013.

5

Density Functional Theory Investigation into the B and Ga Doped Clean and Water Covered γ-Alumina Surfaces

Lihong Cheng,[1] **Tianliang Xu,**[2] **Wenkui Li,**[1] **Zhiqin Chen,**[1] **Jianping Ai,**[1] **Zehua Zhou,**[1] **and Jianwen Liu**[3]

[1] Key Laboratory of Surface Engineering of Jiangxi Province, Jiangxi Science and Technology Normal University, Nanchang, Jiangxi 330013, China
[2] Zhengzhou Institute of Finance and Economics, Zhengzhou, China
[3] National Supercomputing Center in Shenzhen, Shenzhen 518055, China

Correspondence should be addressed to Lihong Cheng; chenglihong001@126.com and Jianwen Liu; liujw@nsccsz.gov.cn

Academic Editor: Tao Wang

The structures and energies of the B and Ga incorporated γ-alumina surface as well as the adsorption of water are investigated using dispersion corrected density functional theory. The results show that the substitution of surface Al atom by B atom is not so favored as Ga atom. The substitution reaction prefers to occur at the tricoordinated A(4) sites. However, the substitution reaction becomes less thermodynamically favored when more Al atoms are substituted by B and Ga atoms on the surface. Moreover, the substitution of bulk Al atoms is not so favored as the Al atoms by B and Ga on the surface. The γ-alumina surface is found to have stronger adsorption ability for water than the B and Ga incorporated surface. The total adsorption energy increases as water coverage increases, while the stepwise adsorption energy decreases. The studies show the coverage of water at 7.5 H_2O/nm^2 (five H_2O molecules per unit cell) can fully cover the active sites and the further water molecule could only be physically adsorbed on the surface.

1. Introduction

The γ-alumina is an important material in chemistry and materials science due to its widespread applications in chemical industry [1–3], ceramics, and semiconductors [4–7]. In order to improve the performance of the material, some heteroatoms were usually chosen to be incorporated into the γ-alumina bulk and surfaces [8, 9]. For example, the Fe atom was usually used to improve the catalytic performance of γ-alumina [10]. Wan et al. investigated the Fe_2O_3/Al_2O_3 catalyst from coprecipitated and spray-dried method with Mösbauer spectroscopy [11] and found the reduction of Fe_2O_3 to FeO. The substitution of surface Al^{3+} by Fe^{3+} in alumina with mixed $(Al_{1-x}Fe_x)_2O_3$ surface formation is also confirmed by transmission Mösbauer spectroscopy. Integral low-energy electron Mösbauer spectroscopy and Fe K-edge X-ray absorption near-edge structure characterization observed the formation of iron nanoclusters from the transformation of γ-$(Al_{1-x}Fe_x)_2O_3$ to α-$(Al_{1-x}Fe_x)_2O_3$ and

the iron distribution on the surface layers and in the cores of grains [12–14]. This field also attracts the interests of theoretical researches. Feng et al. calculated the structures and energies of the Fe promoted γ-alumina surface [10] and found that the incorporation of Fe atom into the γ-alumina surface is possible, while it is thermodynamically not so favored for the Fe substitution for the bulk Al atoms. The adsorption of water on the γ-alumina surface is stronger than that on the Fe_2O_3 covered surface. In addition, the electronic structures also change after the substitution of the Al atoms by the Fe atoms on the surface. Except for the incorporation of Fe into the alumina surface, the B and Ga atoms, which are often used as the trivalent substitution ions for the zeolites [15, 16], are also used in experiments for the preparation of high performance catalysis and semiconductors [8, 9, 17–19]. Kibar et al. prepared nanostructured boron doped alumina catalyst support [20] and found that the morphology of the supports can be modified from cracked surface to nanosphere formation by the introduction of boron. Jansons et al. introduced Ga into the alumina crystal and prepared a new complex luminescence band at about 5 eV [19]. The Ga-related luminescence can be observed under the excitation of X-rays up to 600 K. In order to know the energies as well as the structures of the B and Ga incorporated γ-alumina surface at molecular level, the present work investigated the thermodynamic and structure properties of the B and Ga incorporated γ-alumina surface using dispersion corrected periodic density functional theory. Since γ-alumina is usually prepared and used in atmospheres containing water, the adsorption of water on the B and Ga incorporated γ-alumina is also investigated and compared with the water adsorption on pure γ-alumina surface.

2. Computational Details

The dispersion-corrected periodic density functional as implemented in the Vienna ab initio Simulation Package (VASP) was used for the calculations [21–23]. The DFT-D3 method of Grimme was used to take into account the dispersive interactions as previous work reported that the dispersion correction seriously influences the relative stability order and adsorption energies [24–28]. The exchange and correlation energies were calculated by the generalized gradient approximation (GGA) formulation with the PBE functional [21]. The Kohn-Sham one-electron states were extended in accordance with plane-wave basis sets with a kinetic energy of 400 eV. The projector augmented wave (PAW) method was applied to describe the electron-ion interactions [29–31].

The Brillouin zone was sampled with $5 \times 5 \times 5$ and $3 \times 3 \times 1$ k-points meshes generated by the Monkhorst-Pack algorithm, for the $p(1 \times 1 \times 1)$ γ-alumina cell and $p(1 \times 1)$ γ-alumina (110) surface slab, respectively. The convergence criteria were set to be 10^{-4} eV for the SCF energy, 10^{-3} eV and 0.03 eV/Å for the total energy and the atomic forces, respectively.

The γ-alumina surface was described using the Digne's model [32], and (110) surface was taken into account. As shown in Figure 1, γ-alumina surface was modeled using a $p(1 \times 1)$ supercell with an eight-layer slab, which contains sixteen Al_2O_3 units. A vacuum with thickness of 15 Å was employed to separate each slab from interactions. The top four layers and the adsorbates were fully relaxed, and the bottom four layers were fixed in their bulk position during structure optimization. In order to facilitate the discussions, the surface layer Al and O atoms are indexed with number. The coordination number of each Al atom was expressed by subscript. As has been described in many previous works [33, 34], the Al atoms in bulk γ-alumina are in tetrahedral and octahedron sites. After cleavage, the tetrahedral and octahedron Al atoms expose as the tricoordinated and tetra-coordinated Al, respectively, in the (110) surface. As shown in Figure 1, the $Al(4)_{3c}$ was in tetrahedral site in the bulk, and $Al(1)_{4c}$, $Al(2)_{4c}$, and $Al(3)_{4c}$ were in octahedral sites in the bulk. It could be observed from the top view that $Al(1)$ and $Al(2)$ atoms are in the same chemical environment.

For the substitution of surface Al^{3+} by X^{3+} (X = B and Ga) in reaction (1) the substitution energy is defined as $E_{sub} = nE[Al(OH)_3] + E[Al_{(4-n)}X_nO_6{}^*] - nE[X(OH)_3] - E[Al_4O_6{}^*]$,

where $E[X(OH)_3]$ and $E[Al_4O_6{}^*]$ present the energies of the gas phase $X(OH)_3$ and the isolated oxide surfaces ($*$ presents the 2–8 layers alumina substrate), respectively. The positive substitution energy means substitution reaction is not thermodynamically favored.

$$nX(OH)_3 + Al_4O_6{}^* \longrightarrow nAl(OH)_3 + Al_{(4-n)}X_nO_6{}^*. \quad (1)$$

The adsorption energy of nH_2O (n = 1–6) on the oxide surface was defined as $E_{ads} = E[(H_2O)_n \cdot X_4O_6{}^*] - nE(H_2O) - E(X_4O_6{}^*)$, where $E[(H_2O)_n \cdot X_4O_6{}^*]$, $E(H_2O)$, and $E(X_4O_6{}^*)$ are the total energies of the minima structures of $X_4O_6{}^*$ (X = B, Al, Ga) surface with adsorbed water, gas phase water molecule, and clean $X_4O_6{}^*$ surface, respectively. Following this definition, a more negative value indicates stronger interaction between adsorbed species and the surface.

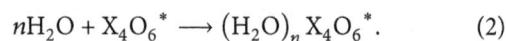

$$nH_2O + X_4O_6{}^* \longrightarrow (H_2O)_n X_4O_6{}^*. \quad (2)$$

3. Results and Discussion

3.1. Substitution of Al by B and Ga on the Surface. Table 1 shows the substitution energies for the substitution of Al by B and Ga on the surface. As indicated by the calculated substitution energies, the substitution of surface Al by B atoms is not thermodynamically favored, since the calculated substitution energies are positive. The tricoordinated Al(4) is the most favored site for one B incorporation with the substitution energy of 1.31 eV. The tetracoordinated Al(1) and Al(3) sites are slightly difficult with the substitution energies of 1.81 and 2.29 eV, respectively.

For two Al atoms substituted by B atoms on the surface, which corresponds to 50% surface, Al were replaced by B; the Al(1,2) are the most favored, with the substitution energy of 3.22 eV, versus Al(1,2,3) for 75% surface Al substitution by B, with the substitution energy of 5.32 eV. The substitution energies positively increase to 7.82 eV as all surface Al atoms were substituted by B atoms.

Figure 2 shows the structures for the B and Ga substituted γ-alumina surface. The corresponding bond distances of the surface layer atoms before and after the substitution are given in Table 1. It is found that the surface Al–O bond distances are in the range of 170–186 pm on the γ-alumina surface. After the substitution of Al by B atoms, the Al–O bond distance almost remains the same. It should be noted that the B–O bond distances (137–167 pm) are much shorter than those of the Al–O bonds, since the radius of B atom is much shorter than that of Al [35]. In addition, the B atom prefers to be tricoordinated, for example, for B(1,2,3,4) in Figure 2, bond c is elongated to 255 pm and became broken.

Since the substitution energies for the substitution of Al by Ga are much less than those for B substitution, the Ga could be more easily incorporated into to the alumina surface than B. Particularly for the Ga(4) structure, the substitution energy is −0.06 eV, which indicates the substitution reaction at Al(4) site by Ga is thermodynamically favored. Similar to that of B substitution, the substitution energy increases as more Al atoms were replaced by Ga atoms, for example, the substitution energies are 0.11, 0.49, and 0.90 eV, respectively,

TABLE 1: Substitution energies (E_{sub}) and the M–O bonds distances (pm) of the surface layer atoms before and after substitution for B and Ga in γ-alumina (110) surface.

Substituted sites	E_{sub} (eV)	M(1)–O, a–d[a]	M(2)–O, e–h[a]	M(3)–O, i–1[a]	M(4)–O, m–o[a]
γ-alumina		174, 185, 173, 182	185, 174, 173, 182	186, 186, 180, 170	180, 171, 171
B(1)	1.81	141, 225, 137, 145	183, 176, 172, 183	184, 187, 179, 169	181, 175, 171
B(3)	2.29	173, 185, 172, 182	185, 173, 172, 182	167, 167, 149, 138	181, 171, 171
B(4)	1.31	173, 183, 177, 185	183, 173, 177, 185	185, 185, 181, 171	149, 139, 139
B(1,2)	3.22	142, 232, 136, 145	232, 142, 136, 145	186, 186, 178, 169	188, 176, 176
B(2,3)	3.94	175, 182, 173, 184	227, 141, 137, 145	147, 218, 145, 136	181, 171, 176
B(2,4)	3.58	175, 191, 174, 183	166, 145, 146, 155	184, 185, 181, 170	147, 138, 142
B(3,4)	3.70	172, 182, 177, 185	182, 172, 177, 185	164, 164, 151, 139	150, 139, 139
B(1,2,3)	5.32	141, 235, 136, 145	235, 141, 136, 145	165, 165, 149, 138	188, 177, 177
B(1,2,4)	6.02	145, 188, 142, 151	188, 145, 142, 151	184, 184, 181, 170	146, 141, 141
B(2,3,4)	5.81	174, 191, 175, 184	167, 145, 147, 155	146, 250, 146, 136	147, 138, 142
B(1,2,3,4)	7.82	146, 151, 146, 161	255, 140, 138, 144	163, 162, 152, 138	145, 139, 142
Ga(1)	0.39	183, 200, 179, 191	184, 174, 172, 182	186, 185, 180, 170	179, 172, 171
Ga(3)	0.17	174, 185, 173, 182	185, 174, 173, 182	200, 200, 186, 175	180, 171, 171
Ga(4)	−0.06	174, 185, 173, 182	185, 174, 173, 182	186, 186, 180, 170	188, 179, 179
Ga(1,2)	0.82	182, 198, 179, 191	198, 182, 179, 191	186, 186, 180, 170	178, 172, 172
Ga(2,3)	0.57	174, 184, 173, 182	200, 183, 180, 191	199, 200, 186, 176	179, 171, 172
Ga(2,4)	0.31	174, 184, 173, 181	200, 183, 180, 191	185, 186, 180, 170	187, 179, 177
Ga(3,4)	0.11	174, 185, 173, 182	185, 174, 173, 182	200, 200, 186, 175	188, 179, 179
Ga(1,2,3)	1.00	183, 198, 179, 191	198, 183, 179, 191	199, 199, 186, 176	178, 172, 172
Ga(1,2,4)	0.72	183, 198, 180, 191	198, 183, 180, 191	186, 186, 179, 170	186, 180, 180
Ga(2,3,4)	0.49	174, 184, 173, 181	200, 183, 181, 191	199, 200, 186, 175	187, 179, 181
Ga(1,2,3,4)	0.90	183, 198, 180, 190	198, 183, 180, 190	199, 199, 186, 176	186, 180, 180

[a]Corresponding to Figure 1(c).

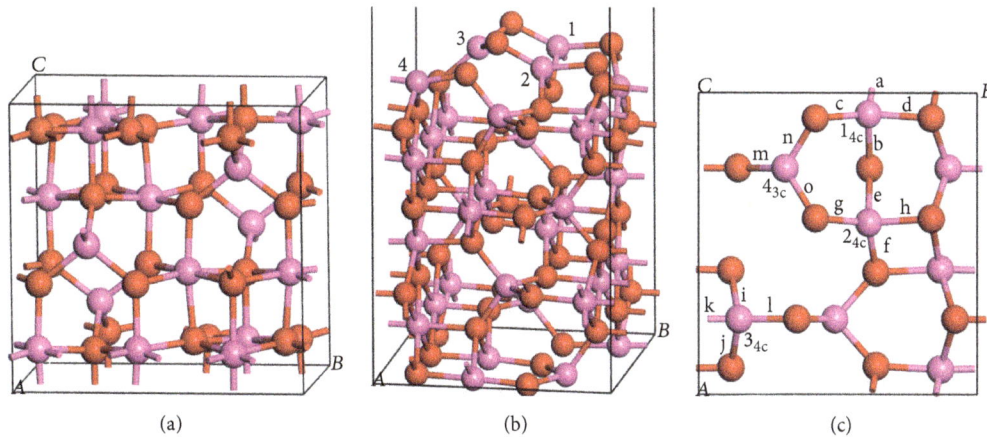

FIGURE 1: Side view of unit cell of γ-Al$_2$O$_3$ (a) and side (b) and top (c) views of the γ-Al$_2$O$_3$(110) surface (Al and O atoms are in rose and red, respectively. Coordination numbers of surface atoms are shown in subscript in (c)).

for Ga(3,4), Ga(1,2,4), and Ga(1,2,3,4). Since the Ga atom radius is larger than that of Al [35], the Ga–O bond distance is much longer than that of Al–O bond distance. In order to map out whether it is possible for B and Ga substitution for the bulk Al atoms of γ-alumina, we also calculated the substitution energies for the substitution of the sublayer hexa- and tetracoordinated Al atom by B and Ga atoms. The calculated substitution energies for B replacing the sublayer hexa- and tetracoordinated Al atoms are 4.34 and 2.89 eV, respectively, versus 0.81 and 0.57 eV for Ga, which are larger than the substitution energies for Al substitution on the surface. It indicates that the substitution reaction should favor happening on

the surface, rather than in the bulk. In addition, the substitution of tetrahedral Al sites is always more thermodynamically favored than the substitution of octahedral Al sites.

As reported in the previous work [10], the substitution of γ-alumina surface Al by Fe atoms is thermodynamically favored, as the substitution energy for the substitution of all surfaces Al by Fe atom is −0.87 eV. It indicates that the Fe should be more easily to be incorporated into the γ-alumina than B and Ga.

3.2. Adsorption of Water Molecules on the X$_2$O$_3$* Surfaces (X = B, Al, Ga). Figure 3 shows the structures and adsorption

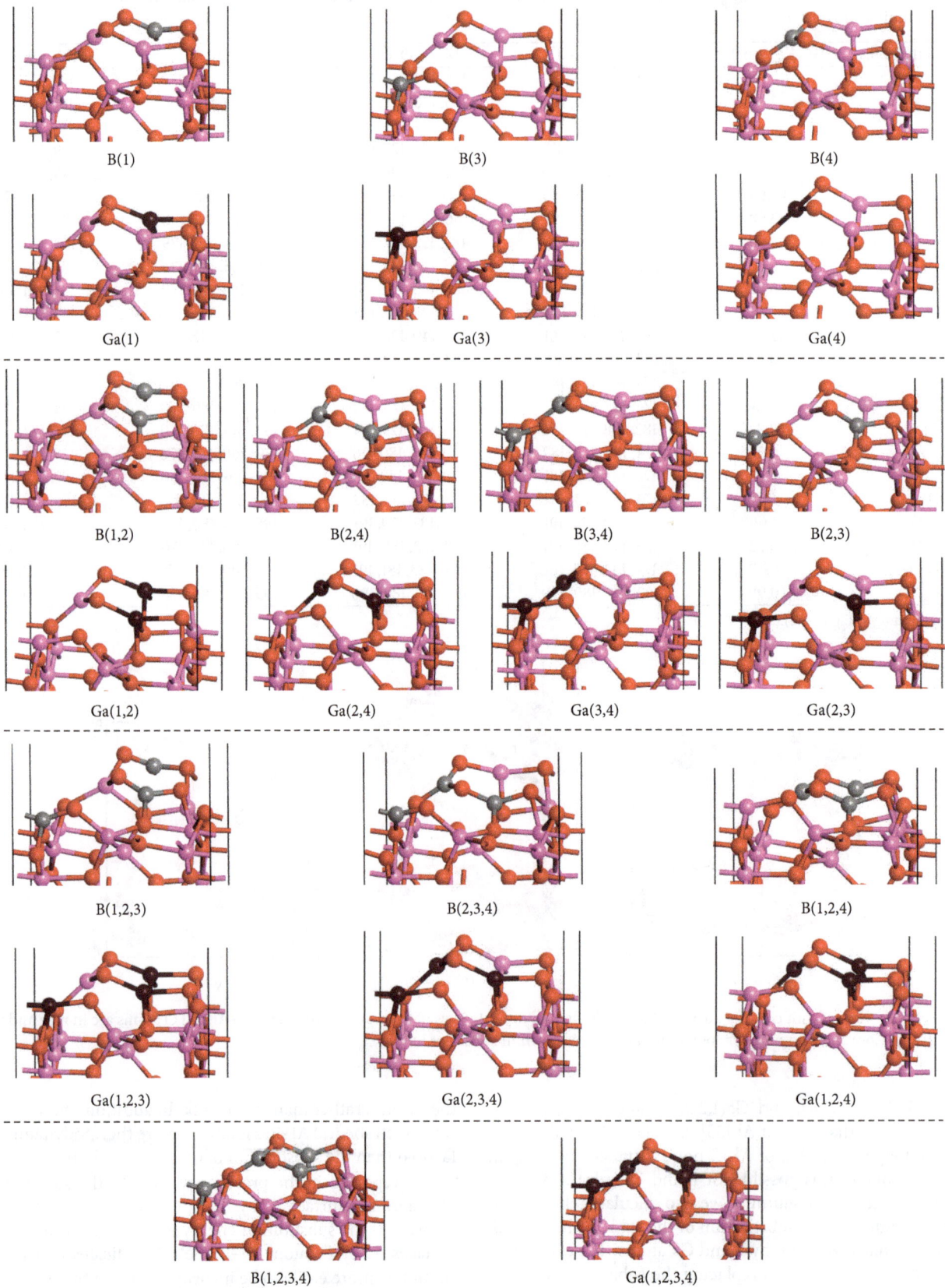

FIGURE 2: Side views of the B and Ga substituted γ-Al$_2$O$_3$(110) surface (B, Al, Ga, and O atoms are in gray, rose, dark, and red, resp.). All figures are indexed corresponding to Table 1.

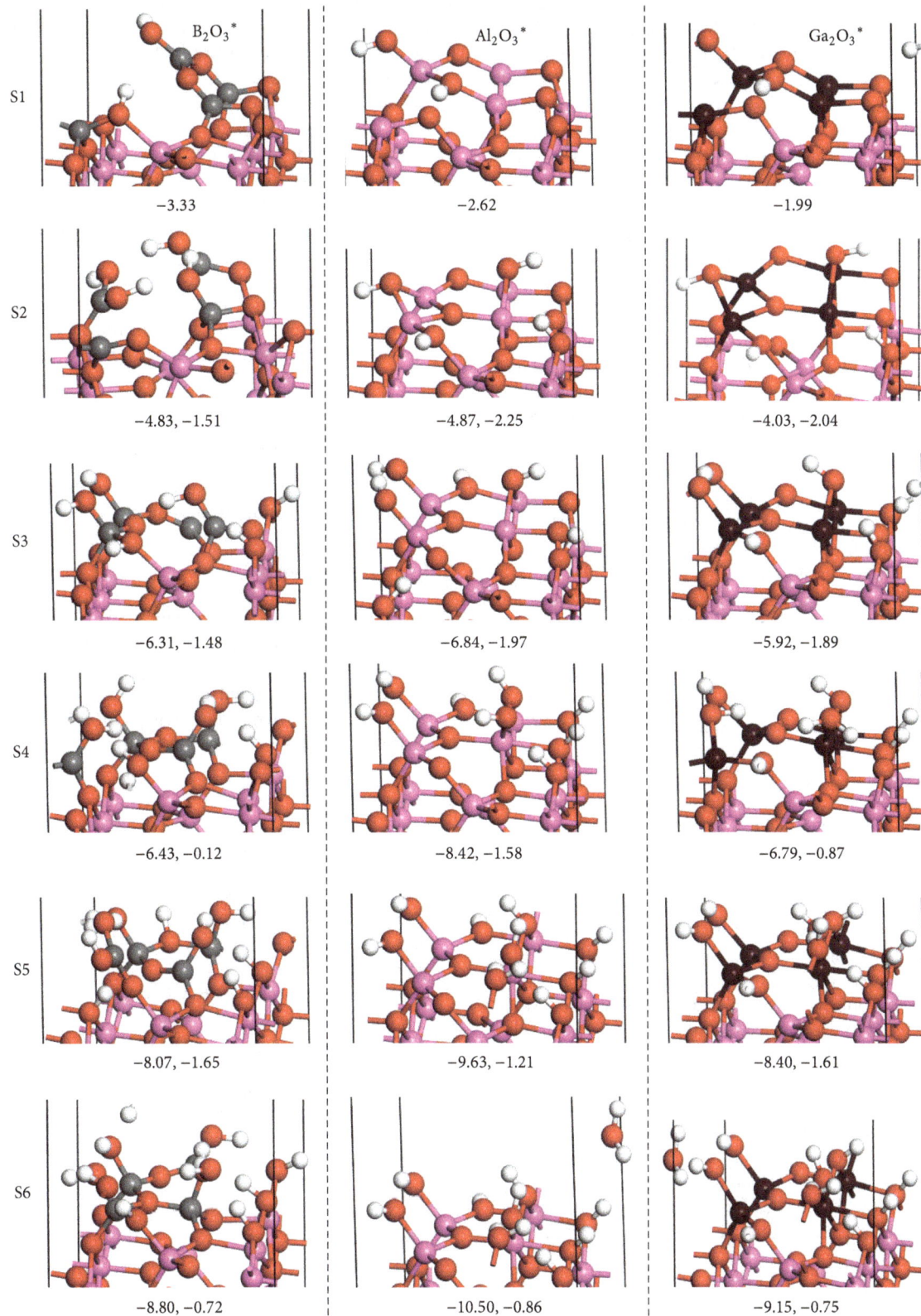

FIGURE 3: Side views for subsequent water adsorption on $B_2O_3{}^*$, $Al_2O_3{}^*$, and $Ga_2O_3{}^*$ surfaces, Sn ($n = 1$–6). The total and (subsequent) adsorption energies are given in eV. (B, Al, Ga, O, and H atoms are in gray, rose, dark, red, and white, resp.).

energies for water adsorption on the $B_2O_3^*$, $Al_2O_3^*$, and $Ga_2O_3^*$ surfaces. It should be noted that our calculated adsorption energies for the water adsorption on the $Al_2O_3^*$ surfaces are −2.62, −4.87, −6.84, −8.42, −9.63, and −10.50 eV, respectively, for 1–6 water molecules adsorbed on the γ-alumina surface. The stepwise adsorption energies for each water molecule are −2.62, −2.25, −1.97, −1.58, −1.21, and −0.86 eV. The previous works reported the adsorption of one water molecule on the γ-alumina (110) surface releases the energy of −2.49 eV [1, 32], which is slightly smaller than the results of present work (−2.62 eV). The reason is that the PBE-D3 methods were used in the present work and dispersion correction effects were taken into consideration. The optimized structures for the adsorption are the same as the previous works. The $Al_2O_3^*$ and $Ga_2O_3^*$ show similar structures with one −OH group on the tricoordinated surface Al/Ga atom, and H atom bonds to the twofold coordinated surface O atoms. It is interesting to see that the surface B–O bond was broken after water adsorption, and the âĂŠOH group from the water bonds to the surface BO_2 in coplanar. The H atom from the water adsorbs onto the surface of O atom forming an in-surface hydroxyl. It leads to a larger adsorption energy for the first water molecule adsorbed on the surface of $B_2O_3^*$ than that of $Al_2O_3^*$ and $Ga_2O_3^*$.

For two water molecules adsorption on the $B_2O_3^*$, $Al_2O_3^*$, and $Ga_2O_3^*$ surfaces, the adsorption energies are −4.83, −4.87, and −4.03 eV, respectively. $Al_2O_3^*$ shows the largest adsorption energy. The second water molecule makes the $B_2O_3^*$, $Al_2O_3^*$, and $Ga_2O_3^*$ surfaces seriously distorted. As shown in Figure 3, the X(3) atom moves to surface X(4) atom for X = Al and Ga, and they share one −OH group from the water and both became tetracoordinated.

As the water coverage increases from one to six water molecules in one $B_2O_3^*$ slab, the adsorption energy increases from −3.33, −4.83, −6.31, −6.43, and −8.07 to −8.80 eV, respectively. In comparison, the adsorption energy increases from −1.99, −4.03, −5.92, −6.79, and −8.40 to −9.15 eV for $Ga_2O_3^*$ surface. Both are smaller than those for the $Al_2O_3^*$ surface. It indicates that the pure γ-alumina surface shows stronger water adsorption than $B_2O_3^*$ and $Ga_2O_3^*$. It should be noted that the subsequent adsorption monotonously decreases for the water adsorption on $Al_2O_3^*$ surface. Since the water adsorption leads to the surface reconstruction, there are the turning points of the stepwise adsorption energy for 4-5 water molecules adsorption for the $B_2O_3^*$ and $Ga_2O_3^*$ slabs. In addition, the stepwise adsorption energy for the sixth water molecule adsorbent on the $B_2O_3^*$, $Al_2O_3^*$, and $Ga_2O_3^*$ surfaces is similar (−0.72, −0.86, and −0.75 eV, resp.). The main reason is that the former five water molecules have totally covered the active sites for water adsorption, and the sixth water molecule could only be physically adsorbed.

4. Conclusions

The dispersion corrected periodic density functional theory was used to investigate the structure and energies for the B and Ga incorporated γ-alumina surface. The results show that the substitution of Al by B is not thermodynamically favored on the surface. However, the substitution of Al by Ga is thermodynamically favored at low coverage on the surface. The substitution reaction prefers to occur at the tricoordinated A(4) sites. The substitution reaction becomes thermodynamically not favored as more and more B and Ga substitutions take place on the surface. The substitutions of Al by B and Ga are not so favored in the bulk as that for the surface.

The adsorption of water molecules on the B and Ga incorporated γ-alumina surface was also investigated and compared to that of the pure γ-alumina surface. It shows that the γ-alumina surface has the strongest adsorption ability for water adsorption. The total adsorption energy increases as water coverage increases from one to six water molecules in each slab, while the stepwise adsorption energy decreases. On the $B_2O_3^*$, $Al_2O_3^*$, and $Ga_2O_3^*$ surfaces, the sixth could only be physically adsorbed on the surface, since the former adsorbed five water molecules (at the coverage of 7.5 H_2O/nm^2) fully covered the active sites for water adsorption.

Conflicts of Interest

The authors declare that there are no conflicts of interest regarding the publication of this paper.

Acknowledgments

This work was supported by the Doctoral Scientific Research Foundation for Dr. Lihong Cheng of Jiangxi Science & Technology Normal University (2017.1-2020.12), Science Foundation of Jiangxi Department of Education (GJJ150827), and National Natural Science Foundation of Jiangxi Province (20151BAB204009).

References

[1] G. Feng, C.-F. Huo, C.-M. Deng et al., "Isopropanol adsorption on γ-Al_2O_3 surfaces: a computational study," *Journal of Molecular Catalysis A: Chemical*, vol. 304, no. 1-2, pp. 58–64, 2009.

[2] L. Huang, Y.-L. Zhu, C.-F. Huo et al., "Mechanistic insight into the heterogeneous catalytic transfer hydrogenation over Cu/Al_2O_3: direct evidence for the assistant role of support," *Journal of Molecular Catalysis A: Chemical*, vol. 288, no. 1-2, pp. 109–115, 2008.

[3] Y. Liu, J. Liu, G. Feng, S. Yin, W. Cen, and Y. Liu, "Interface effects for the hydrogenation of CO_2 on Pt_4/γ-Al_2O_3," *Applied Surface Science*, vol. 386, pp. 196–201, 2016.

[4] B. Ealet, M. H. Elyakhloufi, E. Gillet, and M. Ricci, "Electronic and crystallographic structure of γ-alumina thin films," *Thin Solid Films*, vol. 250, no. 1-2, pp. 92–100, 1994.

[5] J. Libuda, F. Winkelmann, M. Bäumer et al., "Structure and defects of an ordered alumina film on NiAl(110)," *Surface Science*, vol. 318, no. 1-2, pp. 61–73, 1994.

[6] D. W. Susnitzky and C. B. Carter, "The formation of copper aluminate by solid-state reaction," *Journal of Materials Research*, vol. 6, no. 9, pp. 1958–1963, 1991.

[7] G. A. El-Shobaky, N. M. Ghoneim, and E. A. Sultan, "Thermal decomposition of nickel aluminium mixed hydroxides and formation of nickel aluminate spinel," *Thermochimica Acta*, vol. 63, no. 1, pp. 39–49, 1983.

[8] T. Wang, F. Jiang, G. Liu, L. Zeng, Z.-J. Zhao, and J. Gong, "Effects of Ga doping on Pt/CeO_2-Al_2O_3 catalysts for propane dehydrogenation," *AIChE Journal*, vol. 62, no. 12, pp. 4365–4376, 2016.

[9] B. Bonnetot, V. Rakic, T. Yuzhakova et al., "Preparation and characterization of Me_2O_3-CeO_2 (Me = B, Al, Ga, In) mixed oxide catalysts. 2. Preparation by sol-gel method," *Chemistry of Materials*, vol. 20, no. 4, pp. 1585–1596, 2008.

[10] G. Feng, C.-F. Huo, Y.-W. Li, J. Wang, and H. Jiao, "Structures and energies of iron promoted γ-Al_2O_3 surface: a computational study," *Chemical Physics Letters*, vol. 510, no. 4–6, pp. 224–227, 2011.

[11] H.-J. Wan, B.-S. Wu, C.-H. Zhang et al., "Study on Fe-Al_2O_3 interaction over precipitated iron catalyst for Fischer-Tropsch synthesis," *Catalysis Communications*, vol. 8, no. 10, pp. 1538–1545, 2007.

[12] V. G. De Resende, X. Hui, C. Laurent, A. Weibel, E. De Grave, and A. Peigney, "Fe-substituted mullite powders for the in situ synthesis of carbon nanotubes by catalytic chemical vapor deposition," *Journal of Physical Chemistry C*, vol. 113, no. 26, pp. 11239–11245, 2009.

[13] V. G. de Resende, A. Cordier, E. De Grave et al., "Synthesis of γ-$(Al_{1-x}Fe_x)_2O_3$ solid solutions from oxinate precursors and formation of carbon nanotubes from the solid solutions using methane or ethylene as carbon source," *Journal of Materials Research*, vol. 23, no. 11, pp. 3096–3111, 2008.

[14] V. G. de Resende, A. Cordier, E. de Grave et al., "Presence of metallic Fe nanoclusters in α-$(Al,Fe)_2O_3$ solid solutions," *The Journal of Physical Chemistry C*, vol. 112, no. 42, pp. 16256–16263, 2008.

[15] G. Feng, D. Yang, D. Kong, J. Liu, and Z.-H. Lu, "A comparative computational study on the synthesis prescriptions, structures and acid properties of B-, Al- and G-incorporated MTW-type zeolites," *RSC Advances*, vol. 4, no. 89, pp. 47906–47920, 2014.

[16] J. Zhou, J. Teng, L. Ren et al., "Full-crystalline hierarchical monolithic ZSM-5 zeolites as superiorly active and long-lived practical catalysts in methanol-to-hydrocarbons reaction," *Journal of Catalysis*, vol. 340, pp. 166–176, 2016.

[17] C. G. Zuo, A. G. Xiao, Z. H. Zhou et al., "Spectroscopic properties of Ce^{3+}-doped BaO–Gd_2O_3–Al_2O_3–B_2O_3–SiO_2 glasses," *Journal of Non-Crystalline Solids*, vol. 452, pp. 35–39, 2016.

[18] V. A. Silva, P. C. Morais, R. F. Morais, and N. O. Dantas, "Successful Nd^{3+} doping of Li_2O–B_2O_3–Al_2O_3 vitreous system: optical characterization and judd–ofelt spectroscopic calculations," *Brazilian Journal of Physics*, vol. 46, no. 6, pp. 643–648, 2016.

[19] J. L. Jansons, P. A. Kulis, Z. A. Rachko et al., "Luminescence of Ga-doped alpha-Al_2O_3 crystals," *Physica Status Solidi B-Basic Research*, vol. 120, no. 2, pp. 511–518, 1983.

[20] M. E. Kibar, O. Özcan, Y. Dusova-Teke, E. Yonel-Gumruk, and A. N. Akin, "Optimization, modeling and characterization of sol-gel process parameters for the synthesis of nanostructured boron doped alumina catalyst supports," *Microporous and Mesoporous Materials*, vol. 229, pp. 134–144, 2016.

[21] J. P. Perdew, K. Burke, and M. Ernzerhof, "Generalized gradient approximation made simple," *Physical Review Letters*, vol. 77, no. 18, pp. 3865–3868, 1996.

[22] G. Kresse and J. Furthmüller, "Efficiency of ab-initio total energy calculations for metals and semiconductors using a plane-wave basis set," *Computational Materials Science*, vol. 6, no. 1, pp. 15–50, 1996.

[23] G. Kresse and J. Furthmüller, "Efficient iterative schemes for ab initio total-energy calculations using a plane-wave basis set," *Physical Review B - Condensed Matter and Materials Physics*, vol. 54, no. 16, pp. 11169–11186, 1996.

[24] J. Liu, Z.-F. Liu, G. Feng, and D. Kong, "Dimerization of propene catalyzed by Brønsted acid sites inside the main channel of zeolite SAPO-5: a computational study," *The Journal of Physical Chemistry C*, vol. 118, no. 32, pp. 18496–18504, 2014.

[25] G. Feng, Y.-Y. Lian, D. Yang, J. Liu, and D. Kong, "Distribution of Al and adsorption of NH_3 and pyridine in ZSM-12: a computational study," *Canadian Journal of Chemistry*, vol. 91, no. 10, pp. 925–934, 2013.

[26] S. Grimme, A. Hansen, J. G. Brandenburg, and C. Bannwarth, "Dispersion-corrected mean-field electronic structure methods," *Chemical Reviews*, vol. 116, no. 9, pp. 5105–5154, 2016.

[27] S. Grimme, S. Ehrlich, and L. Goerigk, "Effect of the damping function in dispersion corrected density functional theory," *Journal of Computational Chemistry*, vol. 32, no. 7, pp. 1456–1465, 2011.

[28] S. Grimme, J. Antony, S. Ehrlich, and H. Krieg, "A consistent and accurate ab initio parametrization of density functional dispersion correction (DFT-D) for the 94 elements H-Pu," *Journal of Chemical Physics*, vol. 132, no. 15, Article ID 154104, 2010.

[29] G. Kresse and D. Joubert, "From ultrasoft pseudopotentials to the projector augmented-wave method," *Physical Review B*, vol. 59, no. 3, pp. 1758–1775, 1999.

[30] P. E. Blöchl, C. J. Först, and J. Schimpl, "Projector augmented wave method: ab initio molecular dynamics with full wave functions," *Bulletin of Materials Science*, vol. 26, no. 1, pp. 33–41, 2003.

[31] P. E. Blöchl, "Projector augmented-wave method," *Physical Review B*, vol. 50, no. 24, pp. 17953–17979, 1994.

[32] M. Digne, P. Sautet, P. Raybaud, P. Euzen, and H. Toulhoat, "Use of DFT to achieve a rational understanding of acid-basic properties of γ-alumina surfaces," *Journal of Catalysis*, vol. 226, no. 1, pp. 54–68, 2004.

[33] J. Wang, H. Yu, L. Geng et al., "DFT Study of Hg Adsorption on M-substituted Pd(111) and PdM/γ-Al_2O_3(110) (M = Au, Ag, Cu) Surfaces," *Applied Surface Science*, vol. 355, pp. 902–911, 2015.

[34] L. Geng, L. Han, W. Cen et al., "A first-principles study of Hg adsorption on Pd(111) and Pd/γ-Al_2O_3(110) surfaces," *Applied Surface Science*, vol. 321, pp. 30–37, 2014.

[35] B. Cordero, V. Gómez, A. E. Platero-Prats et al., "Covalent radii revisited," *Dalton Transactions*, no. 21, pp. 2832–2838, 2008.

Influence of Surfactant Structure on the Stability of Water-in-Oil Emulsions under High-Temperature High-Salinity Conditions

Abdelhalim I. A. Mohamed,[1] Abdullah S. Sultan,[2] Ibnelwaleed A. Hussein,[3] and Ghaithan A. Al-Muntasheri[4]

[1]*Petroleum Engineering Department, University of Wyoming, Laramie, WY 82071, USA*
[2]*Petroleum Engineering Department and Center for Petroleum & Minerals, King Fahd University of Petroleum & Minerals, Dhahran 31261, Saudi Arabia*
[3]*Gas Processing Center, College of Engineering, Qatar University, P.O. Box 2713, Doha, Qatar*
[4]*EXPEC Advanced Research Center, Saudi Aramco, P.O. Box 62, Dhahran 31311, Saudi Arabia*

Correspondence should be addressed to Abdullah S. Sultan; sultanas@kfupm.edu.sa

Academic Editor: Francisco Javier Deive

Emulsified water-in-oil (W/O) systems are extensively used in the oil industry for water control and acid stimulation. Emulsifiers are commonly utilized to emulsify a water-soluble material to form W/O emulsion. The selection of a particular surfactant for such jobs is critical and certainly expensive. In this work, the impact of surfactant structure on the stability of W/O emulsions is investigated using the hydrophilic-lipophilic balance (HLB) of the surfactant. Different commercial surfactants were evaluated for use as emulsifiers for W/O systems at high-temperature (up to 120°C) high-salinity (221,673 ppm) HTHS conditions. Diverse surfactants were examined including ethoxylates, polyethylene glycols, fluorinated surfactants, and amides. Both commercial Diesel and waste oil are used for the oleic phase to prepare the emulsified system. Waste oil has shown higher stability (less separation) in comparison with Diesel. This work has successfully identified stable emulsified W/O systems that can tolerate HTHS environments using HLB approach. Amine Acetate family shows higher stability in comparison with Glycol Ether family and at even lower concentration. New insights into structure-surfactant stability relationship, beyond the HLB approach, are provided for surfactant selection.

1. Introduction

Emulsions are broadly utilized in different industries such as pharmaceutical [1], hydraulic fluids [2], polymerization [3], paints [4], and food industries [5, 6]. Furthermore, emulsification technology has been extensively applied in the oilfields [7–10]. Usually, the emulsion contains two or more partially or completely immiscible liquids [11], where the dispersed phase exists as droplets suspended in the continuous phase. The interface between hydrophobic and hydrophilic molecules is intrinsically not stable [12]. When two immiscible liquids are stirred, the emulsion is formed [13, 14]. Emulsions are stabilized when a surfactant is added to a two-phase system due to the slowdown of emulsion breaking such as coalescences [15]. The interfacial tension is decreased with the adsorption of more surfactant at the interface, and consequently droplet coalescence is delayed [15, 16]. Stability of the emulsion is determined by different factors such as the nature of the interfacial film, continuous phase viscosity, oil-water-ratio, mixing time, and temperature [17].

2. Emulsification in the Oilfields

The emulsification technique is well documented in oil and gas production literature. For example, it has been employed in drilling fluid formulations [18, 19] and well stimulation treatments [9, 20]. Also, it is used in well productivity enhancement via asphaltene deposition removal [21] drag reduction in multiphase flow [10] and for the control of excessive water production [8, 22]. Emulsification technique was introduced in the oil industry through the use of emulsified acids in 1933. Emulsified acids were invented to address corrosion problems rather than improving the stimulation job [23]. Thus, many researchers comprehensively studied this technique for further understanding of advantages and disadvantages of emulsified acids [20, 24–26]. Moreover, for emulsified acids, there are many reports on the effect of droplet size, water phase volume fraction, and the concentration of the emulsifier on their stability and rheology (see Al-Mutairi et al. [20] and references within). In addition to the well stimulation applications, a new application of emulsification technique in the oilfield is proposed recently as a method for water control with bullhead injection. In this case, the emulsion acts as a relative permeability modifier (RPM). The existence of RPM fluids is well known [27–30]. In a patent, Stavland and Nilsson suggested an injection of the gelant (crosslinked polymer) as an emulsion for RPM field application [22]. In work by Stavland et al. [8], an aqueous polymer gelant is emulsified into an oil and then injected as one component. Eventually, the solution separates into an oil phase and water phase upon reaching the reservoir. Afterward, the water phase gels up in a water-wet pathway of a pore space to reduce permeability to formation brine, and the oil phase remains mobile to secure a path for hydrocarbons to flow [8, 22]. The control of the gel fraction that occupies the porous media leads to the control of RPM, that is, the reduction in relative permeability of the hydrocarbons and formation brine; this is controlled by the water fraction in the emulsion [8]. In a recent publication, our group studied the gelation kinetics of emulsified PAM/PEI system using thermal analysis technique [31].

Undoubtedly, the type of emulsion is critical for those applications. Therefore, the emulsifier, which will be employed to accomplish the emulsification, must be cautiously selected to meet the requirements of those settings, so a fair rate of success could be seized. Nowadays, the biggest difficulty with surfactants, at least from the standpoint of those who have to choose them, is the staggering numbers that are available. Each manufacturer tries to provide one or more of his products that are suitable for every need, which makes the selection process difficult. The large numbers of surfactants available, coupled with the fact that application problems are becoming increasingly difficult, is making the need for a suitable process for the selection of surfactants more and more critical. Selection of surfactants is important for many applications in the oil field such as EOR, stimulation, and water shut-off. Our group has recently conducted a detailed surfactant screening study for chemical EOR purposes [32].

Forming stable emulsion is not straightforward routine. To emulsify two immiscible fluids, a particular emulsifier is necessary to form a specific type of emulsion. Consequently, selecting surfactant (emulsifier) to do the job is a critical subject, and it is certainly a very expensive exercise in terms of both cost and time. There is no systemic procedure in the oil industry for selecting a suitable emulsifying agent for a specific application. The industry mainly relies on the experience and service providers' recommendations. Too often, a series of time-consuming laboratory measurements, such as phase behaviour and interfacial/surface tension, are performed at reservoir condition to select surfactant for an application. Commonly, the selection of an emulsifier is based on (a) the surfactant solubility, (b) controllable separation time (thermal stability), and (c) acceptable environmental consideration for a particular region [8]. This technique is founded on Bancroft's rule, which is an empirical rule grounded on the surfactant solubility [33, 34]; more details are given later. Interestingly, a more robust method such as hydrophobic-lipophilic balance (HLB), which is based on the surfactant chemical structure [35–37], is rarely used in the oilfields at least from surfactant selection point of view. Hence, the objective herein is to relate the surfactant structure to its performance in an attempt to ease the process of selecting a surfactant for emulsified W/O emulsions for applications in high-temperature high-salinity (HTHS) conditions. The performance of the different commercial emulsifiers is evaluated and correlated to their HLB. The usage of both Bancroft's rule and HLB as a selection criterion is investigated. Furthermore, the effects of surfactant chemistry and concentration, temperature, oleic phase composition, and water phase salinity on the emulsion stability are studied.

3. Experimental

In this section, details about materials, equipment, experimental procedures, emulsion preparation, and characterization are detailed.

3.1. Materials. The surfactants used in this study were supplied by Sigma-Aldrich®, AkzoNobel, and Capstone® as presented in Table 1. Sea water and brine formation are used as water phase. Water analysis is given in Table 2. Diesel from local gas stations and refinery waste oil are used for the oleic phase. The Diesel is representative of the oil utilized in the field by the industry for preparing emulsified acids. All salts used in this study are ACS grade.

3.2. Equipment. The emulsions were prepared in a high-performance dispersing instrument (Ultra-Turrax T 50 Basic) provided by VWR International. The homogenizer is equipped with a variable-speed drive with six different speeds in the range 500–10,000 rpm. All emulsions were prepared at room temperature at 2000 rpm mixing speed for 5 minutes. The mixing speed and time were selected following a separate investigation. A conductivity meter is provided by HACH (CDC401 model); the device can handle total dissolved solids in the range 0 to $50 \, mg \cdot L^{-1}$ as NaCl. The meter is used to classify the emulsion type whether it is W/O or O/W. GL-18 high-temperature disposable test tubes and soda-lime-glass (18 × 180 mm) of approximate volume of 32 ml (operational temperature of 180°C) were used. The high-temperature tubes were sealed with a screw cap and a rubber seal case to prevent evaporation. The fact that no evaporation is taking place was assured by comparing the initial and final volumes of the

TABLE 1: Description of the surfactants used in this study.

Surfactant	Type	Mw (g·mol^{-1})	HLB	Weight ratio (%)
Glycolic Acid Ethoxylate-1 Ether ($C_{14}H_{24}O_6$)	AIS*	288.34	12	61.1
Glycolic Acid Ethoxylate-2 Ether ($CH_3(CH_2)_{11-13}(OCH_2 CH_2)_n OCH_2CO_2H$ ($n = 6$))	AIS*	739.20	7.14	35.7
Glycolic Acid Ethoxylate-3 Ether ($C_{55}H_{112}O_8$)	AIS*	901.47	2	9.76
Ethylenediamine-Tetrol ($C_3H_6C_2H_8N_2C_2H_4O)_x$	N/A	3,600	7	35.7
Fluorosurfactant-1	N/A	N/A	N/A	N/A
Fluorosurfactant-2	N/A	N/A	N/A	N/A
Polyethylene Glycol-1 Ether ($C_{58}H_{109}KO_{24}S$)	AIS*	1260	14	69.8
Polyethylene Glycol-2 Ether ($C_{18}H_{35}(OCH_2CH_2)_n OH$, $n \sim 2$)	NIS**	356.58	4	20
Polyethylene Glycol-3 Ether ($C_{16}H_{33}(OCH_2CH_2)_n OH$, $n \sim 2$)	NIS**	330	5	25
Polyethylene Glycol-4 Ether ($C_4H_{10}O_2$)	NIS**	90,12	N/A	N/A
Amine Acetates-1	IS***	N/A	6.8	34
Amine Acetates-2	IS***	200	10.5	52.5
Amine Acetates-3	IS***	263	6.8	34
Ethoxylated Amides-1 (Ethomid-1)	N/A	N/A	4.85	24.3
Ethoxylated Amides-2 (Ethomid-2)	N/A	340–360	5.1	25.5

*AIS = anionic surfactant; **NIS = nonionic surfactant; ***IS = ionic surfactant; Mw = molecular weight; N/A = data not available.

TABLE 2: Chemical analysis of water used in the study.

Ion, ppm	Water type concentration, mg·L^{-1}	
	Brine formation	Sea water
Na$^+$	59,300	18,300
Ca^{2+}	23,400	650
Mg^{2+}	1,510	2,083
SO$_4^-$	110	4,290
Cl$^-$	137,000	32,200
HCO$_3^-$	353	120
Total dissolved solids*	221,673	57,642

*Determined by addition.

Lipophilic (tail)

Hydrophilic
(head)

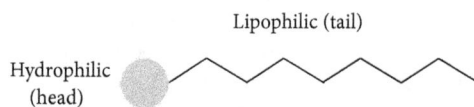

FIGURE 1: Schematic diagram of surfactant (emulsifier) chemical structure.

sample at the end of the experiment. HAAKE FISONS hot oil bath Model N3 is used to study the emulsion separation (separated volume fraction of the phases versus time).

3.3. Experimental Setup and Procedure

3.3.1. Emulsion Preparation.
Several emulsions were prepared systemically to ensure the reproducibility of the results. All emulsions were prepared at room temperature at a fixed mixing speed for 5 minutes. Enough time was given for the emulsifier to mix thoroughly in the external phase. Then, a desired volume of the dispersed phase was slowly added to the continuous (external) phase. It is important to control the addition of the dispersed phase droplets throughout the mixing. It is reported that both the addition rate of the dispersed phase and mixing intensity govern the type of emulsion; a coarse emulsion will be the result of adding the dispersed phase to the continuous phase in one step without intensive mixing. However, a fine emulsion can be produced by adding the dispersed phase in atomized form coupled with intensive mixing [20, 38].

3.3.2. Emulsion Characterization.
Two methods are used for the surfactant selection, namely, Bancroft's rule and HLB

value of a surfactant. Different commercial surfactants with a broad range of properties are selected as shown in Table 1. Conductivity and dilution tests are used simultaneously to identify the type of emulsions.

Bancroft's Rule. The nature of the emulsifying agent controls the emulsion type rather than the oil-water-ratio or the method of emulsion preparation [16, 17, 34]. Bancroft developed one of the first empirical rules to describe the nature of emulsion that could be stabilized by a given emulsifier [33]. The continuous phase is the phase in which an emulsifier is more soluble as stated by Bancroft's rule. Hence, O/W emulsions are formed by the addition of a water-soluble emulsifier. On the other hand, W/O emulsions are formed when an oil-soluble emulsifier is used [33, 34].

Hydrophilic-Lipophilic Balance (HLB) Determination. HLB measures the degree to which a surfactant is hydrophilic or lipophilic. HLB offers an efficient way of picking the suitable surfactant for a specific application as suggested notably by Griffin [35, 36]. A scale of 0 to 20 is proposed. HLB value of 0 represents a completely lipophilic molecule, and a value of 20 accounts for a strongly hydrophilic molecule. The HLB values for W/O emulsifiers are in the range 3.5–6, while those of O/W are in the range 8–18. Wetting agents have HLB values in the range 7–9.

Surfactant HLB value is determined by calculating the contributions of different constituents of the molecular structure (see Figure 1) as described by Griffin [35, 36]. Another method was suggested in 1957 by Davies [37]. It is

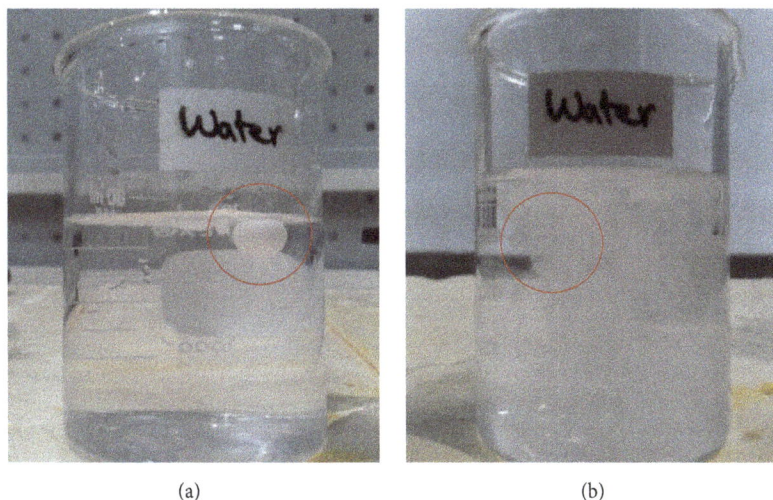

FIGURE 2: Dilution test, (a) Droplet sinking, (W/O) emulsion, and (b) Droplet dispersion, (O/W) emulsion.

based on the chemical groups of the molecule. This method considers the effect of strong and weak hydrophilic groups. However, it requires more information such as numbers of hydrophilic and lipophilic groups in the molecule and values of hydrophilic and lipophilic groups. Consequently, the value of HLB for a particular surfactant has been either provided by the supplier or calculated employing Griffin's method. Also, the weight ratio of the hydrophilic part to the hydrophobic (HLWR) is calculated as shown in Table 1.

Conductivity Test. The type of emulsion has been determined by measuring its conductivity; this is a quantitative method based on the electrical proprieties of the emulsion water phase, which is highly conductive, whereas the oleic phase is nonconductive. It should be pointed out that the external phase dominates the emulsion's conductivity [20, 39, 40]. Thus, O/W emulsion is conductive (i.e., >0.00 μS·cm^{-1}), whereas W/O emulsion is nonconductive (i.e., ~0.00 μS·cm^{-1}). Conductivity measurements for all the fluids used in this study were carried out to set a baseline for the conductivity test. The recorded conductance values at 28°C are as follows: standard NaCl buffer solution 950 μS·cm^{-1}, deionized water 6.71 μS·cm^{-1}, sea water 47.2 × 10^3 μS·cm^{-1}, and field water 1078 μS·cm^{-1}.

Dilution Test. This test identifies the emulsion's external phase by dilution, in which water can be used to dilute O/W emulsion, whereas oil can be used to dilute W/O emulsion. In this test, a droplet of the formed emulsion is dispersed in water and Diesel to see if it spreads or sinks; if the placed droplet disperses, then the external phase is the same as the fluid used for the test. Conversely, if the droplet sinks in the medium, the external phase will be different from the fluid used for the test as shown in Figure 2 [39–41].

4. Surfactant Screening

In this section, two methods for the surfactant screening will be examined: (i) Bancroft's rule empirical based and

(ii) HLB founded on the physiochemical properties of the surfactant.

4.1. Results and Discussion. To investigate the validity of Bancroft's rule, emulsions were prepared using a number of surfactants with different solubility. Each surfactant was dissolved in sea water and Diesel at a time, wherein 2 vol% surfactant concentration was added to an external phase (28 vol%) and then mixed for a one minute at 500 rpm at room temperature. Afterward, the mixing speed was raised to 4000 rpm, and a dispersed phase was added to the solution of a surfactant and an external phase at a specific rate. Then the emulsions were characterized to identify the external and dispersed phases. The conductivity for each emulsion was measured periodically in parallel with the drop test in a span of one hour at room temperature (see Table 3), to make sure there was no inversion taking place.

When surfactants in Table 3 dissolved in sea water and Diesel separately to form the emulsion's continuous phase, various trends were noted. For instance, Fluorosurfactant-1 and Glycolic Acid Ethoxylate-2 Ether were dissolved in the Diesel (oleic phase) to form W/O emulsion, as Bancroft's concept explicitly theorizes. However, the conductivity and dilution tests showed that produced emulsions were O/W as shown in Table 3, which disagrees with Bancroft's rule. Conversely, Polyethylene Glycol-1 Ether did not dissolve in Diesel and formed O/W emulsion when dissolved in sea water, which is in complete agreement with Bancroft's rule. Furthermore, Ethylenediamine-Tetrol and Glycolic Acid Ethoxylate-3 Ether dissolved in Diesel and sea water equally, herein regardless in which phase the surfactants dissolved the resultant emulsions were O/W as in Table 3. Therefore, Bancroft's method for surfactant screening is found inconclusive. Consequently, another approach is proposed for use as a selection tool, which is based on the HLB value for the surfactant.

HLB values were calculated for all surfactants and were used to understand the surfactant behaviour as given in Table 1. The HLB values of most of the surfactants in

TABLE 3: Surfactant characterization based on conductivity and dilution tests.

Surfactant	Conductivity test (μS·cm^{-1})	Dilution test	Observations
*Fluorosurfactant-1	4.5	O/W	Disagrees with Bancroft's rule
*Fluorosurfactant-2	N/A	N/A	The surfactant is not oil-soluble
*Glycolic Acid Ethoxylate-1 Ether	N/A	N/A	The surfactant is not oil-soluble
**Glycolic Acid Ethoxylate-1 Ether	N/A	N/A	No stable emulsion formed
*Glycolic Acid Ethoxylate-2 Ether	1034	O/W	Disagrees with Bancroft's rule
*Glycolic Acid Ethoxylate-3 Ether	20.5×10^3	O/W	Disagrees with Bancroft's rule
**Glycolic Acid Ethoxylate-3 Ether	23.5×10^3	O/W	Agrees with Bancroft's rule
*Ethylenediamine-Tetrol	1072	O/W	Disagrees with Bancroft's rule
**Ethylenediamine-Tetrol	78.2×10^3	N/A	Agrees with Bancroft's rule
*Polyethylene Glycol-1 Ether	N/A	N/A	The surfactant is not oil-soluble
**Polyethylene Glycol-1 Ether	1078	N/A	Agrees with Bancroft's rule

*Surfactant dissolved in Diesel (oleic phase). **Surfactant dissolved in sea water (water phase); 1μS·cm$^{-1} = 1E - 3$ mS·cm$^{-1} = 1E - 6$ mho·cm$^{-1} = 0.640$ ppm (TDS); N/A = data not available.

TABLE 4: New selected surfactant based on the HLB.

Surfactant	HLB	Application based on the HLB	Conductivity test (μS·cm^{-1})	Dilution test
Polyethylene Glycol-2 Ether	4	W/O emulsifier	0.02	W/O
Amine Acetates-2	10.5	O/W emulsifier	3.50×10^3	N/A*
Amine Acetates-3	6.8	W/O emulsifier	0.01	N/A*
Ethoxylated Amides-1	4.85	W/O emulsifier	0.02	W/O
Ethoxylated Amides-2	5.1	W/O emulsifier	0.02	W/O

*N/A = data not available.

Table 3 are in the recommended range for O/W, not W/O emulsion application, which explains why O/W emulsions were formed. Furthermore, HLB and HLWR can explain surfactants solubility as well. Surfactants with HLB values higher than 10 (or HLWR > 50%) are hydrophilic (water-soluble), while surfactants with HLB values less than 10 are lipophilic (oil-soluble). For example, Polyethylene Glycol-1 Ether HLB of 14 and HLWR of 69.8% signifies its affinity to be water-soluble and oil insoluble. Likewise, Glycolic Acid Ethoxylate-1 Ether with HLB of 12 and HLWR of 61.1% was observed earlier (see Table 3). For a surfactant, soluble in both oil and water, this behaviour can be explained by intermediate affinity as shown by HLB and HLWR, Ethylenediamine-Tetrol (HLB of 7 and HLWR of 35.7%). For Glycolic Acid Ethoxylate-3, HLB of 2 and HLWR of 9.76%, those values indicate a dominant lipophilic affinity. However, its solubility in water could be a product of its hydrophilic part strong ionic interaction (anionic surfactant) with sea water molecules.

To further investigate HLB approach, surfactants with HLB values inside the recommended range for W/O and O/W emulsion application were selected as shown in Table 4. The results of the conductivity and dilution tests (at ambient conditions) confirmed that the formed emulsions are in agreement with the predictions based on the HLB values. Thus, this approach is considered to be more reliable for surfactant selection.

4.2. Emulsion Thermal Stability. To investigate the thermal stability of the formed emulsion, a sealed case of high-temperature test tubes was used, by monitoring the separated volume fraction of the oleic and water phases versus time at constant temperature. Such a test can indicate emulsion quality.

In this paper, the possibility of employing commercial surfactants as an alternative emulsifier in forming stable W/O emulsion for HTHS applications in the oilfields is investigated. From an operational point of view, the thermal stability of the emulsified system plays a major role in the success of the placement job. For instance, in emulsified acid, no separation inside well during the injection operation is essential to preventing well tubular's corrosion, such as high bottom-hole static temperature reservoirs (e.g., Thunder Horse 138°C (280°F) and Ursa 121°C (250°F) in Gulf Mexico). The time required for treatment placement is reported around one hour. Thus, the emulsified system must be stable for the time period [9]; any separation will lead to well's metallic-parts corrosion.

Similarly, designed separation time is necessary for the water shut-off applications. For emulsified polymer gel, it is desired to have a controllable separation and gelation time [8, 22, 31]. The time needed for polymer gel placement at high-temperature (≥ 130°C) high-salinity with high Mg^{2+} and Ca^{2+} contents was reported to be about 55 minutes [42, 43] and around one and a half hours for the emulsified gel at similar conditions of high-temperature of 123°C [8]. Consequently, the emulsified gel system should be stable for at least one hour. Deemulsification and gelation should start afterwards. It is preferred that the gelation starts after a complete separation; if not, a weak gel will develop. In case of partial or complete separation during treatment placement, the emulsified gelant

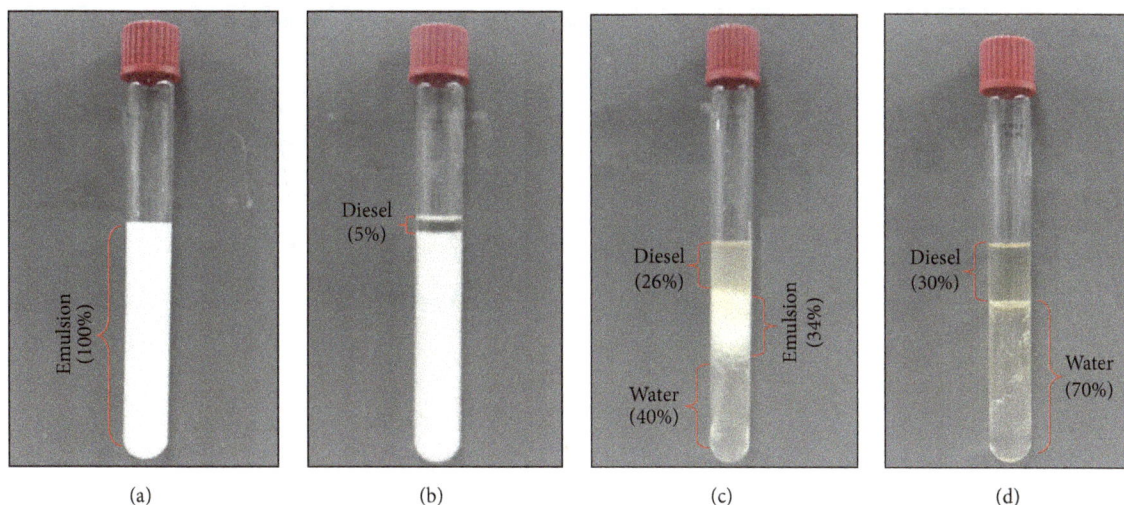

FIGURE 3: Water-in-oil emulsion separation, (a) no separation (100% emulsion), (b) separated volume fraction, zero water phase, and 0.05 oleic phase, (c) separated volume fraction, 0.4 water phase, and 0.26 oleic phase, and (d) complete separation (100%); separated volume fraction is 1 (0.7 water phase, 0.3 oleic phase).

will be exposed to high-temperature which will result in premature gelation inside the well tubular, which is highly undesirable [31]. For field applications, the high-temperature of the near wellbore area can be lowered using a preflush. Literature [41, 44] has shown that this method can reduce the temperature substantially; for example, the injection of 5,000 gallons of water can cool down the near well bore area from 150°C (302°F) to 116°C (240.8°F). Therefore, the thermal stability of all formed emulsions was investigated at 120°C.

In these experiments, no polymer or acid was used in the water phase. This is mainly because the current focus is to develop a surfactant selection criterion, for high-temperature high-salinity conditions, which can be utilized later for emulsified systems. At this phase of the research, the impact of several factors such as surfactant chemistry, temperature, water phase salinity and water-oil-ratio is quite significant. Hence, the focus was on those parameters. It is reported in the literature that the water phase significantly affects the thermal stability and the type of the emulsion. Addition of acid and corrosion inhibitors resulted in a less stable emulsion and, in some cases, lead to emulsion inversion from O/W to W/O [20]. However, addition of the gelant to the water phase slightly affected the stability [8, 31].

Moreover, an increase of salinity of the water phase from 5 to 20,000 ppm led to increase in the stability; increasing the salinity >20,000 ppm has significantly reduced the stability and resulted in emulsion inversion, and similar behaviour was noticed with water-oil ratio [39]. Since the emphasis of the manuscript is on the stability of the emulsion at salinity typical to those found in the oilfields, brine formation and sea water are used. Some of the surfactants screened in this study are successfully used to form stable emulsified PAM/PEI in a recent work by our group [31].

The emulsion stability is presented by using phase changes diagram (volume fraction versus time). This graph illustrates the percentage of the phases at a given time.

Initially, the emulsion is within the sample "homogeneous phase" before it starts to break. The single-phase water and oleic phases at the outset are 0% as the separation is yet to start as shown in Figure 3(a). However, macroemulsions are characteristically thermodynamically unstable; exposing the emulsion to a temperature over time leads to the separation of the emulsion. Then, the percentage of the water and the oleic phases increases as shown in Figures 3(b)–3(d). The volume proportion of the separated oleic and water phases keeps changing until it reaches a plateau or the emulsion is entirely separated. In all emulsions studied herein, the total volume was 30 ml, 70% of which is the water phase, and the rest 30% is the oleic phase (the mixture of emulsifier and the Diesel). Diesel percentage is in the range of 24–29.5%, while the emulsifier percentage is in the range 0.5%–6%. All emulsions were prepared at room temperature at 2000 rpm mixing speed for 5 minutes. Then, the thermal stability was examined in bulk at 120°C (248°F) for 12 hours using the oil heating bath.

4.3. Effect of Surfactant Type and Concentration. To study the influence of emulsifier type and concentration on the thermal stability, four surfactants were selected based on HLB criteria for W/O emulsions. Ethoxylated Amides-2, Amine Acetates-3, Polyethylene Glycol-2 Ether, and Polyethylene Glycol-3 Ether were used to emulsify the water phase, herein brine formation, into Diesel. To test the stability of the emulsion prepared using the surfactants mentioned above, each time three emulsions were prepared using one of the surfactants with different concentrations. The increase in the stability was noticed with the increase in the surfactant concentration. For instance, ~70% separation was observed at 8 and 120 minutes when Ethoxylated Amides-2 increased from 1 to 3 vol% as shown Figure 4, which signifies 93% (or a factor of 15) increase in the separation time. In the case where Amine Acetates-3 used 13% separation is noted at 2 and 330 minutes

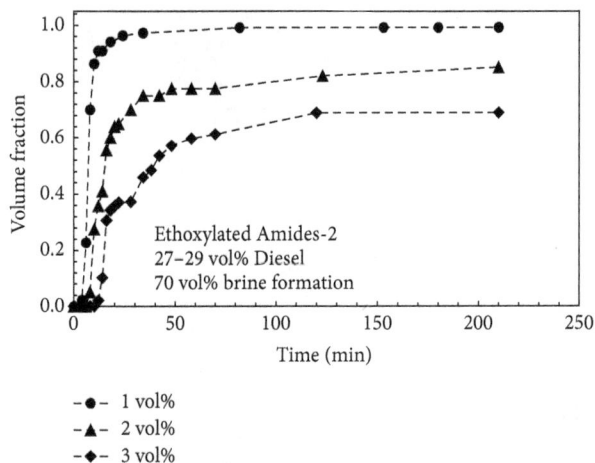

FIGURE 4: Ethoxylated Amides-2 phase behaviour at 120°C (248°F).

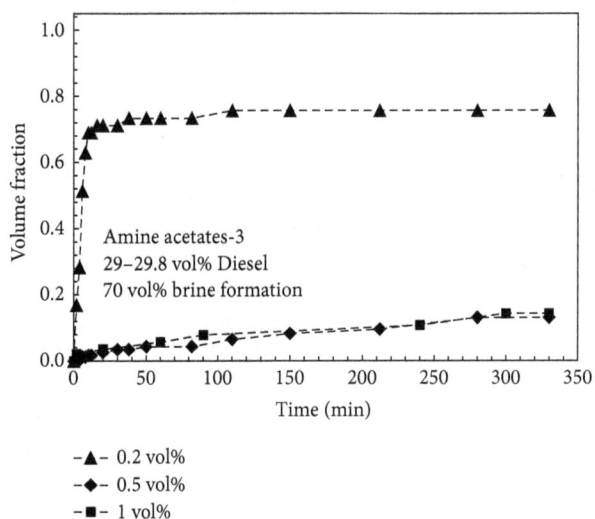

FIGURE 6: Polyethylene Glycol-2 Ether phase behaviour at 120°C (248°F).

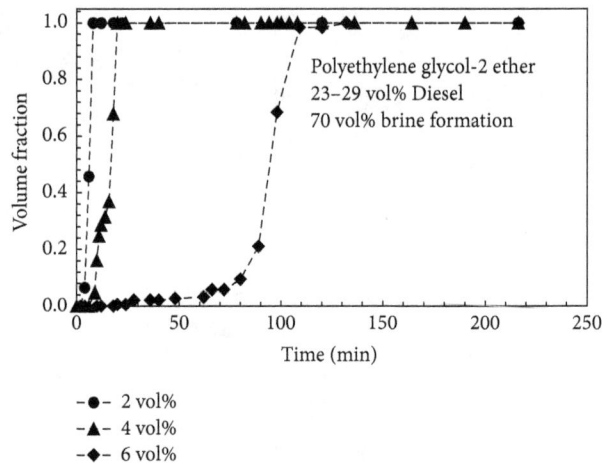

FIGURE 5: Amine Acetates-3 phase behaviour at 120°C (248°F).

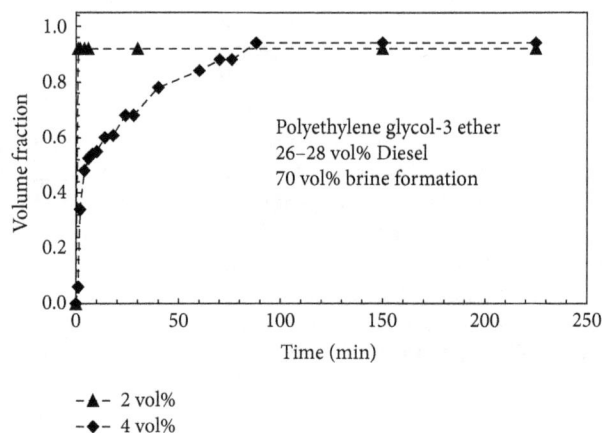

FIGURE 7: Polyethylene Glycol-3 Ether phase behaviour at 120°C (248°F).

when the concentration increased from 0.2 to 0.5 vol% (Figure 5). This indicates 99% increase in the separation time, while further increase in surfactant concentration (up to 1 vol%) did not result in a significant change. When Polyethylene Glycol-2 Ether was utilized, complete separation (100%) took place after 12, 22, and 109 minutes when used at 2, 4, and 6 vol%, respectively (see Figure 6); this represents 89% increase in the separation time (or a factor of 9.1). Also, ~90% separation occurred at 1 and 88 minutes (99% increase), when Polyethylene Glycol-3 Ether concentration increased from 2 to 4 vol% (see Figure 7). When comparing the performance of all the surfactants, three distinct trends were observed. Firstly, with a surfactant with a low concentration and high stability, the use of 0.5 vol% Amine Acetates-3 resulted in 13% separation at 330 minutes. Secondly, surfactants with a high concentration and a low stability, such as Polyethylene Glycol-2 Ether and Glycol-3 Ether, showed complete separation when both were used at a high concentration (~6 vol%). Finally, a surfactant with a moderate concentration and good stability, Ethoxylated Amides-2, showed a 69% separation

after 210 minutes when 3 vol% was used. Generally, with more surfactant adsorbed at the interface, more stability is achieved. Surfactant molecules reduce the interfacial tension and consequently retard droplets coalescence; this agrees with previous reports in the literature [16, 17, 41]. Moreover, surfactants with Amine functional group showed higher stability when compared to the Glycol Ether family group.

4.4. Effect of the Oleic Phase. To investigate the effect of the oleic phase on the emulsion stability, two emulsion samples were prepared, one with Diesel and the other with refinery waste oil. The composition of the waste oil and detailed analysis is given elsewhere (see Figure 5.1 of Sidaoui [45]). The phase separated volume (in Figures 8(a)–8(c)) is calculated as a percentage of the phase (i.e., emulsion, water, and oleic) total volume. For both samples, the water phase (formation brine) separation is almost identical as shown in Figure 8(a). However, the oleic phase separation is different, when waste oil is used instead of Diesel to prepare the emulsion. After 210 minutes in the oil bath, the separated oleic phase decreased from 100% to 9% (91% decrease or a

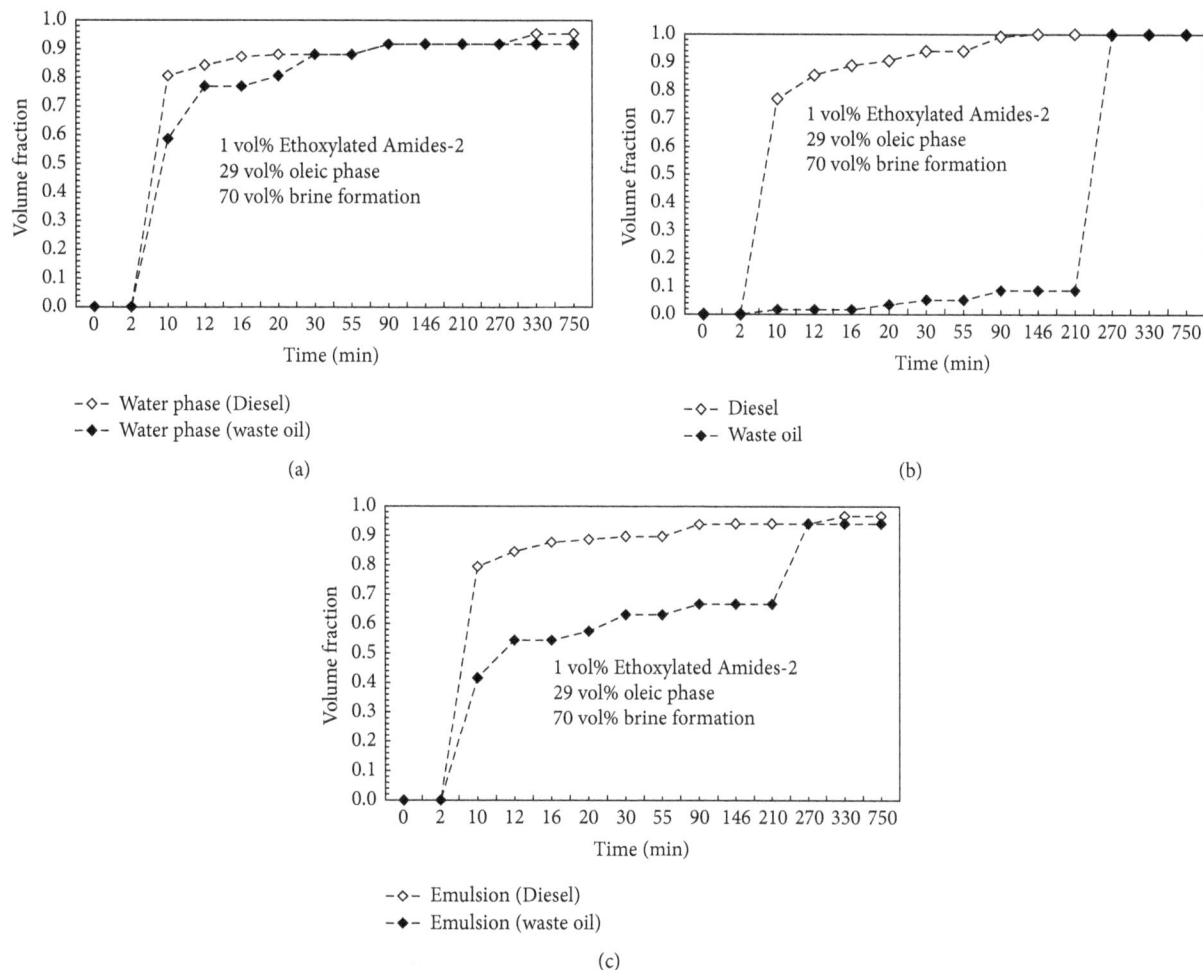

FIGURE 8: Emulsion phase behaviour at 120°C (248°F), (a) water phase, (b) oleic phase, and (c) emulsion phase.

factor of 11) as shown in Figure 8(b). Increase in the stability is noticed for the emulsion prepared with waste oil instead of Diesel with the separated volume decreased from 94% to 67% (28.7% decrease or a factor of 1.4) (see Figure 8(c)). The use of waste oil resulted in more stable emulsion (less separation). This may be because waste oil has resins or asphaltenes, which are natural emulsifiers. Commonly, Diesel and kerosene are used in the oilfields because of availability [8, 21, 40]. This work highlights the possibility of using waste oil as an alternative oleic phase for emulsification purposes, which will result in a reduction of associated cost of using mineral oils.

4.5. Reflections on HLB and Emulsion Stability. By definition, the HLB is a measure of the degree to which a surfactant is hydrophilic or lipophilic. Hence, authors reason that an ideal emulsifier has the hydrophilic part equivalent (equal) to lipophilic portion. This may allow for equal (balanced) distribution of the emulsifier in the water and the oleic phases, which may lead to a more stable emulsion. This hypothesis will be examined by correlating the emulsion stability to HLB and HLWR. It can be seen from Figure 9 that more stability of W/O emulsions is achieved at higher HLB and HLWR. For

instance, when an emulsifier with HLB of 6.8 and HLWR of 34% (Amine Acetate-3) was used at a very low concentration of 0.5 vol%, the emulsion took 310 minutes to break down completely. While in the case of the emulsifier with HLB of 4 and HLWR which equals 20%, the separation time was 20 minutes, although a higher concentration (6 vol%) was used. Also, in Figure 9, it has been noticed that the Polyethylene Glycol-3 Ether with HLB of 5 and HLWR of 25% formed more stable emulsion compared to Polyethylene Glycol-2 Ether with HLB of 4. By comparing the physiochemical properties of the two emulsifiers (see Table 1), the only difference is the H/C ratio which is higher in the case of Polyethylene Glycol-3 Ether (~2.1 for Glycol-3 Ether versus 2.0 for Glycol-2 Ether); this indicates the presence of more unsaturation in the Glycol-2 Ether chain. Here we like to postulate that this behaviour is likely related to the C/H in the surfactant. Higher C/H indicates the presence of more unsaturation, which will influence the polarity and the surfactant may favour one phase (i.e., water or oleic) over the other. Keeping in mind that the water phase contains ions, the oleic phase may contain heavy hydrocarbon/acidic components. Most likely, the reduced stability is due to the unbalanced distribution of the emulsifier molecules between the two phases, because of

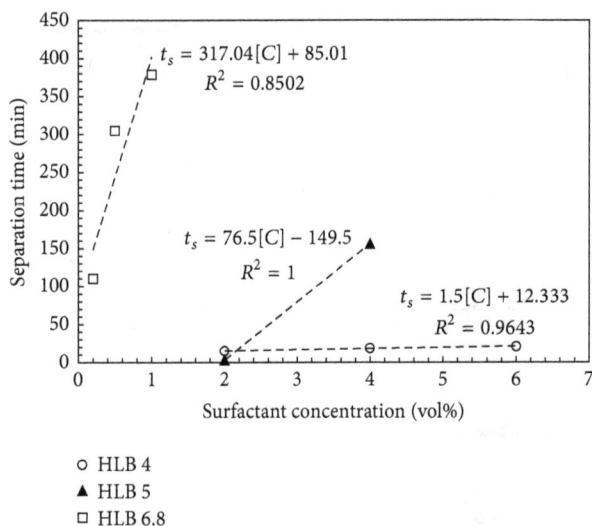

FIGURE 9: Separation time as function of surfactant concentration for different HLB (120°C).

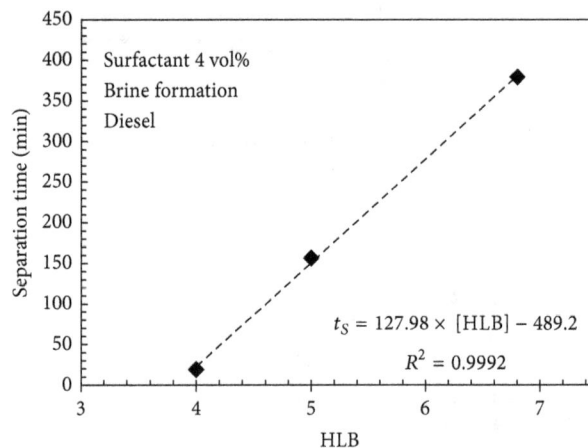

FIGURE 10: Separation time as a function of HLB at 120°C.

the change in polarity. Furthermore, Amine Acetate-3 shows higher stability in comparison with Glycol Ether family and at a lower concentration. The dependency of the separation time to surfactant concentration at constant HLB value is correlated in the form of a linear relationship as follows:

$$t_s = a \times [C] + b, \qquad (1)$$

where t_s is the separation time, [C] is the surfactant concentration, and a and b are constants. Herein, the intercept b is thought to reflect the effect of the HLB as can be seen from Figure 9; the magnitude of b changes as HLB value changes.

Although the constituting functional group for the Amine Acetate is distinct from the two Glycols (i.e., Glycol-2 Ether and Glycol-3 Ether), a trend between the HLB and the stability is evident; the separation time increases with increase in HLB value as shown in Figure 10. The dependence of the separation time on HLB is correlated in linear form:

$$t_s = a \times \text{HLB} - b, \qquad (2)$$

where t_s is the separation time, HLB is the surfactant hydrophilic-lipophilic balance, and a and b are constants. When HLB increased from 4 to 6.8 at a constant concentration, the separation time increased from 19 to 379 minutes (95% increases).

5. Conclusions

Here are the main findings of this article:

(1) Hydrophilic-lipophilic balance is used as a criterion for surfactant selection and was found to work well for many surfactants of different structures. This approach is expected to reduce the time and resources needed for selection of emulsifiers.

(2) The solubility of the surfactant has been postulated to play a vital role in the type of the formed emulsion.

Bancroft's rule states "the phase in which an emulsifier is more soluble constitutes the continuous phase"; this argument was found to not necessarily hold true in all cases. For example, this work showed that when a set of surfactants were dissolved in Diesel W/O emulsions were expected to form as suggested by Bancroft's rule. However, the conductivity and dilution tests indicate that produced emulsions were O/W. To further investigate this observation, the same surfactants were dissolved in water; again, the type of the formed emulsion was O/W, which is not in agreement with Bancroft's rule.

(3) Surfactants with higher HLB values inside the recommended range for W/O emulsions resulted in more stable emulsions. A correlation is provided for the dependency of the emulsion thermal stability on HLB. For surfactants with similar structure and physical properties, this work shows that the surfactant with higher C/H ratio will likely form a less stable emulsion. The authors postulated that this result is probably due to the unbalanced distribution of surfactant because of the difference in polarity.

(4) Surfactants with Amine functional group were found to form more stable W/O emulsion when compared to the Glycol Ether family group.

(5) This work has identified a set of new cost effective emulsifiers, suitable for emulsified water phase (polymeric gels/acids) in the oleic phase. The formed W/O emulsion is very stable at HTHS conditions.

(6) The general trend is that the separation time increases with the increase in emulsifier concentration and vice versa. The increase in stability is likely due to the decrease in interfacial tension with the increase in emulsifier concentration and the consequent delay in the coalescence of the droplets.

(7) Oleic phase composition was found to affect the thermal stability. A higher emulsion stability is achieved when waste oil is used instead of Diesel to prepare the emulsion, and 91% decrease in the separated

oleic phase is recorded when waste oil used instead of Diesel. This behaviour is suggested to be due to the presence of resins or asphaltenes, which are natural emulsifiers, in the waste oil. Usage of waste oil will result in minimizing the associated cost of using mineral oils for the emulsification application in oilfields.

(8) Further investigation is required to understand the effect of the water phase in case of adding materials such as polymer gel and acid loading on the W/O emulsion system.

In general, this work was successful in developing stable emulsified water-in-oil systems that can tolerate high-temperature high-salinity environments using scientific approaches. Also, HLB approach was used for surfactant selection, and new insights into structure-emulsion stability relationships beyond the HLB approach were provided. The findings of this work would open more avenues for future research.

Conflicts of Interest

The authors declare that they have no conflicts of interest.

Acknowledgments

King Abdul-Aziz City for Science and Technology (KACST) is acknowledged for supporting this research through Project no. AT-30-291. KFUPM, Saudi Aramco, and Qatar University are also acknowledged for their support.

References

[1] F. Nielloud and G. Marti-Mestres, *Pharmaceutical Emulsions and Suspensions*, Marcel Dekker Inc, New York, NY, USA, 1st edition, 2000.

[2] M. R. Greaves and J. C. Knoell, "A comparison of the performance of environmentally friendly anhydrous fire resistant hydraulic fluids," *Journal of ASTM International*, vol. 6, no. 10, 2009.

[3] C. D. Anderson and E. S. Daniels, *Emulsion Polymerization and Latex Applications*, Rapra Rev. Rep., Shawbury, England, 4th edition, 2003.

[4] S. A. Osemeahon, AJPAC 2011, 5, 204.

[5] D. J. McClements, "Biopolymers in Food Emulsions," *Modern Biopolymer Science*, pp. 129–166, 2009.

[6] S. Friberg, K. Larsson, and J. Sjoblom, *Food Emulsions*, Marcel Dekker, New York, NY, USA, 3rd edition, 2003.

[7] Q. A. Ahmed, A. Mohamed, R. Salah, and A. Bakhit, "Risk analysis and decision making in relative permeability modifier water shut-off treatment. society of petroleum engineers," *Society of Petroleum Engineers*, North Africa Technical Conference and Exhibition 2010, NATC 2010 - Energy Management in a Challenging Economy.

[8] A. Stavland, K. I. Andersen, B. Sandoey, T. Tjomsland, and A. A. Mebratu, "How to apply a blocking gel system for bullhead selective water shutoff: from laboratory to field," in *Proceedings of the SPE/DOE Symposium on Improved Oil Recovery*, Tulsa, Okla, USA.

[9] P. Nisha and W. Sushant, "A novel emulsified acid system for stimulation of very high-temperature carbonate reservoirs," in *Proceeding of the International Petroleum Technology Conference, IPTC*, Beijing, China, 2013.

[10] M. Al-Yaari, I. A. Hussein, and A. Al-Sarkhi, "Pressure drop reduction of stable water-in-oil emulsions using organoclays," *Applied Clay Science*, vol. 95, pp. 303–309, 2014.

[11] T. F. Tadros and B. Vincent, *Encyclopedia of Emulsion Technology*, Dekker, New York, NY, USA, 1983.

[12] J. Weiss, *Current Protocols in Food Analytical Chemistry*, John Wiley and Sons, 1st edition, 2002.

[13] M. M. Rieger, "Emulsions," in *The Theory and Practice of Industrial Pharmacy*, L. Lachman, H. A. Liberman, J. L. Kanig, and L. Fediger, Eds., pp. 502-503, Philadelphia, PA, USA, 3rd edition, 1986.

[14] S. Tamilvanan, S. R. Senthilkumar, R. Baskar, and T. R. Sekharan, "Manufacturing techniques and excipients used during the formulation of oil-in-water type nanosized emulsions for medical applications," *Journal of Excipients and Food Chemicals*, vol. 1, no. 1, 2010.

[15] P. Becher, *Encyclopedia of Emulsion Technology*, Dekker, New York, NY, USA, 0th edition, 1983.

[16] R. Navneet and I. P. Pandey, *Journal of Industrial Research Technology*, vol. 3, p. 12, 2013.

[17] H. C. Joshi, I. P. Pandey, A. Kumar, and N. Garg, APAC 2012,1, 2167.

[18] C. P. Lawhon, W. M. Evans, and J. P. Simpson, JPT 1967, 19, 943.

[19] A. D. Patel, *SPE Drilling and Completion*, vol. 14, p. 274, 1999.

[20] S. H. Al-Mutairi, H. A. Nasr-Ei-Din, A. D. Hill, and A. D. Al-Aamri, "Effect of droplet size on the reaction kinetics of emulsified acid with calcite," *SPE Journal*, vol. 14, no. 4, pp. 606–616, 2009.

[21] W. A. Fattah and H. A. Nasr-El-Din, *SPE Prod and Oper*, vol. 25, p. 151, 2010.

[22] Nor. 310581, (1999), invs.: A. Stavland, S. Nilsson.

[23] U.S. 1,922,154 (1933), invs.: De Groote.

[24] W. R. Dill, "Reaction times of hydrochloricacetic acid solutions on limestone," *Southwest Regional Meeting of the American Chemical Society, Society of Petroleum Engineers*, pp. 1–3, 1961, Oklahoma City.

[25] P. L. Crenshaw and F. F. Flippen, JPT 1968, 12, 1361.

[26] R. Navarrete, M. Miller, and J. Gordon, "Laboratory and Theoretical Studies for Acid Fracture Stimulation Optimization," in *Proceedings of the SPE Permian Basin Oil and Gas Recovery Conference*, Midland, Texas.

[27] F. N. Schneider and W. W. Owens, SPEJ 1982, 22, 79.

[28] D. D. Sparlin, JPT 1976, 28, 906.

[29] L. J. Kalfayan and J. C. Dawson, "Successful implementation of resurgent Relative Permeability Modifier (RPM) technology in well treatments requires realistic expectations," in *Proceedings of the SPE Annual Technical Conference and Exhibition, Society of Petroleum Engineers*, pp. 26–29, Houston, USA, September 2004.

[30] R. S. Seright, SPEJ 2009, 14, 5.

[31] A. I. A. Mohamed, I. A. Hussein, A. S. Sultan, K. S. M. El-Karsani, and G. A. Al-Muntasheri, "DSC investigation of the gelation kinetics of emulsified PAM/PEI system: Influence of surfactants and retarders," *Journal of Thermal Analysis and Calorimetry*, vol. 122, no. 3, pp. 1117–1123, 2015.

[32] M. S. Kamal, A. S. Sultan, and I. A. Hussein, "Screening of amphoteric and anionic surfactants for cEOR applications using a novel approach," *Colloids and Surfaces A: Physicochemical and Engineering Aspects*, vol. 476, pp. 17–23, 2015.

[33] E. Ruckenstein, "Microemulsions, macroemulsions, and the Bancroft rule," *Langmuir*, vol. 12, no. 26, pp. 6351–6353, 1996.

[34] D. J. McClements, "Lipid-Based Emulsions and Emulsifiers," in *Food Lipids: Chemistry, Nutrition, and Biotechnology*, C. A. Casimir and B. M. David, Eds., pp. 64–83, Taylor and Francis Group and CRC Press, New York, NY, USA, 3rd edition, 2008.

[35] W. C. Griffin, "Classification of surface-active agents by HLB," *Journal of the Society of Cosmetic Chemists*, vol. 1, p. 311, 1949.

[36] W. C. Griffin, "Calculation of HLB values of non-ionic surfactants," *Journal of the Society of Cosmetic Chemists*, vol. 5, p. 249, 1954.

[37] J. T. Davies, "A quantitative kinetic theory of emulsion type, I. Physical chemistry of the emulsifying agent: gas/liquid and liquid/liquid interface," in *Proceedings of the 2nd International Congress of Surface Activity*, pp. 426–438, Butterworths, London, UK, 1957.

[38] A. Sabhapondit, J. R. Vielma Guillen, and C. Prakash, "Laboratory optimization of an emulsified acid blend for stimulation of high-temperature carbonate reservoirs," in *Proceedings of the North Africa Technical Conference and Exhibition 2012: Managing Hydrocarbon Resources in a Changing Environment, NATC 2012*, pp. 352–358, egy, February 2012.

[39] M. Al-Yaari, I. A. Hussein, A. Al-Sarkhi, M. Abbad, and F. Chang, "Pressure drop reduction of stable emulsions: role of aqueous phase salinity," in *Proceedings of the SPE Saudi Arabia Section Technical Symposium and Exhibition 2013*, pp. 19–22, Al-Khobar, Saudi Arabia, May 2013.

[40] M. Al-Yaari, A. Al-Sarkhi, I. A. Hussein, F. Chang, and M. Abbad, "Flow characteristics of surfactant stabilized water-in-oil emulsions," *Chemical Engineering Research and Design*, vol. 92, no. 3, pp. 405–412, 2014.

[41] A. Mohamed, *Development of Emulsified gels for Water control in Oil and Gas Wells*, M.Sc., King Fhad University Petroleum Minerals, Al-Khobar, Saudi Arabia, 2014.

[42] P. Albonico, G. Burrafato, L. A. Di, and T. P. Lockhart, "Effective gelation-delaying additives for cr+3/polymer gels," in *Proceeding of the SPE International Symposium on Oilfield Chemistry, Society of Petroleum Engineers*, pp. 2–5, New Orleans, Louisiana, 1993.

[43] K. S. M. El-Karsani, G. A. Al-Muntasheri, A. S. Sultan, and I. A. Hussein, "Gelation of a water-shutoff gel at high pressure and high temperature: Rheological investigation," *SPE Journal*, vol. 20, no. 5, pp. 1103–1112, 2015.

[44] G. A. Al-Muntasheri, P. L. J. Zitha, and H. A. Nasr-Ei-Din, "A new organic gel system for water control: a computed tomography study," *SPE Journal*, vol. 15, no. 1, pp. 197–207, 2010.

[45] Z. Sidaoui, *A Novel Approach to Formulation of Emulsified Acid Using Waste Oil and Nano-Particles*, M.Sc., King Fhad University Petroleum Minerals, Al-Khobar, Saudi Arabia, 2016.

7

Study of Phase Equilibrium of NaBr + KBr + H$_2$O and NaBr + MgBr$_2$ + H$_2$O at 313.15 K

Qing Chen, Jiping She, and Yang Xiao

College of Energy, Chengdu University of Technology, Chengdu, Sichuan 610059, China

Correspondence should be addressed to Jiping She; 437290779@qq.com

Academic Editor: Christophe Coquelet

The phase equilibrium for the ternary systems NaBr + KBr + H$_2$O and NaBr + MgBr$_2$ + H$_2$O at 313.15 K was investigated by isothermal solution saturation method. The solubilities of salts and the densities of saturated solutions in these ternary systems were determined by chemical methods, while the equilibrium solid phases were analyzed by Schreinermarker wet residues method. Based on the experimental data, phase diagrams and density versus composition diagrams were plotted. The two ternary systems were type of simple common-saturation and without complex salt and solid solution. There are in all two crystalline regions, two univariant curves, and one invariant point in these phase diagrams of two ternary systems at 313.15 K. The equilibrium solid phases in the ternary system NaBr + KBr + H$_2$O are KBr and NaBr·2H$_2$O, and those in the ternary system NaBr + MgBr$_2$ + H$_2$O are NaBr·2H$_2$O and MgBr$_2$·6H$_2$O.

1. Introduction

Phase equilibrium in salt-water systems and phase diagram are the foundation of inorganic chemical production and salt mineral resources exploitation [1–4]. To extract relevant products from the potassium, magnesium, and bromine salt mine, it is essential to investigate the phase equilibrium of NaBr + KBr + H$_2$O and NaBr + MgBr$_2$ + H$_2$O. By now, a number of studies on the Br-bearing phase equilibria have been done, such as quaternary systems KCl–KBr–K$_2$SO$_4$–H$_2$O at 323 K, 348 K, and 373 K [5–7], NaBr–SrBr$_2$–MgBr$_2$–H$_2$O and KBr–SrBr$_2$–MgBr$_2$–H$_2$O at 323 K [8], and quinary system Na$^+$, K$^+$//Cl$^-$, Br$^-$, and SO$_4$$^{2-}$–H$_2$O at 373 K [9]. The two ternary systems NaBr–KBr–H$_2$O and NaBr–MgBr$_2$–H$_2$O also have been reported at 323 K and 348 K [10–12]. However, the data provided is far from enough, so an extensive study at other temperatures needs to be done. The phase equilibrium of NaBr + KBr + H$_2$O and NaBr + MgBr$_2$ + H$_2$O at 313.15 K has not been reported yet. This paper is conducive to fill the blank of data. In this study, the solubility and density of the ternary systems were obtained. The equilibrium solid phases were analyzed, and the crystallization regions were determined. All results can offer fundamental data support for salt mineral resources exploitation and further theoretical studies.

2. Methodology

2.1. Materials and Apparatus. The sources and purity of the chemicals are listed in Table 1. Doubly deionized water (electrical conductivity $\leq 1 \cdot 10^{-4}$ S·m^{-1}) is used in the work. A HZS-H thermostatic water bath shaker is employed to carry out the experiments.

2.2. Experimental Methods. The method of isothermal solution saturation [13–15] was employed to determine the solubility of the ternary systems. The famous Schreinermarks method of moist residues [15–17] was applied to determine the equilibrium solid phase in the experiments.

Based on a fixed ratio and ensuring that one of the components is excessive, the experimental components are added to a series of conical flasks (125 mL) gradually, and the sealed flasks are placed into the oscillator. The oscillator vibrates continuously at 313.15 K (the standard uncertainty of 0.3 K). In a pre-experiment, the liquid phase of the samples

TABLE 1: Purities and suppliers of chemicals.

Chemical	Mass fraction purity	Source
NaBr	≥99.0%	Tianjin Bodi Chemical Holding Co. Ltd., China
KBr	≥99.0%	Tianjin Bodi Chemical Holding Co. Ltd., China
MgBr$_2$·6H$_2$O	≥99.0%	Tianjin Bodi Chemical Holding Co. Ltd., China

is analyzed every 2 days, and it is shown that the phase equilibrium is reached in 10 days. After equilibrium, the oscillation is stopped and the system is allowed to stand for 4 days to make sure that all the suspended crystals settle. The wet residues and liquid phase are transferred to two volumetric flasks, respectively. Simultaneously, some other liquid phases are used to determine density individually. Finally, these samples are quantitatively analyzed by chemical methods.

More details of the experimental method and the procedure are presented in the previous papers [12–14].

2.3. Analysis. The concentration of potassium ion was analyzed by a sodium tetraphenylborate (STPB) hexadecyl trimethyl ammonium bromide (CTAB) titration [18–20] (uncertainty of 0.0058); the concentration of magnesium ion was measured with an EDTA standard solution using the indicator Eriochrome Black-T [21] (uncertainty of 0.0072); the concentration of bromine ion was determined by Mohr's method using a silver nitrate standard solution [21] (uncertainty of 0.0037); and the concentration of sodium was evaluated according to the ion charge balance. The density is measured using a pycnometer (uncertainty of 0.002). Each experimental result is achieved from the average value of three parallel measurements.

3. Results and Discussion

To compare with literature data [22, 23], the experimental data on the solubility for NaBr, KBr, or MgBr$_2$ in pure water at 313.15 K are in good agreement with the literature values, which demonstrates that the experimental devices and methods are feasible.

3.1. Solid-Liquid Phase Equilibrium for NaBr + KBr + H$_2$O. The experimental data were listed in Table 2. The ion concentration values were expressed in mass fraction in the equilibrium solution. The solution densities were given in grams per cubic centimeter. According to the experimental results, the phase diagram was plotted in Figure 1 and the relationship of the solution densities was plotted in Figure 2. In the ternary system NaBr + KBr + H$_2$O at 313.15 K, it contains one invariant point, two univariant curves, and two crystallization regions.

As indicated in Figure 1, A, B, C, and W denote solid NaBr, solid KBr, solid NaBr·2H$_2$O, and H$_2$O, respectively; point S, an invariant point, reflects the cosaturated solution

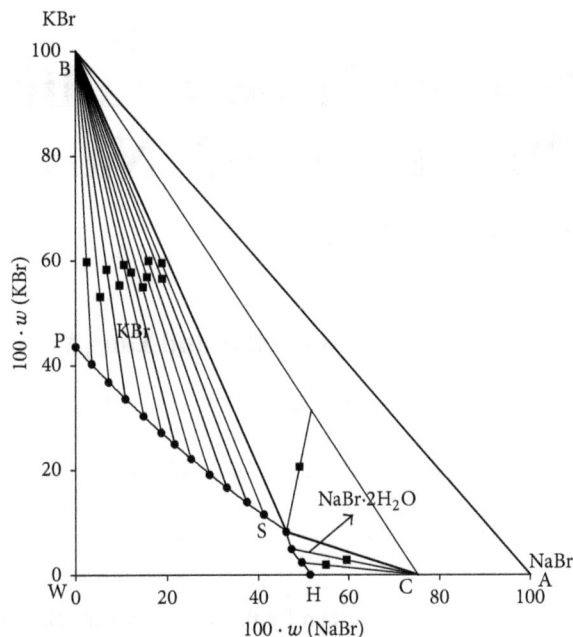

FIGURE 1: Equilibrium phase diagram of the ternary system NaBr + KBr + H$_2$O at 313.15 K. ●, equilibrium liquid phase composition; ■, moist solid phase composition; A, pure solid of NaBr; B, pure solid of KBr; C, pure solid of NaBr·2H$_2$O; W, water; H, solubility of NaBr in water; P, solubility of KBr in water; S, cosaturated point of NaBr·2H$_2$O and KBr.

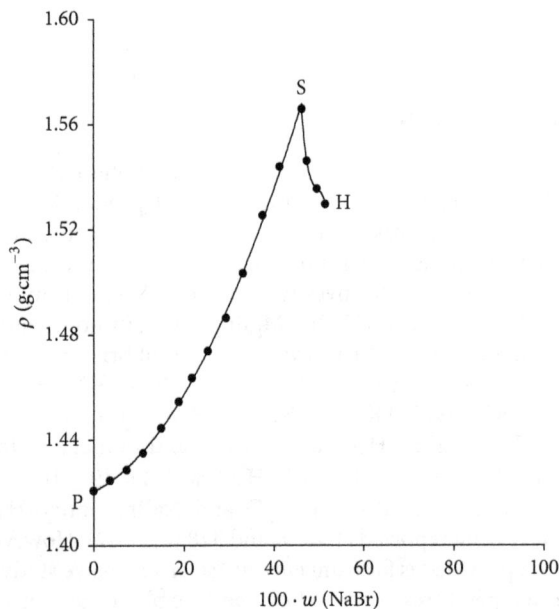

FIGURE 2: Density versus 100w (NaBr) in the ternary system (NaBr + KBr + H$_2$O). H, S, and P have the same meaning as described in Figure 1.

of KBr and NaBr·2H$_2$O at 313.15 K, with w (NaBr) = 0.4612 and w (KBr) = 0.0820; P and H denote the solubility of KBr and NaBr in water at 313.15 K, respectively. Two univariant solubility curves of this ternary system are PS and HS. Curve PS corresponds to the saturated KBr solution and visualizes changes of the KBr concentration with increasing the NaBr

TABLE 2: Mass Fraction Solubility of the ternary NaBr + KBr + H_2O system at temperature = 313.15 K and pressure = 0.1 MPa[a].

Number	Composition of liquid phase, 100w		Composition of wet residue phase, 100w		Densities of liquid phase	Equilibrium solid phase
	$100w_1$[b]	$100w_2$	$100w_1$	$100w_2$	$\rho/(g \cdot cm^{-3})$	
1, P	0.00	43.51	ND[c]	ND	1.4208	KBr
2	3.56	40.23	2.45	59.74	1.4247	KBr
3	7.23	36.77	5.43	53.13	1.4286	KBr
4	10.88	33.61	6.84	58.31	1.4351	KBr
5	14.88	30.33	9.60	55.30	1.4446	KBr
6	18.75	27.21	10.57	59.18	1.4546	KBr
7	21.68	24.98	12.22	57.81	1.4636	KBr
8	25.23	22.18	14.66	54.92	1.4738	KBr
9	29.28	19.08	15.63	56.89	1.4863	KBr
10	33.05	16.62	15.95	59.95	1.5034	KBr
11	37.45	13.87	18.97	56.55	1.5254	KBr
12	41.23	11.48	18.92	59.52	1.5439	KBr
13, S	46.12	8.20	49.09	20.61	1.5658	$NaBr \cdot 2H_2O$ + KBr
14	47.31	4.97	59.53	2.84	1.5462	$NaBr \cdot 2H_2O$
15	49.56	2.36	54.95	1.91	1.5355	$NaBr \cdot 2H_2O$
16, H	51.43	0.00	ND	ND	1.5296	$NaBr \cdot 2H_2O$

[a]Standard uncertainties $u(T)$ = 0.3 K, $u_r(p)$ = 0.05, $u_r(K^+)$ = 0.0058, $u_r(Br^-)$ = 0.0037, and $u_r(\rho)$ = 0.002. [b]w_1, mass fraction of NaBr; w_2, mass fraction of KBr. [c]ND, not determined. H, S, and P have the same meaning as described in Figure 2.

concentration. Curve SH corresponds to the saturated NaBr solution and indicates changes of the NaBr concentration with the KBr concentration increasing in the equilibrating solution. The KBr concentration decreases sharply with increasing the NaBr concentration, which illustrates that NaBr has a strong salting-out effect on KBr.

As indicated in Figure 1, along the curve PS, we connect the composition points of wet residue phase with liquid phase and then extend the intersection of these straight lines which is approximately the equilibrium solid phase for KBr. The same method is utilized to analyze the equilibrium solid phase of SH, and the intersection is $NaBr \cdot 2H_2O$. WPSH denotes unsaturated region at 313.15 K. BPS denotes crystallization region of KBr, while SHC denotes crystallization region of $NaBr \cdot 2H_2O$. Zone BSC represents the mixed crystalline region of KBr + $NaBr \cdot 2H_2O$. It is obvious that the crystalline region of $NaBr \cdot 2H_2O$ is much smaller than that of KBr.

The phase diagrams of the ternary system NaBr + KBr + H_2O at 323 and 348 K have been reported [10]. Apparently, the three phase diagrams have very similar shapes, each of them having an invariant point, two univariant curves, and two crystallization regions. The equilibrium solid phases in the ternary system NaBr + KBr + H_2O are potassium bromide (KBr) and sodium bromide dihydrate ($NaBr \cdot 2H_2O$) at 313 K and 323 K, and those are potassium bromide (KBr) and sodium bromide (NaBr) at 348 K.

Figure 2 indicates the relationship between the mass fraction of NaBr and the density in the solution. With increasing the NaBr concentration, the density first increases

and then the density declines afterwards. At the invariant point S, the density reaches a maximum value.

3.2. Solid-Liquid Phase Equilibrium for NaBr + $MgBr_2$ + H_2O.
The phase equilibrium experimental data is shown in Table 3, and the ternary phase diagram is drawn in Figure 3.

As indicated in Figure 3, A, M, D, C, and W denote solid NaBr, solid $MgBr_2 \cdot 6H_2O$, solid $MgBr_2$, solid $NaBr \cdot 2H_2O$, and H_2O, respectively; point Q, an invariant point, reflects the cosaturated solution of $MgBr_2 \cdot 6H_2O$ and $NaBr \cdot 2H_2O$ at 313.15 K, with w (NaBr) = 0.0418 and w ($MgBr_2$) = 0.4781; N and H represent the solubility of $MgBr_2$ and NaBr in water at 313.15 K, respectively. Two univariant solubility curves of this ternary system are PS and HS. Curve NQ corresponds to the saturated $MgBr_2$ solution and visualizes changes of the $MgBr_2$ concentration with increasing the NaBr concentration. Curve QH corresponds to the saturated NaBr solution and indicates changes of the NaBr concentration with increasing the $MgBr_2$ concentration. The solubility of NaBr decreases sharply with increasing the $MgBr_2$ concentration.

The polarization of ions has a certain effect on the dissolution of ionic crystals. The results show that the ionic dipole intensity in the solution depends on the electric field strength. In this study, the electrolyte concentration increased with the higher solubility of $MgBr_2$ added to the solution; also, the polarity of the solution increases, and the dielectric coefficient of the dielectric medium is reduced, while the ionic electric field strength increases, making it easy to bound more water to its surrounding, so that the reduction of water in the dissolution of other substances leads to enhanced

TABLE 3: Mass Fraction Solubility of the ternary NaBr + MgBr$_2$ + H$_2$O system at temperature = 313.15 K and pressure = 0.1 MPa[a].

Number	Composition of liquid phase, 100w		Composition of wet residue phase, 100w		Densities of liquid phase	Equilibrium solid phase
	100w_1[b]	100w_2	100w_1	100w_2	$\rho/(\text{g·cm}^{-3})$	
1, H	51.43	0.00	ND[c]	ND	1.5296	NaBr·2H$_2$O
2	45.58	4.98	59.07	2.67	1.5391	NaBr·2H$_2$O
3	40.05	10.49	60.65	4.21	1.5496	NaBr·2H$_2$O
4	34.65	15.45	57.84	6.50	1.5663	NaBr·2H$_2$O
5	30.38	20.43	59.93	6.71	1.5843	NaBr·2H$_2$O
6	23.82	26.87	56.46	9.52	1.6049	NaBr·2H$_2$O
7	17.41	32.79	55.09	11.08	1.6242	NaBr·2H$_2$O
8	10.80	38.47	53.04	12.91	1.6488	NaBr·2H$_2$O
9	6.76	43.23	50.73	15.14	1.6723	NaBr·2H$_2$O
10, Q	4.18	47.81	7.78	51.15	1.6846	NaBr·2H$_2$O + MgBr$_2$·6H$_2$O
11	2.72	48.95	2.01	52.78	1.6795	MgBr$_2$·6H$_2$O
12	1.32	50.28	0.98	54.15	1.6705	MgBr$_2$·6H$_2$O
13, N	0.00	51.62	ND	ND	1.6584	MgBr$_2$·6H$_2$O

[a]Standard uncertainties $u(T)$ = 0.3 K, $u_r(p)$ = 0.05, $u_r(\text{Mg}^{2+})$ = 0.0072, $u_r(\text{Br}^-)$ = 0.0037, and $u_r(\rho)$ = 0.002. [b]w_1, mass fraction of NaBr; w_2, mass fraction of MgBr$_2$. [c]ND, not determined. N, Q, and H have the same meaning as described in Figure 4.

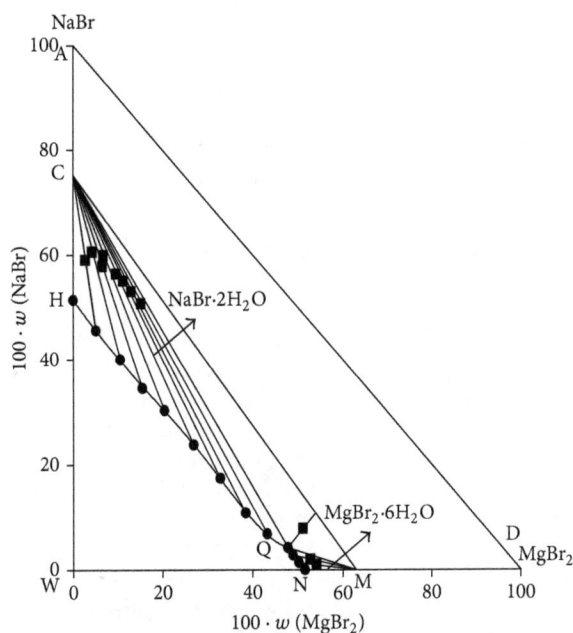

FIGURE 3: Equilibrium phase diagram of the ternary system NaBr + MgBr$_2$ + H$_2$O at 313.15 K. ●, equilibrium liquid phase composition; ■, moist solid phase composition; A, pure solid of NaBr; D, pure solid of MgBr$_2$; C, pure solid of NaBr 2H$_2$O; M, pure solid of MgBr$_2$·6H$_2$O; W, water; H, solubility of NaBr in water; N, solubility of MgBr$_2$ in water; Q, cosaturated point of MgBr$_2$·6H$_2$O and NaBr·2H$_2$O.

+ MgBr$_2$ + H$_2$O. Consequently, curve NQ corresponding equilibrium solid phase is MgBr$_2$·6H$_2$O and curve HQ corresponding equilibrium solid phase is NaBr·2H$_2$O. WNQH denotes unsaturated region at 313.15 K. NQM denotes crystallization region of MgBr$_2$·6H$_2$O, while HQC denotes crystallization region of NaBr·2H$_2$O. Zone MQC denotes the mixed crystalline region of MgBr$_2$·6H$_2$O + NaBr·2H$_2$O. It is obvious that crystallization region of MgBr$_2$·6H$_2$O is much smaller than that of NaBr·2H$_2$O.

The phase diagram of the ternary system NaBr + MgBr$_2$ + H$_2$O has been studied at 323 K and 348 K [11, 12]. Compared with the three phase diagrams at different temperatures, the result shows that the solubility of MgBr$_2$·6H$_2$O is highest at three temperatures. But the numbers of invariant points, crystallization fields, and univariant curves are different. The quaternary systems at 313 K and 348 K are all simple cosaturation type without complex salt and solid solution. They all include one invariant point, two univariant curves, and two crystallization regions (MgBr$_2$·6H$_2$O and NaBr·2H$_2$O at 313 K, MgBr$_2$·6H$_2$O and NaBr at 348 K). The phase diagram at 323 K includes two invariant points, three univariant curves, and three crystallization regions, where the solids are NaBr·2H$_2$O, NaBr, and MgBr$_2$·6H$_2$O, respectively.

Figure 4 indicates the relationship between the mass fraction of MgBr$_2$ and the density in the solution. With an increase of the MgBr$_2$ concentration, the density first increases and then, the density declines afterwards. At the invariant point Q, the density reaches a maximum value.

4. Conclusions

The phase equilibria in the NaBr + KBr + H$_2$O and NaBr + MgBr$_2$ + H$_2$O ternary systems at 313.15 K were investigated. The solubility and density data of the ternary systems were

salting out. In this system, it illustrates that MgBr$_2$ has a strong salting-out effect on NaBr.

In Figure 3, the same method used in Figure 1 is utilized to analyze the equilibrium solid phase of the system NaBr

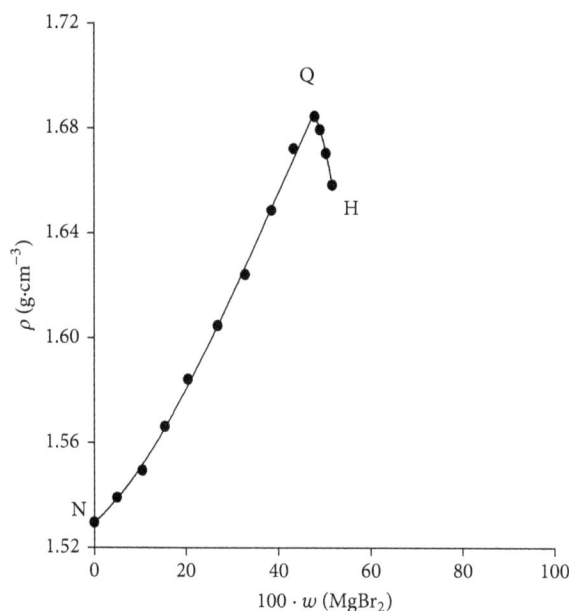

FIGURE 4: Density versus $100w$ (MgBr$_2$) in the ternary system (NaBr + MgBr$_2$ + H$_2$O). N, Q, and H have the same meaning as described in Figure 3.

obtained. The diagrams of density versus composition and the ternary phase diagrams were plotted. The equilibrium solid phases were analyzed and the crystalline regions were determined. In ternary system NaBr + KBr + H$_2$O, the crystalline region of KBr is much larger than that of NaBr·2H$_2$O and NaBr has a strong salting-out effect on KBr. In ternary system NaBr + MgBr$_2$ + H$_2$O, the crystalline region of NaBr·2H$_2$O is much larger than that of MgBr$_2$·6H$_2$O and MgBr$_2$ has a strong salting-out effect on NaBr. There are in all two crystalline regions, one invariant point, and two univariant curves in the ternary phase diagrams. All results can offer fundamental data support for optimizing the processes and further theoretical studies.

Conflicts of Interest

The authors declare that there are no financial conflicts of interest.

References

[1] J. Zhang, X. W. Shi, S. L. Zhao, X. F. Song, and J. G. Yu, "Progress in study on phase equilibria of salt-water systems," *Journal of Chemical Industry and Engineering(China)*, vol. 67, pp. 379–389, 2016.

[2] S. B. Shu, Y. L. Xu, E. X. Xu, and C. W. Xiao, "Study on the occurrence of potassium-rich brine in a geological structure in west Sichuan and the analytical patterns," *China Well and Rock Salt*, vol. 34, pp. 23–26, 2003.

[3] Y. T. Lin and S. X. Cao, "Rare gas field brines rich in potassium and boron of western Sichuan basin," *Geology in China*, vol. 28, pp. 45–47, 2001.

[4] T. L. Deng, H. Zhou, and X. Chen, *The Phase Diagram of Salt-Water System and Its Application*, Chemical Industry Press, Beijing, China, 2013.

[5] D. Wang, S. H. Sang, X. X. Zeng, and H. Y. Ning, "Phase equilibria of quaternary system KCl-KBr-K$_2$SO$_4$-H$_2$O at 323 K," *Petrochemical Technology*, vol. 40, pp. 285–288, 2011.

[6] K. J. Zhang, S. H. Sang, T. Li, and R. Z. Cui, "Liquid–Solid Equilibria in the Quaternary System KCl – KBr – K$_2$SO$_4$ – H$_2$O at 348 K," *Jounal of Chemical and Engineering Data*, vol. 58, pp. 115–117, 2013.

[7] R. Z. Cui, S. H. Sang, and Y. X. Hu, "Solid–Liquid Equilibria in the Quaternary Systems KCl-KBr-K$_2$B$_4$O$_7$-H$_2$O and KCl-KBr-K$_2$SO$_4$-H$_2$O at 373 K," *Jounal of Chemical and Engineering Data*, vol. 58, pp. 477–481, 2013.

[8] Q. Liu, Y. Y. Gao, S. H. Sang, R. Z. Cui, and X. P. Zhang, "Solid–liquid equilibria in the quaternary systems NaBr–SrBr$_2$–MgBr$_2$–H$_2$O and KBr–SrBr$_2$–MgBr$_2$–H$_2$O at 323 K," *Jounal of Chemical and Engineering Data*, vol. 63, pp. 1305–1318, 2016.

[9] R. Z. Cui, S. H. Sang, and Q. Liu, "Solid-liquid equilibria in the quinary system Na$^+$, K$^+$//Cl$^-$, Br$^-$, SO$_4^{2-}$–H$_2$O at 373 K," *Jounal of Chemical and Engineering Data*, vol. 61, pp. 444–449, 2016.

[10] A. Zdanovskii, E. Soloveva, E. Liahovskaia et al., *Khimiizdat*, St. Petersburg, Russia, 1973.

[11] C. Christov and J. Chem, "Study of bromide salts solubility in the (m$_1$NaBr + m$_2$MgBr$_2$)(aq) system at T = 323.15 K, thermodynamic model of solution behavior and solid-liquid equilibria in the (Na + K + Mg + Br + H$_2$O) system to high concentration and temperature," *Journal of Chemical Thermodynamics*, vol. 47, pp. 335–340, 2012.

[12] J. Hu, S. Sang, M. Zhou, and W. Huang, "Phase equilibria in the ternary systems KBr-MgBr$_2$-H$_2$O and NaBr-MgBr$_2$-H$_2$O at 348.15 K," *Fluid Phase Equilibria*, vol. 392, pp. 127–131, 2015.

[13] W. Liu, Y. Xiao, Y. S. Liu, F. X. Zhang, and J. F. Qu, "Phase equilibrium for the ternary system K$_2$SO$_4$ + KCl + H$_2$O in aqueous solution at 303.15 K," *Jounal of Chemical and Engineering Data*, vol. 60, no. 4, pp. 1202–1205, 2015.

[14] X. R. Zhang, Y. S. Ren, P. Li, H. J. Ma, and W. J. Ma, "Solid–liquid equilibrium for the ternary systems (Na$_2$SO$_4$ + NaH$_2$PO$_4$ + H$_2$O) and (Na$_2$SO$_4$ + NaCl + H$_2$O) at 313.15 K and atmospheric pressure," *Jounal of Chemical and Engineering Data*, vol. 59, no. 12, pp. 3969–3974, 2014.

[15] Z. D. Niu and F. Q. Cheng, *The Phase Diagram of Salt-Water System and Its Application*, Tianjin University Press, Tianjin, China, 2002.

[16] H. Schott, "A mathematical extrapolation for the method of wet residues," *Jounal of Chemical and Engineering Data*, vol. 6, pp. 324-324, 1961.

[17] F. A. H. Schreinemakers, "Graphical deductions from the solution isotherms of a double salt and its components," *Zeitschrift Fur Physikalische Chemie-International Journal of Research in Physical Chemistry & Chemical Physics*, vol. 11, pp. 109–765, 1893.

[18] X. B. Si and Y. L. Gao, "Improvement of determination of potassium in fertilizers—gravimetric sodium tetraphenylborate method," *Jounal of Chemical and Engineering Data*, vol. 1, pp. 49-50, 2002.

[19] C. J. Zhao, "Determination of potassium in fertilizers—gravimetric sodium tetraphenylborate method," *China Quality Supervision*, vol. 4, pp. 40-41, 2008.

[20] SN/T 0736.7-1999. Dongying City Agricultural Bureau: China, 1999.

[21] Qinghai Institute of Salt Lakes, *Chinese Academy of Science. Analytical Methods of Brines and Salts*, Chinese Science Press, Beijing, China, 2nd edition, 1988.

[22] A. Seidell, *Solubilities of Inorganic and Metal-Organic Compounds*, American Chemical Society, Washington, DC, USA, 1940.

[23] G. Q. Liu, L. X. Ma, and J. Liu, *Physical Property Data Handbook of Chemistry and Chemical Engineering: Inorganic Volume*, Chemical Industry Press, Beijing, China, 2002.

8

Decay Experiments of Effective N-Removing Microbial Communities in Sequencing Batch Reactors

Chen Lv,[1,2] **Ming Li,**[2] **Shuang Zhong,**[2] **Jianlong Wang,**[1] **and Lei Wu**[2]

[1]*Laboratory of Environmental Technology, Institute of Nuclear and New Energy Technology, Tsinghua University, Beijing 100084, China*
[2]*Key Laboratory of Songliao Aquatic Environment, Ministry of Education, Jilin Jianzhu University, Changchun 130118, China*

Correspondence should be addressed to Jianlong Wang; wangjl@mail.tsinghua.edu.cn

Academic Editor: Claudio Di Iaconi

The temporal changes in the compositions of effective N-removing bacterial communities and the decay coefficients of Anammox were studied within the 120-day decay period under anaerobic or aerobic conditions at 25°C. The maximum nitrogen production rate (MNPR) was determined by measuring the temperature, pH, volatile suspended solids (VSSs), and nitrogen-removal efficiency of the microbial communities during the decay period. The decay coefficients under anaerobic and aerobic conditions at 25°C were determined through equation-based fitting to be $0.031 \, d^{-1}$ and $0.070 \, d^{-1}$, respectively. Through molecular biological means and together with quantitative polymerase chain reaction (qPCR), the proportions of AnAOB in the microbial communities dropped from 48.70% to 3.69% under anaerobic condition and from 48.70% to 1.98% under aerobic condition during the decay period.

1. Introduction

Compared with traditional N-removal processes, the Anammox process is superior with low investment and operation costs, low sludge yield, high processing efficiency, and feasibility to wastewater with low C/N ratio and high ammonia nitrogen [1]. However, Anammox is limited by extremely low cell yield, slow cell growth, environmental sensitivity, and high requirements for temperature, pH, water, and substrate during cultivation [2, 3]. Moreover, the Anammox process is limited by the difficulty in starting, instability after start-up, and difficulty in recovery after destabilization [3–5]. These problems can be overcome if there are abundant favorable bacterial species that can be used for early-phase inoculation or anaphase fed-batch [6]. In current activated sludge models of aerobic degradation, the loss of activity and mass of activated sludge is expressed by only one process called decay [7]. And the decrease in bacterial activity in activated sludge can result from cell death and activity decay [8] and significantly affects the preservation of bacterial species and the decay coefficient [9, 10]. Moreover, the decay coefficient is one of the main variables in the mathematical modeling

that is applied to biological wastewater processing. Correct estimation of the decay constant is a key factor to properly model and better understand the Anammox process; it is also a parameter that is helpful to design and manage an Anammox reactor [11].

The objectives of this study are to accurately measure the decay coefficient of N-removing functional microbes during the whole decay period, and together with quantitative polymerase chain reaction (qPCR), to analyze the compositional changes of functional microbial communities, which were maintained under anaerobic/aerobic conditions and without feeding for about 4 months.

2. Materials and Methods

2.1. Materials. All the microbe samples in the decay experiments were collected from a laboratory small-scale sequencing batch reactor (SBR) fermentation tank (4 L). The reactor ran under controlled conditions (30°C, pH 7.5 ± 0.5) and was blended and stirred at the speed of 80 rpm by the machinery stirrer in the fermentation tank. The reactor had a running period of 8 h, drainage ratio of 50%, hydraulic retention time

(HRT) of 16 h, and volume nitrogen load rate of 750 mgN/L·d. Other devices included a temperature control set installed outside the fermentation tank, the annular aeration line at the bottom, and an online data monitor for measurement of dissolved oxygen (DO), pH, temperature, NH_4^+, and NO_3^-. The composition of the inflow water was 169.7 mg/L KH_2PO_4, 751.1 mg/L $MgSO_4\cdot7H_2O$, 451.6 mg/L $CaCl_2\cdot2H_2O$, 20.0 mg/L EDTA, 5.00 mg/L $FeSO_4\cdot7H_2O$, 0.43 mg/L $ZnSO_4\cdot7H_2O$, 0.24 mg/L $CoCl_2\cdot6H_2O$, 0.99 mg/L $MnCl_2\cdot4H_2O$, 0.25 mg/L $CuSO_4\cdot5H_2O$, 0.22 mg/L $NaMoO_4\cdot2H_2O$, 0.19 mg/L $NiCl_2\cdot6H_2O$, and 0.21 mg/L $NaSeO_4\cdot10H_2O$.

2.2. Experimental Methods. The decay experiments of N-removing functional microbial communities were conducted under anaerobic or aerobic conditions at 25°C. Four groups were conducted, each in triplicate. The microbial communities taken out of the reactor were washed with the substrate-free inflow water. Each time, 250 mL of a sample was placed into a 300 mL sealed bottle, which was put under the corresponding experimental condition. At a certain interval, 25 mL of the sample was collected, pretreated, and measured in terms of $NH_4^+/NO_2^-/NO_3^-$, pH, volatile suspended solids (VSS), specific Anammox activity (SAA), and qPCR. The decay changes of N-removing microbial communities under different conditions were described quantitatively.

SAA was detected using the method from Buys [12]. This method was first applied into denitrifying bacteria, but it was also feasible to measurement of low biomass and low gas production, so it was used in this study. The principle of bacterial activity test is that an appropriate ratio of NH_4^+ and NO_2^- was added to the substrate, and then N_2 production was detected. The anaerobic condition was realized by the ventilation of nitrogen gas into the reactor. Under the anaerobic condition, the substrate (NH_4^+ and NO_2^-, each 5 mmol) was added. According to (1), if the functional microbial community was active, the generated gas should be N_2.

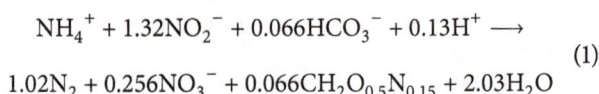

$$NH_4^+ + 1.32NO_2^- + 0.066HCO_3^- + 0.13H^+ \longrightarrow \\ 1.02N_2 + 0.256NO_3^- + 0.066CH_2O_{0.5}N_{0.15} + 2.03H_2O \tag{1}$$

SAA can be used to measure the newly added air pressure in the test bottle (unit: mV), or, namely, the N_2 production from the reaction. Then we could determine the nitrogen production rate n as follows:

$$PV = nRT, \tag{2}$$

where P is the overpressure (higher than normal pressure); V is the space volume in the top; R is the ideal gas constant (=0.0821 atm·mL/K·mmol); T is the temperature.

According to (2), we could determine the maximum nitrogen production rate (MNPR), which is proportional by K to the biomass solid concentration $X_{anx}(t)$. Thus, we could estimate the decay coefficient b_{AN} as follows:

$$MNPR(t) = k * X_{anx}(t)$$

$$\frac{dx_{anx}}{d_t} = -b_{AN} * X_{anx} \tag{3}$$

$$X_{anx}(t) = X_{anx(t=0)}e^{-b_{AN}*t}.$$

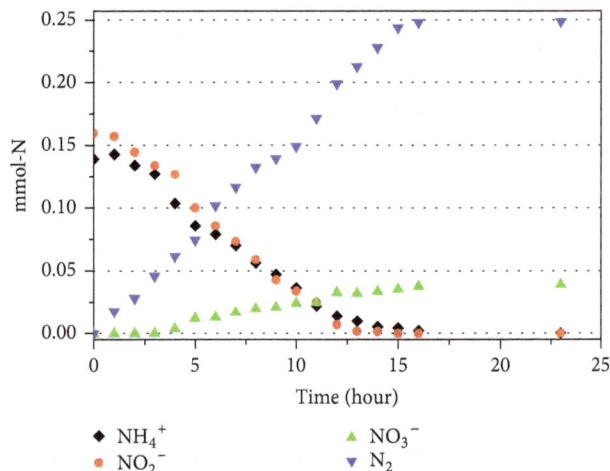

FIGURE 1: Initial N-removing performance of functional microbial communities.

2.3. Analytical Methods. VSS was measured by a standard method; pH was measured by a PHM210 device. N-containing particle concentration was detected by a colorimetric kit after filtration by a 0.45 μm acetic acid fibrin injector (Merck KGaA, Darmstadt, Germany).

In DNA extraction, a microbe sample after frozen drying at −50°C was weighed, and total DNA from each activated sludge sample was extracted using an MP soil DNA rapid extraction kit (Bio101, Vista, CA, USA) according to the manual. The qPCR amplification of 16rRNA functional genes was conducted with the following primers: 1055f/1391r (EUB) [13, 14], CTO 189fA/B/RT1r (AOB) [15], Nspra-675f/746r (NOB) [16], and Amx809f/1066r (AnAOB) [17, 18].

3. Results and Discussion

3.1. Initial N-Removing Performances of Functional Microbial Communities. The initial activity of each functional microbial community was measured. The results were atmospheric pressure = 18 mV and initial MNPR = 2.63 mLN$_2$-N/L·d. The air pressure peaked within 24 h, while the N_2 production during the whole reaction increased with time, which proves the initial microbes were highly active.

The N_2 production was measured simultaneously with sampling. The measurements of N-containing particles were converted to nitrogen molar concentrations (Figure 1). Clearly, at the 16th hour, under the action of the functional microbial communities, the newly added NH_4^+ and NO_2^- (totally 0.3 mmol-N) almost all reacted, forming 0.04 mmol-N NO_3^- and 0.25 mmol-N N_2, and the change of molar concentrations obeyed (1), which proves that the initial functional microbial communities had high N-removing ability.

3.2. Variation of VSS with Time. During the 120-day decay period, the functional microbial communities were sampled, and it was found the VSSs significantly declined under all test conditions (Figure 2). Under the anaerobic condition, VSSs dropped at a significant rate, which was slower than under the aerobic condition. The VSSs under the anaerobic condition in

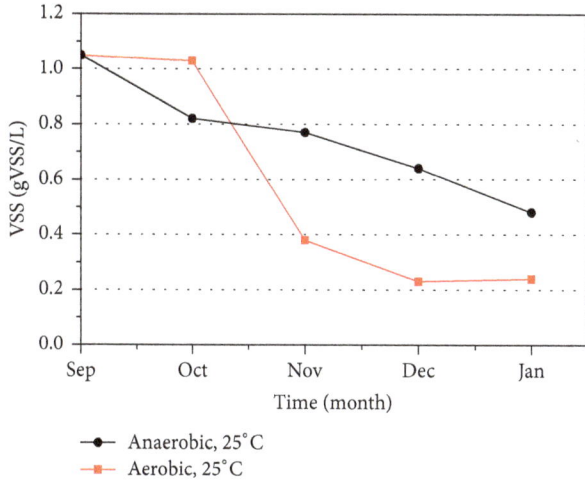

FIGURE 2: Variation of VSS during 120 days.

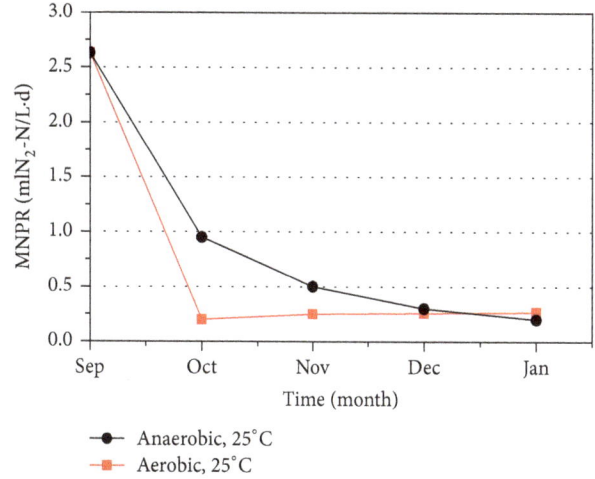

FIGURE 3: Variation of MNPR with time.

Oct, Nov, Dec, and Jan were 0.82, 0.77, 0.64, and 0.48 gVSS/L, respectively. Under the aerobic condition, the VSS in Oct did not change significantly, and was 1.03 gVSS/L. Nevertheless, the color changed significantly and turned light grey, which indirectly indicates the reduction of microbial activity. The VSS in Nov declined severely from the previous month and was 0.38 gVSS/L. The VSSs in Dec and Jan were 0.23 and 0.24 gVSS/L, respectively, indicating the microbes almost all decayed in the third and fourth months.

3.3. Variation of MNPR with Time.

In a reactor under continuous stirring, the equation derived from the mass balance under stable conditions represents the relationship between X_{AN} and b_{AN}:

$$X_{AN} = Y_{AN} \cdot \frac{\left(NH_4^+{}_{\_IN} - NH_4^+{}_{\_OUT}\right) \cdot SRT}{HRT \times \left(1 + b_{AN} \cdot SRT\right)}$$

$$SRT \longrightarrow \infty \quad (4)$$

$$X_{AN} \longrightarrow Y_{AN} \cdot \frac{\left(NH_4^+{}_{\_IN} - NH_4^+{}_{\_OUT}\right)}{HRT \times b_{AN}}.$$

Thus, when b_{AN} and other bioreactor parameters [hydraulic retention time (HRT), sludge retention time (SRT), nitrogen-removal efficiency, and nitrogen load] are known, the Anammox biomass concentration can be easily estimated [11]. In the following equation, the relevant parameters were cited from an Anammox reaction equation [19], and the MNPR (mLN$_2$-N/L·d) was associated with active biomass (X_{AN}):

$$MNPR(t) = \left(\frac{\mu_{max} \cdot 2.04 \cdot 22.4}{Y_{AN} \cdot 14 \cdot 2}\right) \cdot X_{ANX}(t)$$

$$= k \cdot X_{ANX}(t=0) \cdot e^{-b_{AN} \cdot t}, \quad (5)$$

where b_{AN} is the decay coefficient (d^{-1}), μ_{max} is the maximum growth rate (d^{-1}), X_{AN} is the concentration of Anammox organisms (mgCOD L^{-1}), Y_{AN} is the Anammox growth yield

(mg COD mg NH$_4$-N^{-1}), and k is the maximum specific nitrogen gas production rate (mLN$_2$-N L^{-1} d^{-1} mg COD^{-1}).

The SAAs during the 120-day decay period were measured and used to determine the MNPR according to (5). As shown in Figure 3, the MNPR (mLN$_2$-N/L·d) is 2.63 at first and then declines significantly with time. Under the anaerobic condition, the MNPRs in Oct, Nov, Dec, and Jan are 0.95, 0.5, 0.3, and 0.2 mLN$_2$-N/L·d, respectively. Under the aerobic condition, MNPRs decline significantly and are 0.25, 0.27, and 0.26 mLN$_2$-N/L·d, in the first three months, respectively. In the fourth month, nearly no air pressure could be detected, which indicates the inactivity of the functional microbial community. The above results suggest that MNPRs under the anaerobic condition decline regularly and in a gradient way with the prolonging of time. MNPR under the aerobic condition drops rapidly in the first month but does not change severely in the following three months, indicating that the O$_2$ concentration largely affects the decaying process.

3.4. Decay Coefficient b_{AN} of Functional Microbes.

The decay coefficient (b_{AN}) is commonly used in mathematical modeling of biological wastewater processing. Specifically, for the majority of Anammox process reactors, the balanced concentration of active biomass is largely dependent on b_{AN}.

b_{AN} can be determined by fitting NMPR according to (5) on software (Figure 4). b_{AN} of the microbes under the anaerobic and aerobic conditions at 25°C are 0.031 d^{-1} and 0.070 d^{-1}, respectively, indicating, at the same temperature, b_{AN} under the aerobic condition is two times larger than the anaerobic condition. Strous et al. studied the effects of oxygen on Anammox by using SBR [19]. In a reactor, the anaerobic and aerobic conditions were alternated, and Anammox reaction occurred only after the stop of oxygen supply but not during oxygen supply. Thus, the experiments prove that oxygen supply could inhibit the activity of Anammox, which can be restored after oxygen supply. Nevertheless, the inhibitory effect of oxygen concentrations on the Anammox activity should be further studied. When the oxygen concentration was 0.5%–2.0% of the saturated oxygen concentration in air,

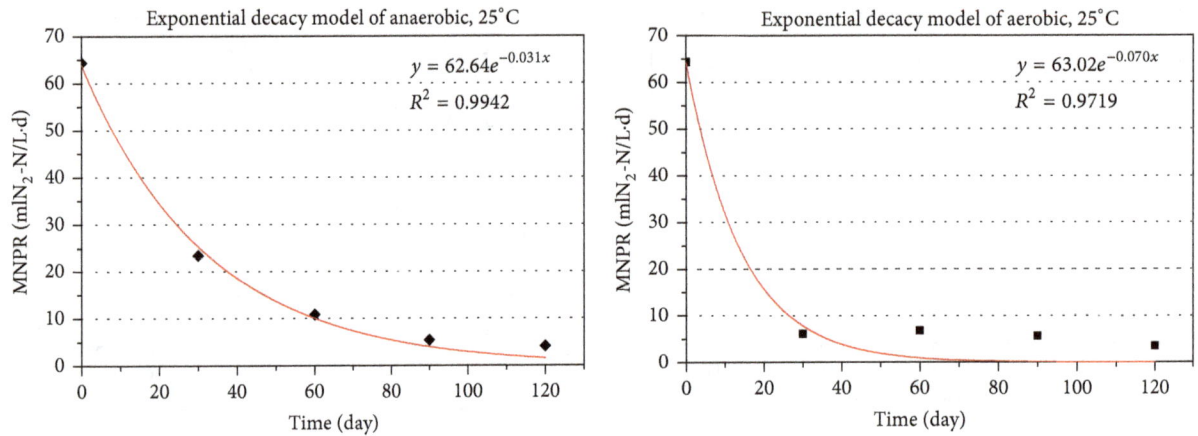

FIGURE 4: Fitting curves of NMPR during the decay period under different conditions.

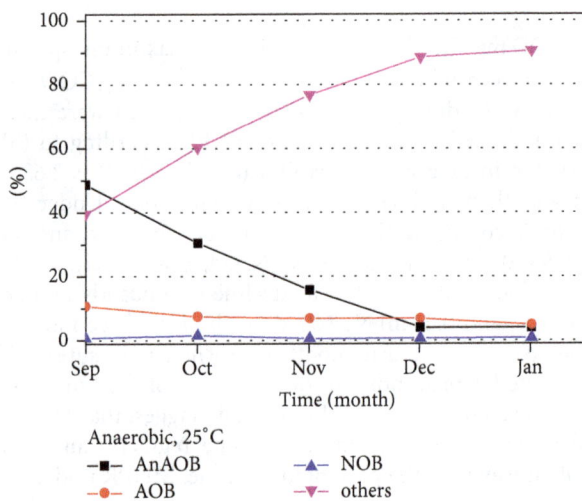

FIGURE 5: Proportion variation of species in the total functional microbial communities under anaerobic condition.

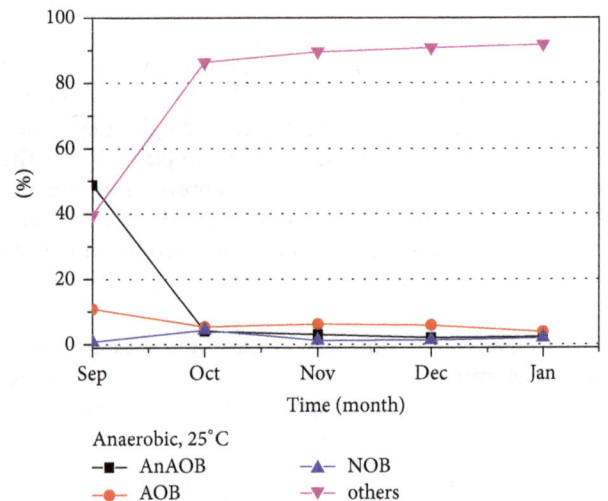

FIGURE 6: Proportion variation of species in the total functional microbial communities under aerobic condition.

the activity of Anammox was completely inhibited, indicating that the inhibitory oxygen concentration on the activity of Anammox is 0.5% of saturated oxygen concentration in air. Through our experiments, we not only determined the decay coefficient but also validated that again the oxygen concentration could inhibit the activity of Anammox.

3.5. Molecular Biology Analysis. Fluorescence qPCR can be used to detect the copy number of a gene in an unknown sample. Namely, a known concentration of the gene was diluted to a series of gradient concentrations, which were detected through one qPCR trial. The results determined from this gradient of concentrations were used to plot a standard curve, from which we could deduce the concentration of an unknown sample. The detected microbes were divided into four classes: AOB, NOB, AnAOB, and others. Based on qPCR, the measured data were used to estimate the proportion of each bacterial species in the overall biomass. When the proportion declines, the decaying speed

of this species surpasses the average rate of other species, and vice versa. The functional microbial communities were initially composed of AOB (10.80%), AnAOB (48.70%), NOB (0.87%), and others (39.63%).

Under the anaerobic condition, the functional microbial communities consisted of AnAOB (30.50%) and others (60.37%) (Oct); AnAOB (15.70%) and others (76.86%) (Nov); AnAOB (4.04%) and others (88.44%) (Dec); AnAOB (3.92%) and others (90.53%) (Jan) (Figure 5). Under the aerobic condition, the functional microbial communities consisted of AnAOB (3.98%) and others (96.31%) (Oct); AnAOB (2.98%) and others (89.51%) (Nov); AnAOB (1.98%) and others (90.80%) (Dec); AnAOB (2.30%) and others (91.79%) (Jan) (Figure 6). Clearly, under the aerobic condition, the decaying rate of AnAOB surpassed those of other species throughout the experiments, indicating that oxygen is extremely unfavorable for the survival of AnAOB and the decay of AnAOB is severe under aerobic conditions. The results of qPCR are consistent with the results of VSS, MNPR, and b_{AN}.

4. Conclusions

Correct evaluation of decay coefficient helps to better understand the Anammox process. The decay of functional microbial communities in 120-day experiments was monitored. MNPR was determined by measuring the temperature, pH, VSS, and nitrogen-removal efficiency of microbial communities during the decay period. The decay coefficients under anaerobic and aerobic conditions at 25°C were determined through fitting to be $0.031\,\mathrm{d}^{-1}$ and $0.070\,\mathrm{d}^{-1}$, respectively.

Through molecular biology means, AnAOB communities decreased faster under aerobic than anaerobic condition. The proportion of AnAOB in the microbial community is proportional to the N-removing efficiency.

Conflicts of Interest

The authors declare that they have no conflicts of interest.

References

[1] S. Tomar and S. K. Gupta, "Investigating the role of co-substrate-substrate ratio and filter media on the performance of anammox hybrid reactor treating nitrogen rich wastewater," *Journal of Bioscience and Bioengineering*, vol. 121, no. 3, pp. 310–316, 2016.

[2] Z. Lei and Z. Ping, "Metabolism of anaerobic ammonium oxidation (anammox) bacteria," *Bulletin of Science, Technology and Society*, vol. 26, no. 6, pp. 931–937, 2010.

[3] W. Cai-hua, Z. Ping, C. Jing, and C. Ting-ting, "Preservation of ANAMMOX bacteria," *China Environmental Science*, vol. 33, no. 8, pp. 1474–1482, 2013.

[4] A. G. Mutlu, A. K. Vangsgaard, G. Sin, and B. F. Smets, "An operational protocol for facilitating start-up of single-stage autotrophic nitrogen-removing reactors based on process stoichiometry," *Water Science and Technology*, vol. 68, no. 3, pp. 514–521, 2013.

[5] B.-S. Xing, Q. Guo, X.-Y. Jiang et al., "Influence of preservation temperature on the characteristics of anaerobic ammonium oxidation (anammox) granular sludge," *Applied Microbiology and Biotechnology*, vol. 100, no. 10, pp. 4637–4649, 2016.

[6] W. Cai-hua, Z. Ping, C. Jing, and C. Ting-ting, "Effects of intermittent starvation on preservation characteristics of ANAMMOX bacteria," *Acta Scientiae Circumstantiae*, vol. 33, no. 1, pp. 36–43, 2013.

[7] M. Friedrich and I. Takács, "A new interpretation of endogenous respiration profiles for the evaluation of the endogenous decay rate of heterotrophic biomass in activated sludge," *Water Research*, vol. 47, no. 15, pp. 5639–5646, 2013.

[8] X. Hao, Q. Wang, X. Zhang, Y. Cao, and C. M. V. Mark Loosdrecht, "Experimental evaluation of decrease in bacterial activity due to cell death and activity decay in activated sludge," *Water Research*, vol. 43, no. 14, pp. 3604–3612, 2009.

[9] X. Hao, Q. Wang, Y. Cao, and M. C. M. Van Loosdrecht, "Experimental evaluation of decrease in the activities of polyphosphate/glycogen-accumulating organisms due to cell death and activity decay in activated sludge," *Biotechnology and Bioengineering*, vol. 106, no. 3, pp. 399–407, 2010.

[10] R. Manser, W. Gujer, and H. Siegrist, "Decay processes of nitrifying bacteria in biological wastewater treatment systems," *Water Research*, vol. 40, no. 12, pp. 2416–2426, 2006.

[11] D. Scaglione, S. Caffaz, E. Bettazzi, and C. Lubello, "Experimental determination of Anammox decay coefficient," *Journal of Chemical Technology and Biotechnology*, vol. 84, no. 8, pp. 1250–1254, 2009.

[12] B. R. Buys, A. Mosquera-Corral, M. Sanchez, and R. Mendez, "Development and application of a denitrification test based on gas production," *Water Science and Technology*, pp. 41-12, 2000.

[13] H. Qianyi, Z. Ping, and K. Da, "Taxonomy, characteristics, and biotechniques used for the analysis of anaerobic ammonium oxidation bacteria," *Chinese Journal of Applied and Environmental Biology*, vol. 02, pp. 384–391, 2017.

[14] W. Yinhua, M. A. Yuexin, L. Changfa, Z. Xuehui, and Z. Ying, "The bacterial diversity in boifilms associated with media of anammox bioreactors," *Journal of Ocean University of China*, vol. 26, no. 6, pp. 500–506, 2011.

[15] P. Fangfang, Z. Peng, M. Hang, L. Kaihong, and P. Yangfang, "Comparing the primer specificity for betaproteobacterial ammonia-oxidizing bacteria in recirculation aquaculture systems," *Acta Microbiologica Sinica*, vol. 51, no. 10, pp. 1342–1350, 2011.

[16] T. Katipoglu-Yazan, C. Merlin, M.-N. Pons, E. Ubay-Cokgor, and D. Orhon, "Chronic impact of sulfamethoxazole on the metabolic activity and composition of enriched nitrifying microbial culture," *Water Research*, vol. 100, pp. 546–555, 2016.

[17] A. Daverey, S.-H. Su, Y.-T. Huang, S.-S. Chen, S. Sung, and J.-G. Lin, "Partial nitrification and anammox process: A method for high strength optoelectronic industrial wastewater treatment," *Water Research*, vol. 47, no. 9, pp. 2929–2937, 2013.

[18] A. Daverey, Y.-C. Chen, K. Dutta, Y.-T. Huang, and J.-G. Lin, "Start-up of simultaneous partial nitrification, anammox and denitrification (SNAD) process in sequencing batch biofilm reactor using novel biomass carriers," *Bioresource Technology*, vol. 190, pp. 480–486, 2015.

[19] M. Strous, J. J. Heijnen, J. G. Kuenen, and M. S. M. Jetten, "The sequencing batch reactor as a powerful tool for the study of slowly growing anaerobic ammonium-oxidizing microorganisms," *Applied Microbiology and Biotechnology*, vol. 50, no. 5, pp. 589–596, 1998.

9

Amperometric Formaldehyde Sensor Based on a Pd Nanocrystal Modified C/Co$_2$P Electrode

Huan Wang,[1] **Yaodan Chi,**[1,2] **Xiaohong Gao,**[1] **Sa Lv,**[1] **Xuefeng Chu,**[1] **Chao Wang,**[1] **Lu Zhou,**[1] **and Xiaotian Yang**[1]

[1]*Jilin Provincial Key Laboratory of Architectural Electricity & Comprehensive Energy Saving, Department of Materials Science, Jilin Jianzhu University, Changchun 130118, China*
[2]*College of Instrumentation & Electrical Engineering, Jilin University, Changchun 130012, China*

Correspondence should be addressed to Xiaotian Yang; hanyxt@163.com

Academic Editor: C. R. Raj

Ultrafine Pd nanocrystals were grown on the cobalt phosphide (Co$_2$P) decorated Vulcan XC-72 carbon (C/Co$_2$P), which is realized by first implementing the corresponding metal precursor and then the further chemical reduction process. The as-synthesized C/Co$_2$P/Pd composite was further constructed to form a gas permeable electrode. This electrode can be applied for formaldehyde (HCHO) detection. The results demonstrate that the Co$_2$P nanocrystal can significantly improve the sensing performance of the C/Co$_2$P/Pd electrode for catalytic oxidation of HCHO, which is considered to be attributed to the effective electron transfer from Co$_2$P to Pd in the C/Co$_2$P/Pd composites. Furthermore, the assembled C/Co$_2$P/Pd sensor exhibits high sensitivity of 617 nA/ppm and good selectivity toward various interfering gases such as NO$_2$, NO, SO$_2$, CO$_2$, and CO. It also shows the excellent linear response that the correlation coefficient is 0.994 in the concentration range of 1–10 ppm. Therefore, the proposed cost-effective C/Co$_2$P/Pd nanocomposite, which owns advantages such as high activity and good stability, has the potential to be applied as an effective electrocatalyst for amperometric HCHO detection.

1. Introduction

Formaldehyde, which is a colorless and toxic volatile organic compound (VOC), is an essential material that is applied in many aspects including the chemical factories and the residence decorating field. It is also well known that the formaldehyde can significantly increase the probability to get cancer for people [1]. Therefore, the detection of formaldehyde gas is significantly important. Until now, numerous technologies have been employed, such as spectrophotometry [2], gas chromatography (GC) [3], high-performance liquid chromatography [4], ion chromatography [5], polarography [6], and integrated sensor [7]. Among the above-mentioned techniques, the gas sensors have been widely reported by many research groups due to the advantages of high sensitivity, fast response, high stability, small size, and cost-effective features. In addition, these small packaged sensors can be mass fabricated easily in order to satisfy the large demand in many industrial fields. Generally, there are two types of formaldehyde gas sensor: (1) semiconductors formaldehyde gas sensor based on metal oxide such as SnO$_2$ [8, 9], ZnO [10], In$_2$O$_3$ [11, 12], and NiO [13] and (2) amperometric formaldehyde gas sensor made from noble metal nanomaterial, such as platinum or gold [14, 15]. In contrast with semiconductor gas sensor, the amperometric formaldehyde gas sensors have numerous advantages such as lower power consumption, excellent linear response, high sensitivity, and reasonable selectivity at room temperature.

Due to their high catalytic activity and excellent chemical stability, Pt-based catalysts are generally used in amperometric gas sensor. However, the scarcity and high cost of platinum greatly hinder its application. Much effort has been devoted to developing the platinum-free electrocatalysts, among which palladium-containing catalysts have been proven to be an effective candidate [16, 17]. The C/Pd catalyst is a good choice, due to its low cost and competitive intrinsic electrocatalytic

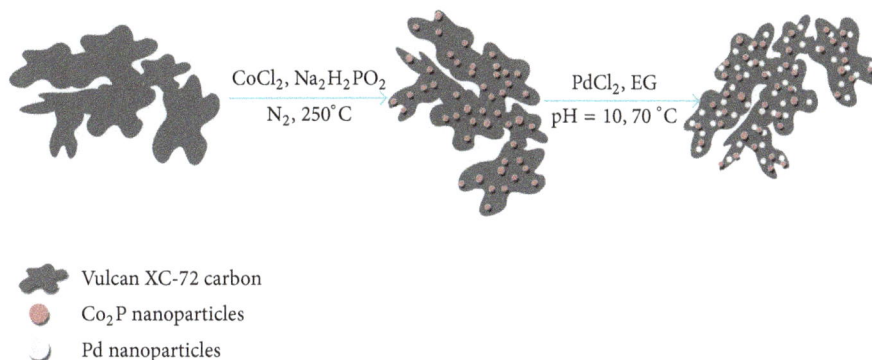

FIGURE 1: Synthesis procedure of Pd nanoparticle-modified C/Co$_2$P composites.

activity [18]. Furthermore, it is reported that the addition of another element, such as Fe, Co, Ni, N, or P, can enhance the catalytic activity of the C/Pd catalyst [19–22]. Unfortunately, the catalytic performance decreases rapidly by the dissolution or instability of the addition elements. It is well known that the transition metal phosphide nanoparticles such as Ni$_2$P, CoP, and FeP can significantly improve the catalytic activity and stability of catalyst [23–25]. But only a few works were reported about metal phosphide nanoparticles decorated Pd/C hybrids.

In this work, an amperometric formaldehyde gas sensor was fabricated by using the C/Co$_2$P/Pd catalyst as active electrode in H$_2$SO$_4$ electrolyte. The obtained C/Co$_2$P/Pd sensor exhibits better sensing performance for HCHO detection than that of the C/Co$_2$P and C/Pd sensor.

2. Experiment Section

2.1. Materials. Vulcan XC-72 carbon powder (99.9%) was purchased from Cabot Co. (USA). Sodium hypophosphite (99%), sodium hydroxide (99.9%), cobalt nitrate hexahydrate (99.99%), ethylene glycol (99.99%), ethanol (99.99%), and hydrochloric acid (37 wt.%) were purchased from Beijing Beihua Chemicals Co., Ltd. Nafion solution (5%) was purchased from DuPont Co. Palladium chloride (99.9%) was purchased from Tianjin Guangfu Chemicals Co., Ltd. Highly purified nitrogen (≥99.99%) and oxygen (≥99.99%) were supplied by Changchun Juyang Co., Ltd. Formaldehyde, nitric oxide, nitrogen dioxide, sulfur dioxide, ammonia, and carbon monoxide were supplied by Dalian Special Gases Co., Ltd.

2.2. Preparation of Co$_2$P Modified Pd/C Composites. Pd modified C/Co$_2$P composite was synthesized by solid phase reaction and ethylene glycol reduction according to a reported method [18, 23, 26, 27] and the corresponding schematic illustration is shown in Figure 1. Typically, 0.5 g Vulcan XC-72 was added to 20 mL aqueous solution containing 0.664 g CoCl$_2$. Then the mixed solution was stirred overnight followed by drying at 120°C for 5 h. The resulting Vulcan XC-72 containing CoCl$_2$ and 1.46 g NaH$_2$PO$_2$·H$_2$O was mechanically mixed in a quartz boat at room temperature. The precursor was directly heated to 250°C and was kept for 1 h under N$_2$ flowing condition. Then it was cooled to room

temperature. The resulting C/Co$_2$P composite was passivated in a 1.0 mol% O$_2$/N$_2$ mixture for 3 h. The obtained sample was washed with deionized water and was dried at 80°C for 24 h.

The 41.6 mg PdCl$_2$ and 100 mg C/Co$_2$P composites were dispersed into ethylene glycol (EG, 60 ml) followed by sonicating for 1 h under ambient condition. Then the solution was further stirred for 5 h. Its pH value was adjusted to 10 by 1 M NaOH solution. Then the mixture was placed in water bath and was heated at 70°C for 3 h with vigorous stirring. After reaction, its pH value was adjusted to 3 by adding hydrochloric acid and then incubated overnight. The resulting product was collected by vacuum filtration and was washed with hot deionized water and ethanol, and the resulting C/Co$_2$P/Pd composites were dried at 80°C for 12 h. For comparison, the C/Pd composite was prepared under the same condition.

2.3. Preparation of Modified Electrodes. The glass carbon electrodes (GCE, d = 3 mm) surfaces were carefully polished to a mirror using alumina slurries with different diameter (1.0, 0.3, and 0.05 mm) and then thoroughly cleaned ultrasonically with ethanol and deionized water. Subsequently, 5 μL of homogeneous C/Co$_2$P, C/Pd, and C/Co$_2$P/Pd dispersion (5 mg mL^{-1}) was dropped onto the surface of GCEs, respectively. After being dried at room temperature for 24 h, three modified GCEs were obtained and denoted as C/Co$_2$P/GCE, C/Pd/GCE, and C/Co$_2$P/Pd/GCE. Each working of C/Pd and C/Co$_2$P/Pd electrodes contained ca. 15 μg/cm^2 of Pd.

2.4. Fabrication and Measurement of Gas Sensor. Amperometric HCHO sensor was fabricated by the method depicted in Figure 2. Circular working electrode (diameter: 1 cm, area: 0.785 cm^2) was prepared as follows: 2.0 mg of catalyst was dispersed in ethanol (0.25 mL) containing 1 μL of 0.5 wt% Nafion ethanol solution to form homogenous suspension by sonication. The mixture was sprayed onto the porous PTFE membrane (PM 28Y, Porex) and was dried at room temperature. The Pd density of the C/Co$_2$P/Pd electrode was 0.667 mg/cm^2. The platinum wire counter electrode and Ag/AgCl (in 3 M KCl) reference electrode were then inserted into the sensor housing. The electrolyte is 0.5 M H$_2$SO$_4$. The distances among the three electrodes are all equal to 6 cm.

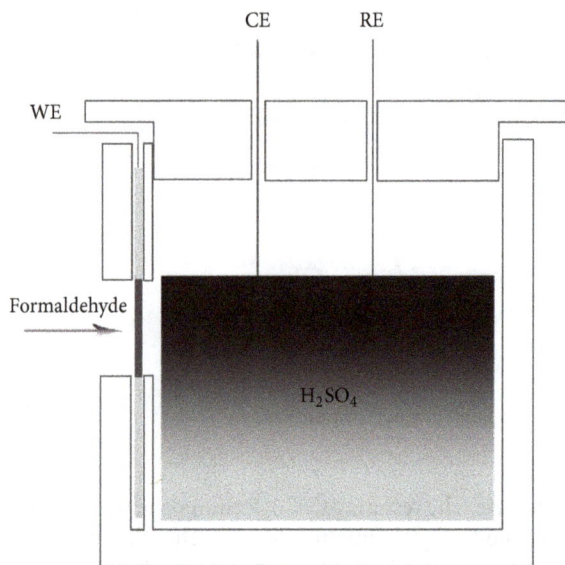

FIGURE 2: Schematic illustration of formaldehyde gas sensor.

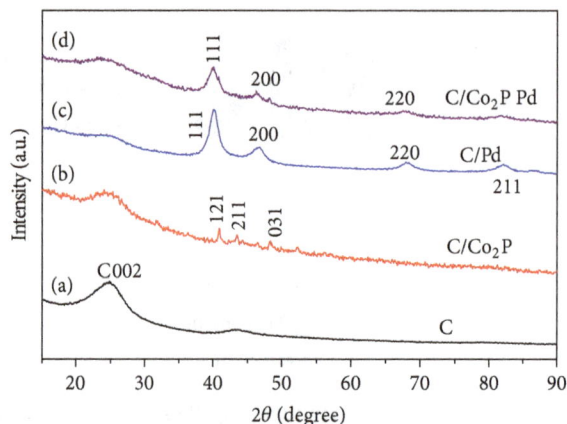

FIGURE 3: X-ray diffraction patterns for the Vulcan XC-72 carbon (a), C/Co_2P composite (b), C/Pd composite (c), and the C/Co_2P/Pd composite (d).

2.5. Characterizations. Transmission electron microscope (TEM), high-resolution TEM (HRTEM) (TEM, JOEL TEM-2010), scanning electron microscope (SEM, Phenom ProX), and energy-dispersive X-ray (EDX, Phenom ProX) were employed to inspect the morphologies and composition of samples. X-ray photoelectron spectroscopy (XPS) analysis was performed on an ESCALAB-MKII X-ray photoelectron spectrometer (VG Co.) with Al Ka X-ray radiation as the X-ray source for excitation. Cyclic voltammetric (CV) and the gas sensing properties were performed by a CHI660C electrochemical workstation (CHI). A conventional three-electrode cell comprises a modified GCE or prepared membrane as working electrode, a platinum wire as the auxiliary electrode, and Ag/AgCl (in 3 M KCl) as reference electrode. All the gas sensors were polarized in 0.5 M H_2SO_4 solution for 24 h at 0.75 V prior to amperometric measurements. Amperometric responses were measured to various gases of HCHO, NO_2, NO, SO_2, CO_2, and CO at the applied potential of 0.75 V (versus Ag/AgCl). The standard gases of HCHO, NO_2, NO, SO_2, CO_2, and CO were purchased from Dalian Special Gases Co., Ltd. The detection gas was mixed by using a mass flow controller which was linked to one N_2 chamber and a HCHO standard chamber. Different gas concentration can be obtained by modulating the mass flow controller digitally.

3. Results and Discussion

3.1. Morphology and Component Characterization. Figure 3 shows the XRD patterns of the Vulcan XC-72 carbon (a), C/Co_2P composites (b), C/Pd composite (c), and the C/Co_2P/Pd composite (d), wherein a (002) peak is observed at 24.5° from the XC-72 carbon. After solid phase reaction, the Co_2P nanoparticles are loaded on the XC-72 carbon. The major peaks at 40.7° (121), 40.9° (201), 43.2° (211), and 48.7° (031) of the C/Co_2P composites are well matched with standard Joint Committee on Powder Diffraction Standards (JCPDS) card number 32-0306 for orthorhombic Co_2P [28, 29]. The XRD

pattern of C/Pd clearly shows the peaks of metallic state Pd (JCPDS file: 65-2867) (Figure 3(c)). The XRD pattern of the C/Co_2P/Pd composite (Figure 3(d)) indicates the successful deposition of Pd nanoparticles by comparison with that of C/Pd (Figure 3(c)). However, the peaks of Co_2P are not observed in the C/Co_2P/Pd composite, which may be due to the Pd nanoparticles coated on the C/Co_2P supporters. The presence of Co_2P was confirmed by EDX and element distribution maps, and the results are similar to the previous report [23].

Morphology and element distribution of the C/Co_2P/Pd composites were characterized by TEM, HRTEM, SEM, EDX, and elemental mapping techniques. Figures 4(a) and 4(b) show the typical TEM and HRTEM images of the C/Co_2P/Pd composites. As shown in Figure 4(b), a nanocrystal particle was indicated by a red circle having a finger lattice of 0.21 nm, which is corresponding to the Co_2P (211) lattice [28]. In addition, the distances among the adjacent lattice fringes in the crystalline regions, which are marked by green circles, are 0.23 nm. This value is agrees well with the d spacing Pd (111) plane that the literature value is 0.229 nm (JCPDS number 65-6174). In order to further illustrate the surface chemical composition of the C/Co_2P/Pd composites, XPS measurements were performed from Figures 4(c)–4(f).

As shown in Figure 4(c), the survey XPS spectrum confirms the presence of C, O, Pd, Co, and P in the Pd/Co_2P/C composites. Figure 4(d) shows the XPS spectrum of Pd_{3d} in C/Co_2P/Pd composites. Two typical peaks at around 335.2 eV and 340.4 eV were corresponding to $Pd_{3d5/2}$ and $Pd_{3d3/2}$, respectively, which are located at 335.8 and 341.2 eV for the C/Pd composite (Figure S1) [30]. Such significant shift indicates a partial electron transfer from Co_2P to Pd [23]. This would increase the electron density of Pd. The unobvious peaks of P_{2p} and Co_{2p} are ascribed to Co_2P in the C/Co_2P/Pd composites, indicating the low content of the Co and P element on the surface of the C/Co_2P/Pd composites.

Figure 5 shows SEM image (a), element mapping images (b)–(f), and EDX spectrum (g) of the C/Co_2P/Pd composites. SEM image indicates that the obtained C/Co_2P/Pd

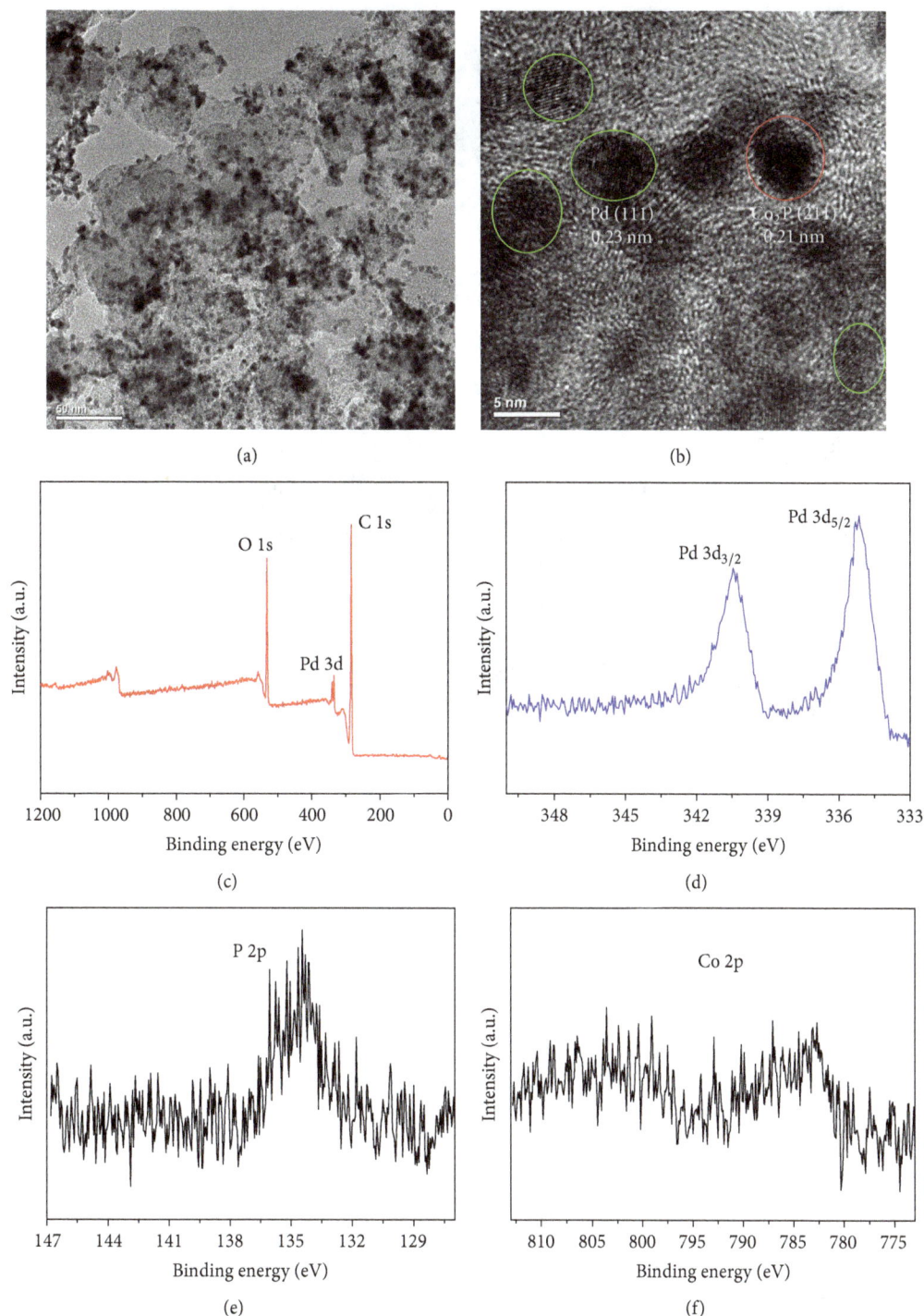

FIGURE 4: The TEM (a), HRTEM (b), and the survey, Pd 3d, P 2p, and Co 2p XPS spectra (c)–(f) of the $C/Co_2P/Pd$ composites.

composites are assembled of nanoparticles. As shown in Figures 5(b)–5(f), the element distribution of the $C/Co_2P/Pd$ composites was investigated by the element mapping, which clearly reveals the existence of C, Pd, O, Co, and P element in the $C/Co_2P/Pd$ composites. This result is coincident with the XPS spectra of the $C/Co_2P/Pd$ composites. Furthermore, EDX spectrum confirms the presence of C, O, Pd, Co, and P in the $Pd/Co_2P/C$ composite.

3.2. Electrochemical Properties of the $C/Co_2P/Pd$ Composites Electrodes. Figure 6(a) shows the cyclic voltammetry (CV) curves of the $C/Co_2P/GCE$ (A), $C/Pd/GCE$ (B), and $C/Co_2P/Pd/GCE$ (C) in 0.5 M H_2SO_4 solution in potential range of −0.2–1.0 V. As shown in Figure 6(A), no current could be observed on the $C/Co_2P/GCE$. It can be seen that the CV curves of the $C/Pd/GCE$ exhibited the characteristic Pd electrochemical reactions in H_2SO_4 solution,

(a)

(b)   (c)   (d)   (e)   (f)

(g)

FIGURE 5: SEM image (a); element mapping images (b)–(f) for C, Pd, O, Co, and P in the $C/Co_2P/Pd$ composites and EDX spectrum of the $C/Co_2P/Pd$ composites.

including reversible H_{upd} region (-0.2 V–0.1 V) and Pd oxidation/reduction regions. Interestingly, compared with curve B, peak current of curve C is significantly increased, which is attributed to the strong electronic interaction between Co_2P and Pd. Figure 6(b) is the CV curve of 5 mM formaldehyde in 0.5 M H_2SO_4 solution at the $C/Co_2P/Pd/GCE$. In the anodic sweep from 0.2 to 0.9 V, the oxide peak appears at 0.75 V attributed to the oxidation of CO_{ads} intermediate to CO_2 on Pd surface during the formaldehyde oxidation. In the successive cathodic sweep, the current peak observed at 0.38 V was due to the reduction of palladium oxide [31].

In order to compare the catalytic activity of the $C/Co_2P/$ GCE, C/Pd/GCE, and $C/Co_2P/Pd/GCE$, CVs of 50 mM formaldehyde in 0.5 M H_2SO_4 solution at the three electrodes were measured, which are shown in Figure 7(a). It is found

that the $C/Co_2P/Pd/GCE$ exhibits the relative negative oxidation potential and large current response for a given formaldehyde concentration, indicating that the $C/Co_2P/Pd/GCE$ has much higher catalytic activity toward the oxidation of formaldehyde than $C/Co_2P/GCE$ and C/Pd/GCE. Figure 7(b) shows CV responses of the $C/Co_2P/Pd/GCE$ toward formaldehyde with different concentrations. It is clearly shown that the peak current of formaldehyde increases with the rise of formaldehyde concentration from 0 to 50 mM.

Figure 8(a) shows the CVs with different scan rates at the $C/Co_2P/Pd/GCE$ in the same solution. The result reveals that as the scan rate increased, the forward oxidation peak currents increased and their peak potentials shifted slightly toward positive directions. In the case of backward potential sweep, increasing in the sweeping rate can cause the peak

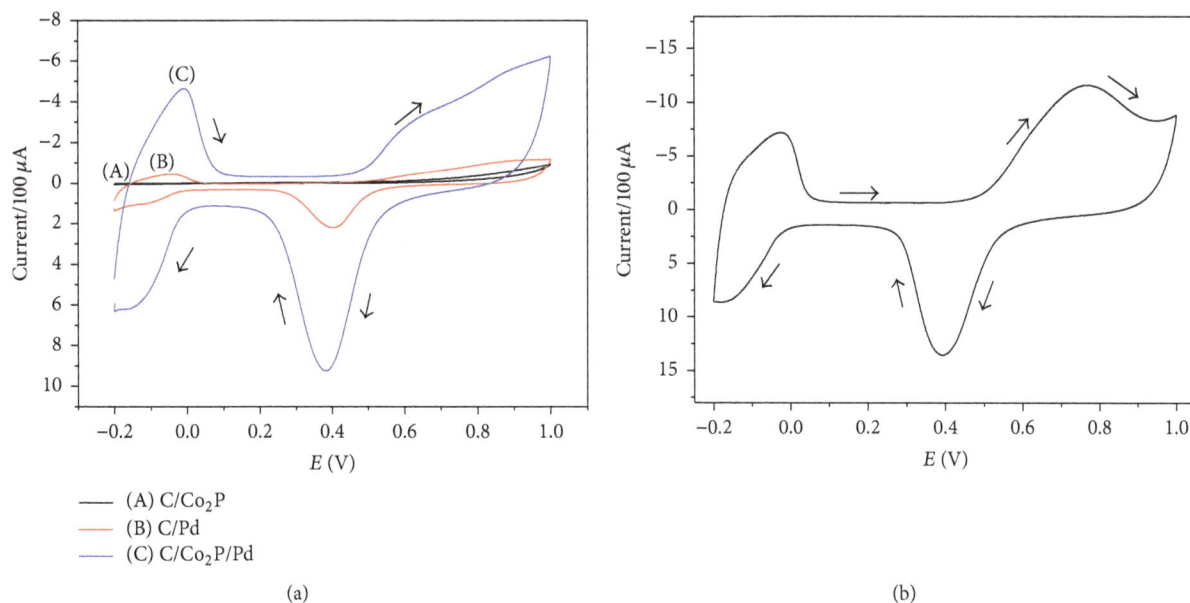

FIGURE 6: (a) CVs of the $C/Co_2P/GCE$ (A), $C/Pd/GCE$ (B), and $C/Co_2P/Pd/GCE$ (C) in 0.5 M H_2SO_4 solution. Scan rate: 50 mV/s. (b) CV of the $C/Co_2P/Pd/GCE$ in 0.5 M H_2SO_4 solution containing 5 mM formaldehyde from −0.2 V to 1.0 V at 50 mV/s.

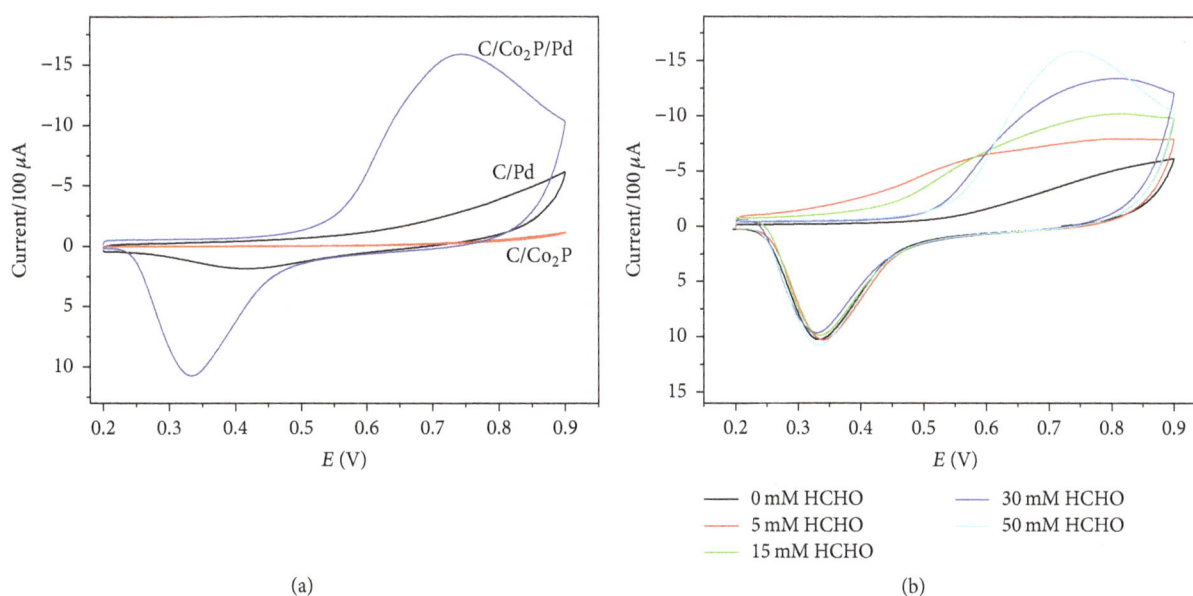

FIGURE 7: (a) CVs of the different electrodes in 0.5 M H_2SO_4 solution containing 50 mM formaldehyde from 0.2 V to 0.9 V at 50 mV/s. (b) CVs of the $C/Co_2P/Pd/GCE$ in 0.5 M H_2SO_4 solution with different concentration of formaldehyde from 0.2 V to 0.9 V at scan rate of 50 mV/s.

potential to shift negatively and peak currents increased. The peak current toward formaldehyde oxidation as function of the square root of the scan rate ($v^{1/2}$) in the range of 10 to 150 mV s^{-1} is shown in Figure 8(b). A good linear relationship with correlation coefficient of 0.9985 between current response and square root of the scan rate ($v^{1/2}$) indicates a diffusion-controlled electron transfer process.

3.3. Properties of the Amperometric Gas Sensor. Figure 9(a) presents the amperometric current response of the C/Pd

(black line), C/Co_2P composite (blue line), and $C/Co_2P/Pd$ (red line) sensor. It clearly shows that the $C/Co_2P/Pd$ sensor exhibits a higher current response than that of the C/Pd and C/Co_2P sensor. The enhanced activity of $C/Co_2P/Pd$ may be attributed to the electron transfer from Co_2P to Pd, which was demonstrated by XPS (Figure S1 in Supplementary Material available online at https://doi.org/10.1155/2017/2346895) and previous work [23]. Typical dynamic current response for the $C/Co_2P/Pd$ sensor was conducted under different formaldehyde concentrations (Figure 9(b)). It is seen that the peak

(a)

(b)

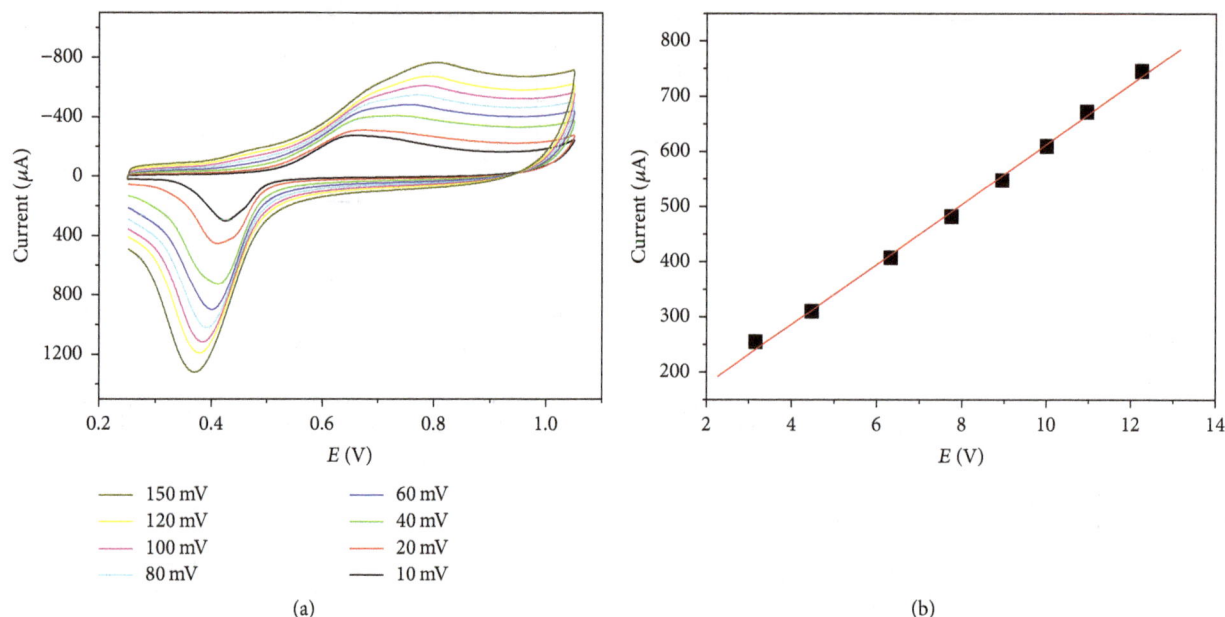

FIGURE 8: (a) CVs of the C/Co$_2$P/Pd/GCE in 5 mM formaldehyde of 0.5 M H$_2$SO$_4$ solution with different scan rates (10, 20, 40, 60, 80, 100, 120, and 150 mV/s). (b) The relation of the oxidation current of formaldehyde with square root of scan rate.

current increases linearly with the concentrations of HCHO from 1 to 10 ppm. The calibration curve (I (μA) = 3.57 + 5.70C (ppm); R = 0.994) was obtained from the inset of Figure 8(b), which indicates that the reaction on the C/Co$_2$P/Pd sensor was limited by the diffusion rate of HCHO gas [32]. On the basis of Fick's diffusion law, the limiting current I_{lim} was shown by the following equation [33]:

$$I_{\text{lim}} = k\,[\text{HCOH}]_{\text{gas}}, \qquad (1)$$

where the current is directly proportional to the gaseous concentration. It further indicates that the response is under the diffusion control at steady state.

Figure 9(c) is the enlarged parts of Figure 8(a). It shows that when the C/Co$_2$P/Pd sensor is exposed to HCHO gas, its current increases quickly, and it will quickly recover in the absence of HCHO. The response and recovery time are 3.8 s and 23.1 s, respectively, which are much shorter than those of the C/Pd and C/Co$_2$P sensor. The sensitivity of the C/Co$_2$P/Pd sensor is 617 nA/ppm. The background noise level is about 50 nA, which is confirmed by a previous report [34]. The limit of detection (LOD) of the C/Co$_2$P/Pd sensor is about 0.25 ppm (S/N = 3). The reproducibility of the HCHO gas sensor was evaluated in 10 ppm HCHO/N$_2$ mixture gas at six sensors prepared under the same condition. The RSD (n = 10) of the peak current is 3.25%. To evaluate the operational reproducibility of the C/Co$_2$P/Pd sensor, i-t curves were determined by 10 successive measurements in 10 ppm formaldehyde flow. The obtained RSD is 2.56%, which indicates that the C/Co$_2$P/Pd sensor can be used for routine analysis of HCHO in clinical use. The stability of the electrochemical sensor was also studied. After continuous 30 times' measurements, the current response almost does not decrease. The amperometric response with long time test for

the C/Co$_2$P/Pd sensor was studied under continuous 10 ppm formaldehyde flow. No obvious decrease is observed in 400 s, indicating the good stability of the C/Co$_2$P/Pd sensor (Figure S2).

In order to investigate the selectivity of the C/Co$_2$P/Pd gas sensor, the amperometric measurements were performed under various testing gases such as HCHO, NO$_2$, NO, SO$_2$, CO$_2$, and CO. Figure 9(d) shows the current response of the C/Co$_2$P/Pd sensor upon exposure to 10 ppm HCHO, 10 ppm NO$_2$, 10 ppm NO, 10 ppm SO$_2$, 5000 ppm CO$_2$, and 50 ppm CO. The detection selectivity, defined as the relative ratio of the response current of 10 ppm HCHO to that of 10 ppm NO$_2$, 10 ppm NO, 10 ppm SO$_2$, 5000 ppm CO$_2$, and 50 ppm CO, was 9.3, 10.0, 33.5, 13.2, and 40.5, respectively. The as-prepared C/Co$_2$P/Pd sensor exhibits excellent selectivity for HCHO detection [33].

4. Conclusions

The C/Co$_2$P/Pd hybrids were successfully synthesized through a two-step method and further applied for fabricating a sensor for detecting HCHO. The results show that the C/Co$_2$P/Pd based sensor has high sensitivity of 617 nA/ppm and its LOD can be determined as 0.25 ppm. Meanwhile, it also owns a good linear response in the concentration range of 1–10 ppm. The good performance of C/Co$_2$P/Pd gas sensor can be analytically attributed to the fast electron transfer from Co$_2$P to Pd nanocrystal. The proposed work provides an alternative way for developing the cost-effective catalyst for fabricating HCHO gas sensor.

Conflicts of Interest

The authors declare that they have no conflicts of interest.

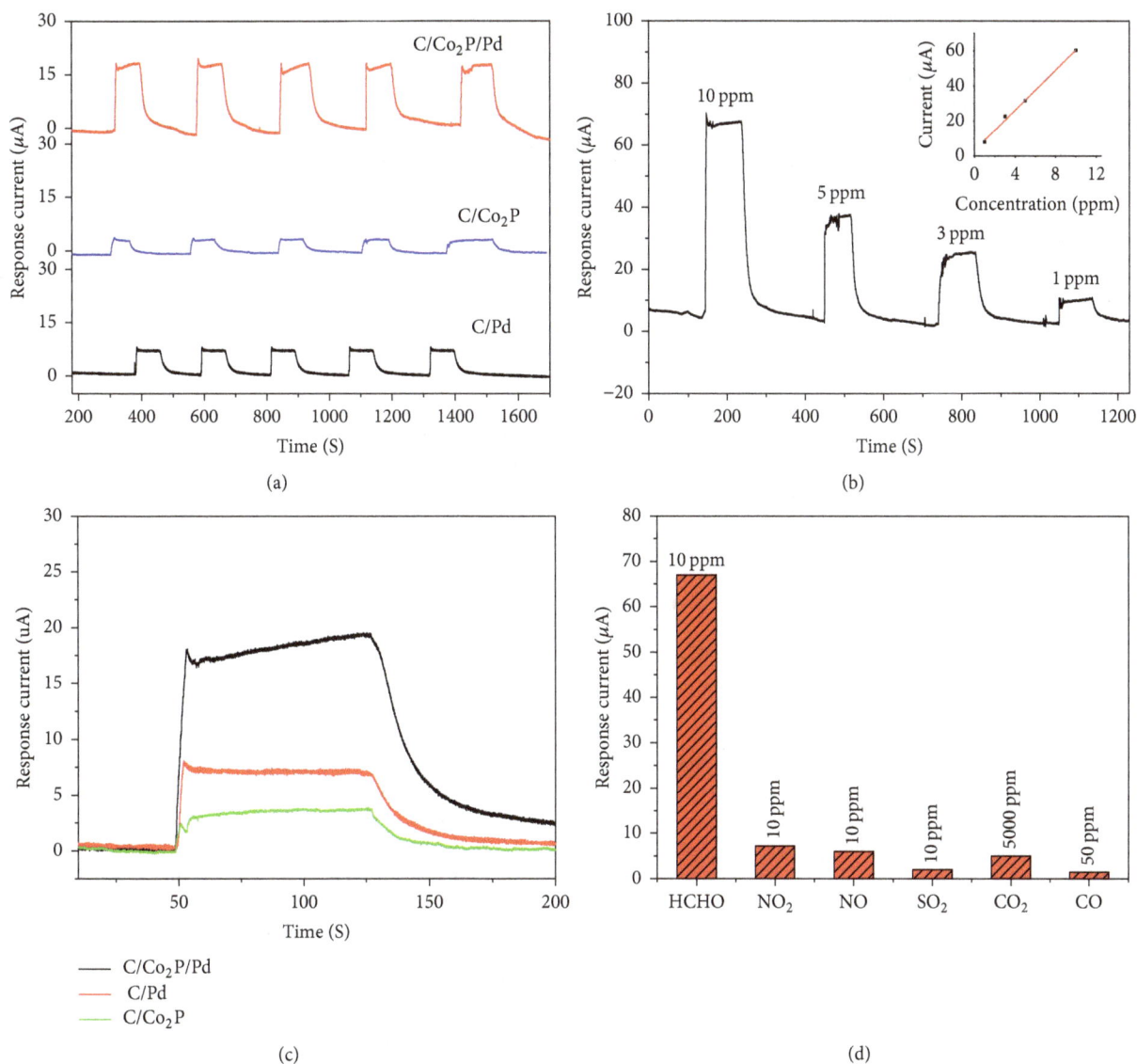

FIGURE 9: (a) Amperometric current responses of the C/Pd (black line), C/Co$_2$P composite (blue line), and C/Co$_2$P/Pd (red line) composites based gas sensors to 3 ppm formaldehyde. (b) Typical dynamic current response curves of C/Co$_2$P/Pd sensor to formaldehyde ranging from 1 ppm to 10 ppm (inset: calibration curve). (c) Enlarged part of the current responses to 3 ppm HCHO gas of the C/Pd (red line), C/Co$_2$P composite (green line), and C/Co$_2$P/Pd (black line) electrode based gas sensors. (d) Bar plot of current response of the C/Co$_2$P/Pd composites based formaldehyde sensor to different interfering gases under exposure to 10 ppm formaldehyde, 10 ppm NO, 10 ppm NO$_2$, 10 ppm SO$_2$, 5000 ppm CO$_2$, and 50 ppm CO. Applied potential to working electrode was +0.75 V (versus Ag/AgCl).

Acknowledgments

This work was supported by the National Natural Science Foundation of China (nos. 51272089, 51672103, and 21575136), National Key R&D Program of Strategic Advanced Electronic Materials (2016YFB0401103), and the Science and Technology Development Project of Jilin Province, China (nos. 20160204069GX and 20170520169JH).

References

[1] P.-R. Chung, C.-T. Tzeng, M.-T. Ke, and C.-Y. Lee, "Formaldehyde gas sensors: a review," *Sensors*, vol. 13, no. 4, pp. 4468–4484, 2013.

[2] G. R. Mohlmann, "Formaldehyde detection in air by laser induced fluorescence," *Applied Spectroscopy*, vol. 39, no. 1, pp. 98–101, 1985.

[3] T. Dumas, "Determination of formaldehyde in air by gas chromatography," *Journal of Chromatography A*, vol. 247, no. 2, pp. 289–295, 1982.

[4] B. Mann and M. L. Grayeski, "New chemiluminescent derivatizing agent for the analysis of aldehydes and ketones by high-performance liquid chromatography with peroxyoxalate chemiluminescence," *Journal of Chromatography A*, vol. 386, pp. 149–157, 1987.

[5] J. M. Lorrain, C. R. Fortune, and B. Dellinger, "Sampling and ion chromatographic determination of formaldehyde and

acetaldehyde," *Analytical Chemistry*, vol. 53, no. 8, pp. 1302–1305, 1981.

[6] J. C. Septon and J. C. Ku, "Workplace air sampling and polarographic determination of formaldehyde," *AIHAJ*, vol. 43, no. 11, pp. 845–852.

[7] B. J. Privett, J. H. Shin, and M. H. Schoenfisch, "Electrochemical sensors," *Analytical Chemistry*, vol. 82, no. 12, pp. 4723–4741, 2010.

[8] B. Kim, Y. Lu, A. Hannon, M. Meyyappan, and J. Li, "Low temperature Pd/SnO2 sensor for carbon monoxide detection," *Sensors and Actuators B: Chemical*, vol. 177, pp. 770–775, 2013.

[9] S. Tian, X. Ding, D. Zeng, J. Wu, S. Zhang, and C. Xie, "A low temperature gas sensor based on Pd-functionalized mesoporous SnO_2 fibers for detecting trace formaldehyde," *RSC Advances*, vol. 3, no. 29, pp. 11823–11831, 2013.

[10] L. Han, D. Wang, Y. Lu, T. Jiang, B. Liu, and Y. Lin, "Visible-light-assisted HCHO gas sensing based on Fe-doped flowerlike ZnO at room temperature," *The Journal of Physical Chemistry C*, vol. 115, no. 46, pp. 22939–22944, 2011.

[11] C. Dong, X. Liu, B. Han, S. Deng, X. Xiao, and Y. Wang, "Non-aqueous synthesis of Ag-functionalized In2O3/ZnO nanocomposites for highly sensitive formaldehyde sensor," *Sensors and Actuators B: Chemical*, vol. 224, pp. 193–200, 2016.

[12] X. Lai, D. Wang, N. Han et al., "Ordered arrays of bead-chain-like In_2O_3 nanorods and their enhanced sensing performance for formaldehyde," *Chemistry of Materials*, vol. 22, no. 10, pp. 3033–3042, 2010.

[13] C. Dong, Q. Li, G. Chen, X. Xiao, and Y. Wang, "Enhanced formaldehyde sensing performance of 3D hierarchical porous structure Pt-functionalized NiO via a facile solution combustion synthesis," *Sensors and Actuators B: Chemical*, vol. 220, pp. 171–179, 2015.

[14] C.-H. Chou, J.-L. Chang, and J.-M. Zen, "Effective analysis of gaseous formaldehyde based on a platinum-deposited screen-printed edge band ultramicroelectrode coated with Nafion as solid polymer electrolyte," *Sensors and Actuators B: Chemical*, vol. 147, no. 2, pp. 669–675, 2010.

[15] S. Liu, K. Hua, Y. Su, X. Lu, C. Li, and Y. J. Wang, "Application of gold hollow nanospheres in amperometric formaldehyde gas sensor," *Chinese Journal of Analytical Chemistry*, vol. 37, pp. 1092–1096, 2009.

[16] M. Yuan, A. Liu, M. Zhao et al., "Bimetallic PdCu nanoparticle decorated three-dimensional graphene hydrogel for non-enzymatic amperometric glucose sensor," *Sensors and Actuators B: Chemical*, vol. 190, pp. 707–714, 2014.

[17] Y. Jiang, Y. Lu, F. Li, T. Wu, L. Niu, and W. Chen, "Facile electrochemical codeposition of "clean" graphene-Pd nanocomposite as an anode catalyst for formic acid electrooxidation," *Electrochemistry Communications*, vol. 19, no. 1, pp. 21–24, 2012.

[18] M. Chen, Z.-B. Wang, K. Zhou, and Y.-Y. Chu, "Synthesis of Pd/C catalyst by modified polyol process for formic acid electrooxidation," *Fuel Cells*, vol. 10, no. 6, pp. 1171–1175, 2010.

[19] B. M. Leonard, Q. Zhou, D. Wu, and F. J. Disalvo, "Facile synthesis of PtNi intermetallic nanoparticles: Influence of reducing agent and precursors on electrocatalytic activity," *Chemistry of Materials*, vol. 23, no. 5, pp. 1136–1146, 2011.

[20] L. Zhang, Y. Tang, J. Bao, T. Lu, and C. Li, "A carbon-supported Pd-P catalyst as the anodic catalyst in a direct formic acid fuel cell," *Journal of Power Sources*, vol. 162, no. 1, pp. 177–179, 2006.

[21] J. Jia, R. Wang, H. Wang et al., "A novel structural design of CNx-Fe3O4 as support to immobilize Pd for catalytic oxidation of formic acid," *Catalysis Communications*, vol. 16, no. 1, pp. 60–63, 2011.

[22] X. Wang and Y. Xia, "Electrocatalytic performance of PdCo-C catalyst for formic acid oxidation," *Electrochemistry Communications*, vol. 10, no. 10, pp. 1644–1646, 2008.

[23] J. Chang, L. Feng, C. Liu, W. Xing, and X. Hu, "An effective Pd-Ni2P/C anode catalyst for direct formic acid fuel cells," *Angewandte Chemie International Edition*, vol. 53, no. 1, pp. 122–126, 2014.

[24] Z. Zhang, J. Hao, W. Yang, and J. Tang, "Defect-Rich CoP/Nitrogen-Doped Carbon Composites Derived from a Metal-Organic Framework: High-Performance Electrocatalysts for the Hydrogen Evolution Reaction," *ChemCatChem*, vol. 7, no. 13, pp. 1920–1925, 2015.

[25] J. J. Podestá, R. C. V. Piatti, and A. J. Arvia, "The influence of iridium, ruthenium and palladium on the electrochemical behaviour of Co-P and Ni-Co-P base amorphous alloys for water electrolysis in KOH aqueous solutions," *International Journal of Hydrogen Energy*, vol. 20, no. 2, pp. 111–122, 1995.

[26] X. Yang, A.-Y. Lu, Y. Zhu et al., "Rugae-like FeP nanocrystal assembly on a carbon cloth: an exceptionally efficient and stable cathode for hydrogen evolution," *Nanoscale*, vol. 7, no. 25, pp. 10974–10981, 2015.

[27] L. Song, S. Zhang, and Q. Wei, "A new route for synthesizing nickel phosphide catalysts with high hydrodesulfurization activity based on sodium dihydrogenphosphite," *Catalysis Communications*, vol. 12, no. 12, pp. 1157–1160, 2011.

[28] B. Tian, Z. Li, W. Zhen, and G. Lu, "Uniformly Sized (112) Facet Co2P on Graphene for Highly Effective Photocatalytic Hydrogen Evolution," *The Journal of Physical Chemistry C*, vol. 120, no. 12, pp. 6409–6415, 2016.

[29] M. Zuo, L. Pan, T. Sun, D. Zhao, Z. Wang, and H. Geng, "Controlled synthesis of Co2P particles with various morphologies using an in-situ melt reaction," *Materials and Corrosion*, vol. 90, pp. 858–866, 2016.

[30] M. S. Ahmed and S. Jeon, "Highly active graphene-supported NixPd100-x binary alloyed catalysts for electro-oxidation of ethanol in an alkaline media," *ACS Catalysis*, vol. 4, no. 6, pp. 1830–1837, 2014.

[31] Q. Hao and Y. Zhang, "Electrochemical degradation of formaldehyde with a novel Pd/GO modified graphite electrode," *International Journal of Electrochemical Science*, vol. 11, no. 2, pp. 1496–1511, 2016.

[32] D.-D. La, C. K. Kim, T. S. Jun et al., "Pt nanoparticle-supported multiwall carbon nanotube electrodes for amperometric hydrogen detection," *Sensors and Actuators B: Chemical*, vol. 155, no. 1, pp. 191–198, 2011.

[33] J. R. Stetter and J. Li, "Amperometric gas sensors - A review," *Chemical Reviews*, vol. 108, no. 2, pp. 352–366, 2008.

[34] M. Rashid, T.-S. Jun, Y. Jung, and Y. S. Kim, "Bimetallic core-shell Ag@Pt nanoparticle-decorated MWNT electrodes for amperometric H2 sensors and direct methanol fuel cells," *Sensors and Actuators B: Chemical*, vol. 208, pp. 7–13, 2015.

Synthesis and Characterization of Ag-Modified V$_2$O$_5$ Photocatalytic Materials

Dora Alicia Solis-Casados,[1] **Luis Escobar-Alarcon,**[2]
Antonia Infantes-Molina,[3] **Tatyana Klimova,**[4]
Lizbeth Serrato-Garcia,[1] **Enrique Rodriguez-Castellon,**[3]
Susana Hernandez-Lopez,[5] **and Alejandro Dorazco-Gonzalez**[1]

[1]*Centro Conjunto de Investigación en Química Sustentable UAEM-UNAM, Km 14.5 Carretera Toluca-Atlacomulco,*
Unidad San Cayetano, 50200 Toluca, MEX, Mexico

[2]*Departamento de Física, Instituto Nacional de Investigaciones Nucleares, P.O. Box 18-1027, 11801 Mexico City, Mexico*

[3]*Departamento de Quimica Inorganica, Facultad de Ciencias, Universidad de Malaga, 29071 Malaga, Spain*

[4]*Departamento de Ingenieria Quimica, UNAM, Mexico City, Mexico*

[5]*Facultad de Química, Universidad Autónoma del Estado de México, Paseo Colon esq Paseo Tollocan Col Nueva la Moderna, 50000*
Toluca, MEX, Mexico

Correspondence should be addressed to Dora Alicia Solis-Casados; solis_casados@yahoo.com.mx

Academic Editor: Julie J. M. Mesa

V$_2$O$_5$ powders modified with different theoretical silver contents (1, 5, 10, 15, and 20 wt% as Ag$_2$O) were obtained with acicular morphologies observed by scanning electron microscopy (SEM). Shcherbinaite crystalline phase is transformed into the Ag$_{0.33}$V$_2$O$_5$ crystalline one with the incorporation and increase in silver content as was suggested by X-ray diffraction (XRD) and X-ray photoelectron spectroscopy (XPS) analysis. With further increase in silver contents the Ag$_2$O phase appears. Catalysts were active in photocatalytic degradation of malachite green dye under simulated solar light, which is one of the most remarkable facts of this work. It was found that V$_2$O$_5$-20Ag was the most active catalytic formulation and its activity was attributed to the mixture of coupled semiconductors that promotes the slight decrease in the rate of the electron-hole pair recombination.

1. Introduction

In recent years, wastewaters from domestic and industrial uses have contributed to the environmental problem because they arrive to the soil and aquifers mantles polluting clean water. In order to address this issue, the effort of many researchers from several scientific disciplines around the world [1, 2] has been focused on wastewater remediation. Heavy metals such as mercury, iron, cadmium, and chromium are included between the most dangerous pollutants in wastewaters as well as some organic compounds such as the phenols, dyes, pesticides, pharmaceutical, and fertilizers and in some cases solvents. All of them are extremely toxic to the humans and also living organisms, even if they are in

a low concentration [3]. Several researchers have proposed solutions to reduce the most dangerous and toxic pollutants contained in wastewaters and also to improve water quality modifying the chemical processes, proposing new absorbent materials and some of the advanced oxidation processes as photocatalysis, which includes developing a photocatalytic material. The remotion of the most resilient organic compounds is among the most important topics concerning wastewater remediation and widely studied nowadays. In so many cases, one of the most extended methods is the photocatalytic degradation, whereby the catalyst is activated by using light. It is noteworthy that degradation of the resilient organic compounds dissolved in water occurs in a natural way by the photolysis process by using the cheapest source energy such as that provided by the sun. The main disadvantage is its very low efficiency since the organic molecules degrade slowly, taking days and even months to achieve the complete mineralization of organic compounds into water and CO_2. The photocatalytic process increases the decomposition rate of organic compounds present in wastewaters. The photocatalyst employed is commonly a semiconductor material with desirable characteristics such as photoactivity, being chemically and biologically inert, photostability, nontoxicity, and low cost [4]. A photocatalyst could be employed in its pure, mixed, or doped form. An example of this is the TiO_2 in its anatase crystalline phase with a band gap energy around 3.2 eV and the rutile crystalline phase with a band gap energy of 3.0 eV [5, 6]. It has been reported that mixing different ratios of anatase : rutile results in higher catalytic activities, which can be attributed to the synergistic effect between both phases, as occurs in coupled semiconductors [7]. Titania in its anatase crystalline phase is in disadvantage for generating the electron-hole pair if the excitation source is sunlight [8, 9]. It is the reason to improve the photocatalytic performance of TiO_2 by doping and modification with metals and nonmetals [10, 11]. It is well known that, in order to obtain better catalytic performance, it is not enough that the photocatalytic material has a low band gap energy to be active under sunlight. Some unstable materials with reduced band gap energy are Fe_2O_3 (2.3 eV), GaP (2.23 eV), and GaAs (1.4 eV) that are not so good as photocatalysts to degrade organic compounds in aqueous solutions [12]. It has been reported before by some researchers that one of the materials with low band gap energy, of around 2.8 eV, and some stability in aqueous solutions is the V_2O_5-based photocatalysts [13], potentially active under irradiation with visible light and investigated in the last years [14–17]. The synthesis of V_2O_5 has been reported by using techniques as the hydrothermal synthesis [18], sol-gel technique [19], thermal decomposition of several precursors as ammonium metavanadate (NH_4VO_3) [20], flame-spray pyrolysis [21, 22] magnetron sputtering, electron-beam evaporation, and pulsed laser deposition [23, 24], obtaining several morphologies as nanobelts [18, 19], nanowires [18], nanoribbons, nanopowders [18, 21], and also thin films [19, 21–23]. Thin films and powders have been obtained with different textural and structural properties related in some cases with their photocatalytic activity [24–26]. It should be considered that the request of doped or coupled a V_2O_5 catalyst to another semiconductor to retarding the recombination of the electron-hole pair, improving catalytic activity [27], is due to the disadvantage of short migration distances for excited electron-hole pairs which increases the recombination rate and decreasing the photocatalytic activity. Other researchers to solve this fact in photocatalysts with low band gap energy have reported that adding a small amount of noble metals (such as Pt, Ag, and Pd) [28–30] could retard the recombination of the electron-hole pair and increase the photocatalytic activity of the V_2O_5. In this research work, the synthesis of Ag-modified V_2O_5 photocatalytic materials through the surfactant assisted technique by using the non-ionic surfactant molecule polyoxyethylene lauryl ether (Brij L23) is proposed. The main purpose was to obtain catalytic formulations based in V_2O_5-xAg that could be photoactive to degrade organic compounds contained in water under simulated solar light.

2. Experimental

2.1. Synthesis of V_2O_5 and Ag-Modified V_2O_5 Photocatalysts. V_2O_5 powders from Fermont were treated by the surfactant assisted technique. This technique uses a micellar solution prepared using polyoxyethylene (23) lauryl ether ($C_{12}E_{23}$, Brij L23 30% w/v solution from Sigma Aldrich), dibutyl ether ($[CH_3(CH_2)_3]_2O$, DBE-reagent plus ≥99% from Sigma Aldrich), and $AgNO_3$ (ACS Reagent, ≥99.0% from Sigma Aldrich). To obtain V_2O_5 photocatalyst, the commercial V_2O_5 was added to the micellar solution previously prepared with 30% Brij L23, 10% DBE, and 60% of water, solvothermally pressurized in a reactor at 60°C for 12 hours, with slow stirring. The obtained solids were filtered, washed with deionized water, and then thermally treated, raising temperature from room temperature to 400°C, at a heating rate of 5°C/min; afterwards, the temperature was isothermally maintained at 400°C for 3 hours in an air convection oven to eliminate the surfactant Brij L23; this sample was called hereinafter V_2O_5. The Ag-modified V_2O_5 photocatalysts were prepared by following the same synthesis procedure. Thus, the required amounts of V_2O_5 and $AgNO_3$ precursors were added to the micellar solution of Brij L23 to obtain a theoretical content of 1, 5, 10, 15, and 20 wt.% of Ag_2O. Photocatalysts will be referred hereinafter as V_2O_5-xAg, where x represents the percentage in weight of Ag_2O (1, 5, 10, 15, and 20 wt% Ag_2O) and for simplicity the Ag_2O content in some parts of the manuscript has been referred as the Ag load.

2.2. Physicochemical Characterization of V_2O_5-xAg. The temperature of surfactant elimination to obtain the photocatalytic formulation was determined from thermogravimetry and differential scanning calorimeter analysis (TGA-DSC), performed by using a Netzsch, STA 449 F3 Jupiter equipment. The weight loss and the heat flow during decomposition were measured in flowing air (20 mL/min) and heating from room temperature to 600°C at a heating rate of 3°C/min. Molecular structure was studied through the infrared spectra (IR) acquired in a Bruker tensor 27 IR spectrometer with ATR accessory. Surface morphology was observed with scanning electron microscopy (SEM) by using a JEOL JSM 6510 LV microscope; additionally, compositional analysis was carried out with an acceleration voltage of 15 kV, using an EDS probe coupled to the same microscope. Raman spectra (RS) were acquired using an HR LabRam 800 system equipped with an Olympus BX40 confocal microscope; a Nd:YAG laser beam (532 nm) was focused by a 100x objective onto the sample surface. A cooled CCD camera was used to record the spectra, usually averaged for 100 accumulations of 10 seconds in order to improve the signal-noise ratio. All spectra were calibrated using the 521 cm^{-1} line of monocrystalline silicon. X-ray diffraction (XRD) patterns were obtained with a X'Pert PRO MPD Philips diffractometer (PANanalytical), using monochromatic CuKα radiation ($\lambda = 1.5406$ Å). The Kα radiation was selected with a Ge (1 1 1) primary monochromator. The X-ray tube was set at 45 kV and 40 mA. X-ray diffraction (XRD) measurements were also performed on a

Rigaku D/MAX III B diffractometer with a copper target. X-ray line broadening analysis (XLBA) was performed using computer software supplied by Rigaku after measurement in the step scan mode. The textural properties (S_{BET}, Vp, and dp) were measured from the nitrogen adsorption-desorption isotherms at $-196°C$ by an automatic ASAP 2020 system from Micromeritics. Prior to the measurements, samples were outgassed at 200°C and 10^{-4} mbar overnight. Surface areas were determined by using the Brunauer-Emmett-Teller (BET) equation and a nitrogen diatomic molecule cross section of 16.2 $Å^2$. The total pore volume was calculated from the adsorption isotherm at $P/P_0 = 0.996$. The optical band gap energy (E_g) values were determined using the Kubelka-Munk method; this was done by transforming the reflectance spectra of the samples with different Ag contents to the Kubelka-Munk function, $F(R)$, and then plotting $(F(R)E)^{1/2}$ versus E. The values of Eg were obtained by a linear fit at linear segments of the curve, determining its intersection with the photon energy axis [32]. It was mentioned in [33] that there are some discrepancies in opinion about the Kubelka-Munk method, but it is used with the sufficient accuracy required in photocatalytic applications. With the maximum absorbance in the UV-Vis spectra, it was observed the excitation wavelength of each sample to emit photoluminescence; this maximum has a slight shift; thus emission spectra were obtained with an excitation source fixed at $\lambda = 492$ nm in a FluoroMax-4, HORIBA, Jobin Yvon fluorometer. The intensity of the emission spectrum seems to be related to the rate of electron-hole recombination [34]. X-ray photoelectron spectra (XPS) were collected using a Physical Electronics PHI 5700 spectrometer with nonmonochromatic Mg Kα radiation (300 W, 15 kV, and 1256.6 eV) with a multichannel detector. Survey spectra were recorded from 0 to 1000 eV at constant pass energy of 100 eV, onto 720 μm diameter analysis area; narrow spectra of samples were recorded in the constant pass energy mode at 20 eV, using a 720 μm diameter analysis area and 10 scans at least. Charge correction was adjusted by using the carbon signal (C 1s at 285 eV). A PHI ACCESS ESCA-V6.0 F software package was used for acquisition and data analysis. A Shirley type background was subtracted from the spectra and then they were fitted using Gaussian-Lorentzian curves to determine the binding energies of the different element core levels more accurately. NIST database was used to identify the corresponding element to the measured binding energies.

2.3. Photocatalytic Activity Tests. The photocatalytic activity of the V$_2$O$_5$-xAg photocatalysts was evaluated through the degradation reaction of a model molecule as the malachite green (MG) dye ($C_{23}H_{26}N_2O$) using an aqueous solution of 10 μmol/l. For this purpose, 0.05 g of catalyst was added to a batch reaction system and stirred in a dark room until reaching the adsorption equilibrium. Catalysts were activated by illumination with a solar simulator SF-150B class ABA from Sciencetech, emitting 6% of UV radiation. The light source was placed at a height of 40 cm from the solution surface. The degradation reaction was monitored by the decrease in intensity of the characteristic absorption band of the MG peaking at 619 nm in the UV-Vis spectra. Reaction was followed by 3 hours taking aliquots every 15

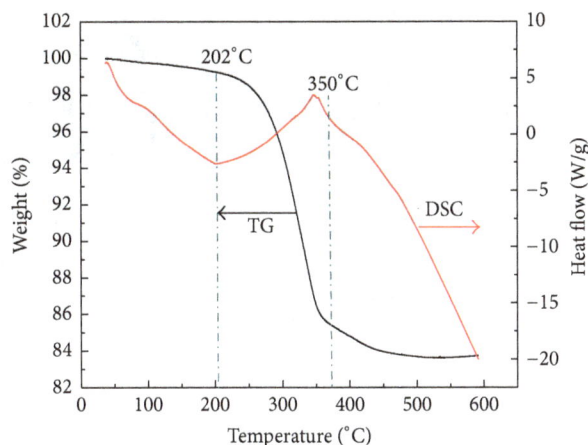

FIGURE 1: TGA-DSC curves of V$_2$O$_5$ catalyst precursor.

minutes. The obtained absorbances at each reaction time were correlated to MG concentrations through a calibration curve traced previously. The kinetic trace at 619 nm is related mainly to the MG concentration change. This trace could be fitted to a pseudo-first-order expression (exponential curve) with acceptable precision to determine the rate constant (k_{app}). Total organic carbon (TOC) was determined by the combustion method, for each solution at the end of the corresponding reaction as ppm of carbon. The mineralization degree was calculated taking as the percent from $[(T_{initial} - T_{final})/T_{initial}]$, where T_{final} and $T_{initial}$ are the TOC values at the end of the reaction and of the original solution, respectively.

3. Results and Discussion

3.1. Surfactant Elimination

3.1.1. Thermal Analysis and Infrared Spectroscopy of V$_2$O$_5$ Precursors. From TGA-DSC curves shown in Figure 1, the temperature to eliminate the surfactant Brij L23 from V$_2$O$_5$ catalyst precursor was determined. The TGA curve shows a main weight loss starting at 200°C and finishing at 450°C, accompanied with an exothermic peak (DSC curve) at 350°C that can be attributed to the thermal decomposition of the surfactant [35]. The incorporation and further increase of Ag promotes a slight decrease in the exothermic peak temperature until 340°C. Therefore, the thermal treatment was carried out at 400°C for 3 hours, in order to achieve the elimination of the surfactant and to favor the thermal decomposition of the AgNO$_3$ into Ag$_2$O and also to induce the crystallization of the catalytic formulation. To confirm if the organic material was removed after the thermal treatment, IR spectra of V$_2$O$_5$-xAg photocatalysts, before and after the thermal treatment, were compared; the results will be discussed later.

3.2. Surface Properties

3.2.1. Scanning Electron Microscopy. The morphology of the V$_2$O$_5$-xAg photocatalysts was observed from the SEM images shown in Figures 2(a)–2(d). It is clear that silver

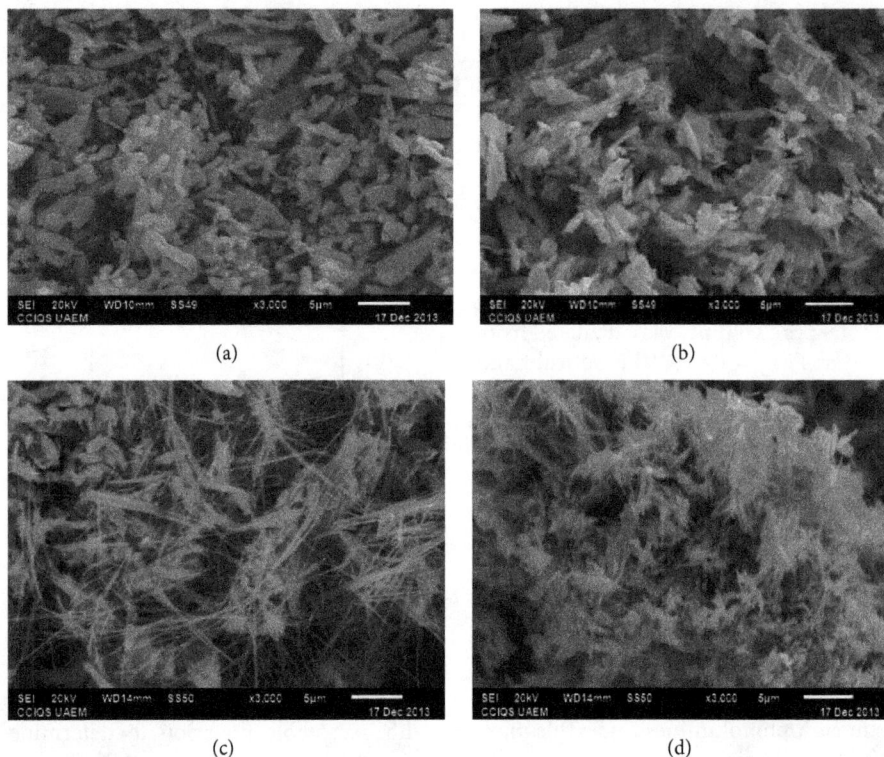

(a)

(b)

(c)

(d)

FIGURE 2: Scanning electron microscopy images at 3000x of (a) V_2O_5, (b) V_2O_5-5Ag, (c) V_2O_5-10Ag, and (d) V_2O_5-20Ag photocatalysts.

incorporation into the photocatalytic formulation changes morphology, obtaining acicular forms, which begin with the morphology bar-type with lengths about 2-3 microns as can be seen in Figures 2(a) and 2(b) for V_2O_5 and V_2O_5-5Ag photocatalysts. Further increase in silver content in the catalytic formulation promotes changes in the morphology, obtaining wires as can be seen in Figures 2(c) and 2(d) for V_2O_5-10Ag and some sharped bars in V_2O_5-20Ag photocatalysts. The catalytic performance observed for each formulation could be associated with the morphology as some changes are expected in textural properties related to morphology. It seems that nucleation process generates crystallites with different morphologies when silver is incorporated into the photocatalytic formulation. That process could be attributed to the solvothermal synthesis procedure mainly and to the increase in Ag_2O load to form different crystalline phases that have different crystallographic data. Elemental atomic composition % in the catalytic surface was determined by the EDS analysis, values shown in Table 1. The atomic % of Ag content increases as was expected with the increase in the theoretical Ag_2O content.

3.2.2. N_2-Physisorption Measurements.

Textural properties, such as the specific surface area and total pore volume, were studied using the N_2 physisorption technique. This was done to estimate the effect of the changes observed in the morphology of the V_2O_5-xAg photocatalysts, on their surface properties. The specific surface area values, calculated using the BET model, are included in Table 2. These values increase with the Ag_2O content, from 5.3 to 14.6 m^2/g, which

TABLE 1: Comparison of atomic percent of silver in V_2O_5-xAg photocatalysts measured by XPS and EDS.

	Ag content (at.%) (XPS)	Ag content (at.%) (SEM)
V_2O_5	—	—
V_2O_5-1Ag	1.5	1.1
V_2O_5-5Ag	2.9	1.3
V_2O_5-10Ag	4.5	4.3
V_2O_5-15Ag	4.5	5.1
V_2O_5-20Ag	6.8	8.1

represent an increase close to 300%. These data indicate that Ag_2O incorporation increases the specific surface area and the porosity compared with pure V_2O_5. This result could be explained in terms of a higher amount of empty spaces between the formed particles with narrow dimensions as was observed in the SEM images. N_2-physisorption isotherms are shown in Figure 3, which according to IUPAC are type II, characteristic of low porous solids presenting meso- and macroporosity. An increase in the nitrogen adsorption volume with the increase of the silver content in the catalytic formulation is clear from the isotherms. A small hysteresis loop is observed, attributed to the empty spaces between particles with irregular forms and sizes.

3.3. Crystalline and Molecular Structures

3.3.1. X-Ray Powder Diffraction.

Figure 4 shows the diffractograms of the V_2O_5-xAg photocatalysts, where the main

TABLE 2: Specific surface area (S_{BET}) and pore volume (V_p) of pure V_2O_5 and V_2O_5-xAg photocatalysts.

Sample	S_{BET} (m^2/g)	V_p (cm^3/g)
V_2O_5*	3.3	0.005
V_2O_5**	3.4	0.010
V_2O_5-1Ag	5.3	0.020
V_2O_5-5Ag	11.0	0.023
V_2O_5-10Ag	10.6	0.030
V_2O_5-15Ag	11.9	0.025
V_2O_5-20Ag	14.6	0.026

*Commercial.
**Synthesized.

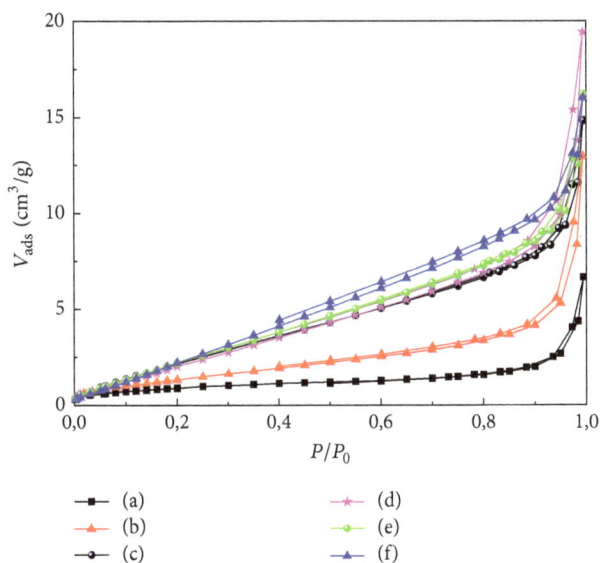

FIGURE 3: N_2 adsorption-desorption isotherms of (a) V_2O_5, (b) V_2O_5-1Ag, (c) V_2O_5-5Ag, (d) V_2O_5-10Ag, (e) V_2O_5-15Ag, and (f) V_2O_5-20Ag.

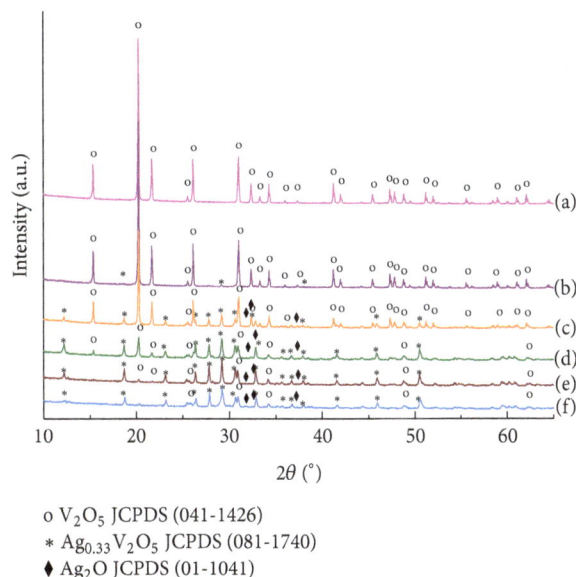

o V_2O_5 JCPDS (041-1426)
* $Ag_{0.33}V_2O_5$ JCPDS (081-1740)
♦ Ag_2O JCPDS (01-1041)

FIGURE 4: XRD diffraction patterns of (a) V_2O_5, (b) V_2O_5-1Ag, (c) V_2O_5-5Ag, (d) V_2O_5-10Ag, (e) V_2O_5-15Ag, and (f) V_2O_5-20Ag photocatalysts.

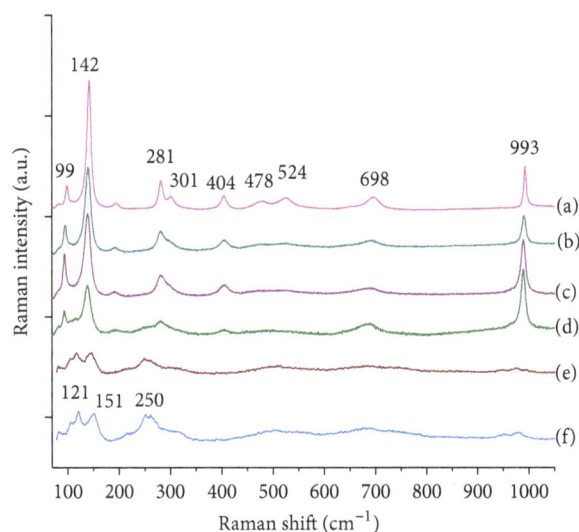

FIGURE 5: Raman spectra of samples (a) V_2O_5, (b) V_2O_5-1Ag, (c) V_2O_5-5Ag, (d) V_2O_5-10Ag, (e) V_2O_5-15Ag, and (f) V_2O_5-20Ag.

crystalline phases such as the Shcherbinaite V_2O_5, the $Ag_{0.33}$ V_2O_5 bronze, and the Ag_2O oxide are observed. The intensity in the diffraction lines changes with the increase of the Ag load in the photocatalytic formulation. The diffractogram of the synthesized V_2O_5 catalyst (Figure 4(a)) shows diffraction lines at 2θ = 15.3, 20.2, 21.7, 26.1, 31.0, 32.3, 34.3, 41.2, and 47.3° that characterize unambiguously the Shcherbinaite crystalline phase of V_2O_5 (JCPDS 41-1426). After Ag incorporation, other diffraction lines become noticeable at 2θ = 29.2, 27.8, 30.6, 30.8, and 32.8° in the samples with the V_2O_5-1Ag and V_2O_5-5Ag catalysts. These new signals can be assigned to the called silver vanadium bronzes, specifically to the $Ag_{0.33}V_2O_5$ crystalline structure (JCPDS 81-1740). Further increase in silver load, samples V_2O_5-10Ag and V_2O_5-20Ag, gives rise to the appearance of new diffraction signals at 2θ = 32.9, 38.1, 54.9, and 65.7°, attributed to the Ag_2O crystalline phase (JCPDS 01-1041). These diffraction lines are very weak probably due to a dilution or dispersion effects and to a small crystallite size under 40 Å. It is clear that V_2O_5 crystalline phase almost disappears at the highest silver loads.

3.3.2. Raman Spectroscopy. To corroborate the microcrystalline structure observed by XRD, Raman spectra of unmodified and Ag-modified V_2O_5 photocatalysts were recorded. It can be seen in the spectra shown in Figure 5 that the Ag incorporation modifies some signals of the Raman spectra with respect to the synthesized V_2O_5 catalyst. The characteristic peaks of the V_2O_5, located at 99, 193, 281, 301, 404, 478, 524, 698, and 993 cm^{-1}, decrease and almost disappear by raising the amount of Ag_2O added during the synthesis along with the appearance of new peaks. The main peak located at 142 cm^{-1}, characteristic of the skeleton bend vibration, shifts from 142 cm^{-1} in V_2O_5 to 138 cm^{-1} in V_2O_5-10Ag photocatalyst, probably due to the distortion of the V_2O_5

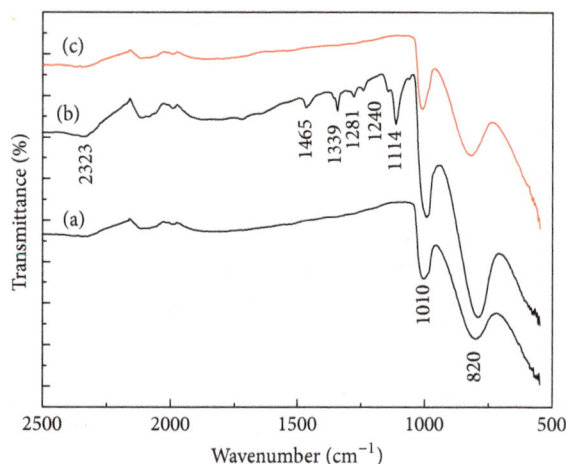

FIGURE 6: Infrared spectra of commercial V_2O_5 (a), V_2O_5 obtained through the surfactant assisted technique before calcination (b), and synthesized V_2O_5 after calcination (c).

structure caused by the Ag incorporation into the V_2O_5 lattice. New Raman bands at 121, 151, and 250 cm^{-1} appear for samples with the highest Ag_2O loads and V_2O_5-15Ag and V_2O_5-20Ag samples, which could be attributed to the formation of the $Ag_{0.33}V_2O_5$ bronze. It is noteworthy that no signals associated with Ag_2O were observed, contrary with the XRD results; this could be attributed to the small Raman section of this compound.

3.3.3. Infrared Spectroscopy. Infrared spectra were recorded with two main purposes: firstly, to corroborate the surfactant elimination from the photocatalytic preparation, as can be seen in Figure 6; secondly, to observe the bond vibration when silver is added at different Ag_2O loads. IR spectra of V_2O_5, commercial and as-synthetized, before and after thermal treatment, were compared. The IR spectrum of the commercial sample shows two bands at 1010 cm^{-1} and 820 cm^{-1} ascribed to the coupled vibration between V=O and V–O–V stretching vibrations [36, 37]. The infrared spectrum of the synthesized V_2O_5 before the thermal treatment also presents IR absorption bands characteristics of the surfactant at 1114, 1240, 1281, 1339, and 1465 cm^{-1} [38], attributed to the C-C and C-H bending in the alkyl groups of the surfactant Brij L23. After thermal treatment, these bands disappear confirming the surfactant elimination. IR spectra of the different V_2O_5-xAg photocatalysts are shown in Figure 7. The incorporation and further increase of the Ag_2O load promote the shift of the band at 1010 cm^{-1} to 1000 cm^{-1} and the disappearance of the band located at 820 cm^{-1} when the theoretical Ag_2O loading is higher than 5 wt.%. The observed shift has been assigned to disordered vacancies in the V_2O_5 lattice [39] due to the presence of Ag. The spectrum of the V_2O_5-1Ag sample shows bond vibrations located at 983 cm^{-1} and 970 cm^{-1}; the last one remains for samples with higher Ag_2O loads. The band at 983 cm^{-1} is related to the decrease of the bond strength of V=O stretch vibration by the incorporation of silver into the photocatalytic formulation as observed for Na-bronzes [40]. New infrared bands, at 917,

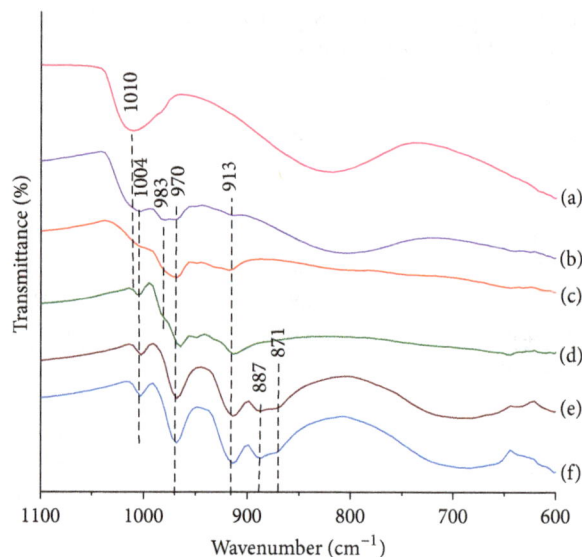

FIGURE 7: IR spectra of (a) V_2O_5, (b) V_2O_5-1Ag, (c) V_2O_5-5Ag, (d) V_2O_5-10Ag, (e) V_2O_5-15Ag, and (f) V_2O_5-20Ag photocatalysts.

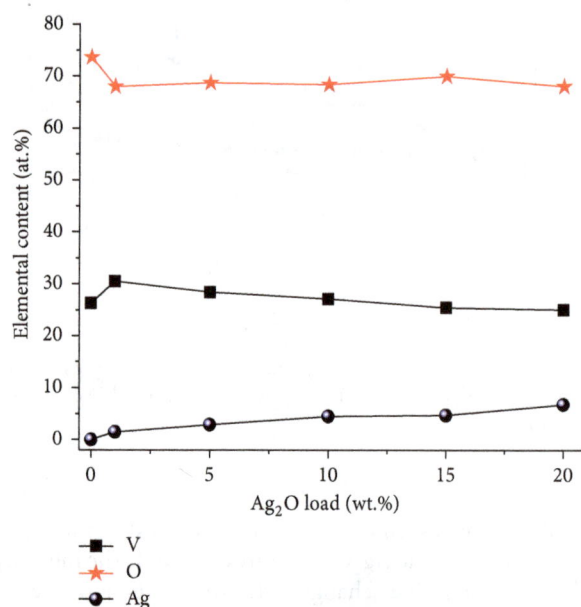

FIGURE 8: Elemental atomic contents for the V_2O_5-xAg photocatalysts.

890 and 871 cm^{-1} and a broad band at 700 cm^{-1}, appear with the increase in Ag content. These could be ascribed to the presence of the $Ag_{0.33}V_2O_5$ bronze identified by XRD.

3.3.4. X-Ray Photoelectron Spectroscopy. Figure 8 shows the corresponding atomic contents of O, V, and Ag, determined from XPS, as a function of the Ag_2O load. From this figure, it is important to remark that the silver atomic content increases as the theoretical Ag_2O load increases. The spectrum region corresponding to V $2p_{3/2}$ has been only considered for the analysis. The corresponding V $2p_{3/2}$ core level spectra of vanadium pentoxide are depicted in Figure 9(a). The fitted

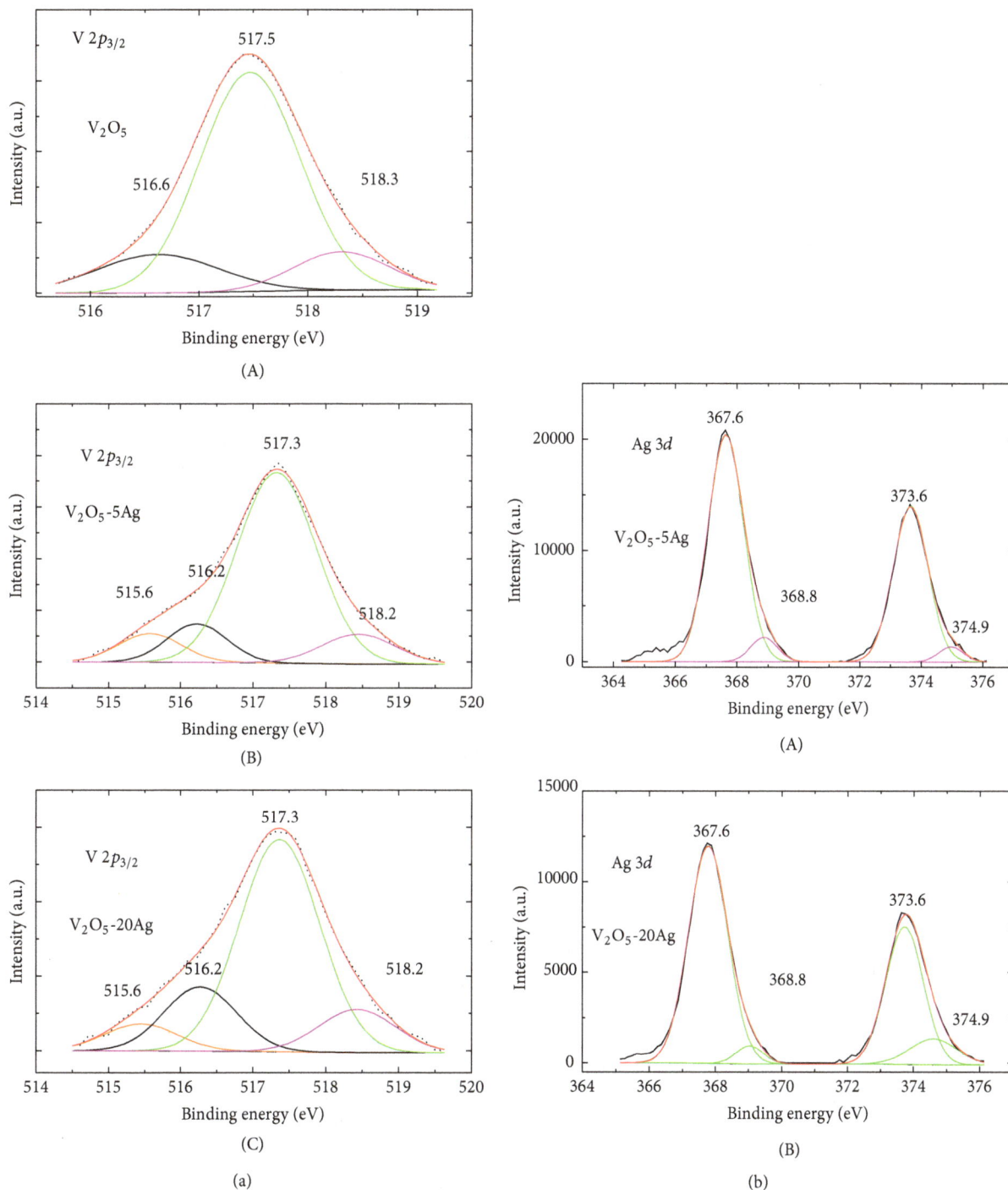

FIGURE 9: Core level spectra of (a) V $2p_{3/2}$ of (A) V_2O_5, (B) V_2O_5-5Ag, and (C) V_2O_5-20Ag and of (b) Ag $3d_{5/2}$ for (A) V_2O_5-5Ag and (B) V_2O_5-20Ag.

spectrum of the synthesized V_2O_5 shows binding energies at 515.6 and 517.5 eV. The peak at 515.6 eV is associated with V^{4+} species and the contribution at 517.5 eV corresponds to V^{5+} indicating the presence of V_2O_5 and confirming the Shcherbinaite crystalline structure observed by XRD. The incorporation of different Ag_2O loads into the photocatalytic formulation promotes the appearance of a new peak at binding energy of 516.2 eV, ascribed to the formation of

the Ag-vanadium bronze ($Ag_{0.33}V_2O_5$) observed from XRD. Figures 9(a)(A)–(C) seem to indicate that the contribution at 516.2 eV increases with the Ag_2O loads. These results indicate that on the photocatalyst surfaces coexist mixed-valence vanadium oxides [41]. Concerning the fitting of the Ag $3d$ signal (Figure 9(b)), the contribution of two doublets is observed in all cases. One of them, centered at 367.7 and 373.6 eV, is attributed to the presence of Ag^+ species in the

TABLE 3: Band gap energy (E_g).

Catalyst	E_g (eV)	Wavelength (nm)
V_2O_5	2.3	564
V_2O_5-1Ag	2.2	566
V_2O_5-5Ag	1.6	765
V_2O_5-10Ag	1.4	867
V_2O_5-15Ag	1.2	1000
V_2O_5-20Ag	1.2	1033

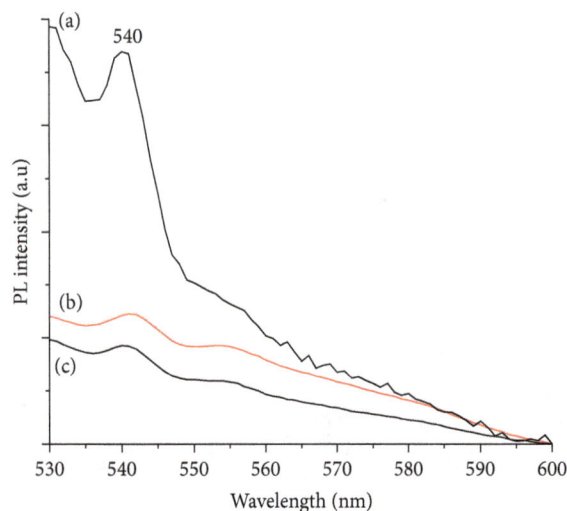

FIGURE 10: Photoluminescence spectra of (a) commercial V_2O_5, (b) synthesized V_2O_5, and (c) V_2O_5-20Ag.

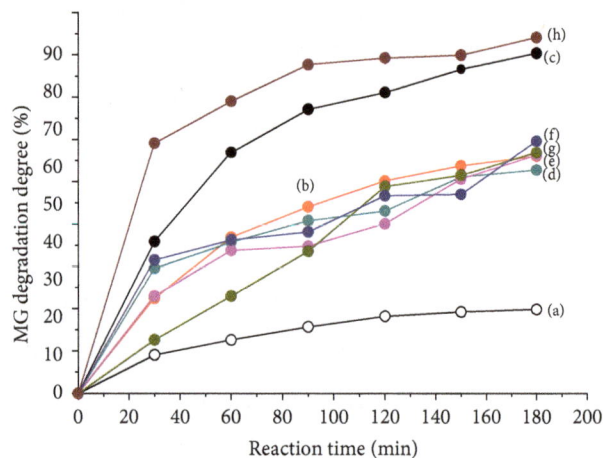

FIGURE 11: Photocatalytic degradation of malachite green dye during the first 180 minutes of reaction time (a) without catalyst (photolysis) and the photocatalytic process using (b) commercial V_2O_5, (c) synthesized V_2O_5, (d) V_2O_5-1Ag, (e) V_2O_5-5Ag, (f) V_2O_5-10Ag, (g) V_2O_5-15Ag, and (h) V_2O_5-20Ag photocatalysts.

$Ag_{0.33}V_2O_5$ oxide, while the other, centered at 368.8 and 374.9 eV, could be due to the presence of the Ag^+ species in Ag_2O, in agreement with the XRD results [41]. Finally, it should be mentioned that the O $1s$ spectra decrease slightly their intensity with the increase in atomic silver content and the characteristic peak is at 530.2 eV.

3.4. Band Gap Energy

3.4.1. Diffuse Reflectance Spectroscopy.
It is well known that pentavalent vanadium has no d electrons and hence d-d transitions are not possible. Therefore, the observed bands in the electronic absorption spectrum are ascribed to charge transfer bands. The reflectance spectrum was processed with the Kubelka-Munk function and the optical band gap energy (E_g) was obtained as can be seen in Table 3. Results reveal that the increase in the Ag_2O load in photocatalytic formulation promotes narrowing of the band gap from 2.3 eV to values as low as 1.2 eV, remarking that this low band gap value makes this material potentially active under solar light.

3.4.2. Photoluminescence Spectroscopy.
Figure 10 shows the photoluminescence spectra of the commercial and synthesized V_2O_5 as well as the V_2O_5-20Ag samples. The PL spectrum of the commercial V_2O_5 is characterized by an intense band peaking at 543. In contrast, the PL emission of the V_2O_5 obtained by the surfactant assisted technique shows

the same band but with a lower intensity, almost three times less intensity. This can be interpreted as a higher electron-hole recombination rate in the commercial sample as the PL intensity is related directly to the recombination rate. An additional PL intensity decrease is also seen in the spectrum of the V_2O_5-20Ag photocatalyst, indicating that this sample exhibits the lowest recombination rate.

3.5. Photocatalytic Activity.
Photocatalytic activity was evaluated in the degradation of the malachite green dye. Figure 11 shows that the photodegradation of the MG dye in the photolysis process is very low as can be seen in Figure 11(a). The use of the commercial V_2O_5 sample degrades 56% of the initial MG concentration after 180 min as is shown in Figure 11(b). The synthesized V_2O_5 catalyst (Figure 11(c)) reached a higher photocatalytic activity than the commercial sample, close to 80% of MG degradation. The Ag incorporation and further increase of the silver load in the photocatalyst decrease the photocatalytic activity in the range of 52–60% of MG degradation (Figures 11(d)–11(g)); however, the V_2O_5-20Ag photocatalyst exhibits the highest MG degradation reaching 84% (Figure 11(h)). It is worth noting that this is the best result found in this work, in which a photocatalyst that degrades organic molecules, as the MG dye, takes advantage of the cheapest illumination source as the sunlight. Moreover, a catalyst with high insolubility in water was obtained. This photocatalytic activity obtained was correlated with the acicular morphology of this material, which enhances the contact area in the reaction system. Also, the mixture of several semiconductors such as the V_2O_5, the $Ag_{0.33}V_2O_5$, and the small fraction of Ag_2O works as coupled conduction bands with electron transfer between them and decreases the recombination rate of the electron-hole pair, as was suggested by the photoluminescence results. Table 4 shows the reaction rate constant values k_{app} (min^{-1}) obtained from the fitting of the MG concentration at the early reaction times. Assuming a

TABLE 4: Kinetic rate constant (k_{app}) for photocatalytic formulations as a function of the silver load, determined using a nonlinear least squares data treatment [31].

Photocatalyst	% of degradation (UV-Vis)	% of degradation (TOC)	k_{app} (min^{-1})
V_2O_5	80.4	69.9	0.0081 ± 0.0001
V_2O_5-1Ag	52.8	67.5	0.0049 ± 0.0008
V_2O_5-5Ag	56.3	54.3	0.0056 ± 0.0007
V_2O_5-10Ag	56.9	56.0	0.0081 ± 0.0009
V_2O_5-15Ag	59.7	73.9	0.0053 ± 0.0001
V_2O_5-20Ag	84.2	74.4	0.0082 ± 0.0002

pseudo-first-order expression, a nonlinear least square fitting was used with an acceptable precision [31]. This reaction rate constant values agrees well with the degradation degree results. Additionally, the mineralization degree was followed by the quantification of the total organic carbon (TOC) through the reaction time; both results are quite similar as can be seen in Table 4, indicating that the photodegradation process follows the mineralization route of the organic dye tested.

4. Conclusions

Ag-modified photocatalysts were obtained in an easy way with acicular morphologies and enhanced specific surface areas. The changes in morphology and textural properties are associated with the appearance of new phases such as $Ag_{0.33}V_2O_5$ as well as Ag_2O coexisting at higher Ag loads as was suggested by XRD and XPS. These new phases have a strong influence in the photocatalytic response of these systems. It was observed that Ag incorporation into the photocatalytic formulation narrows the band gap energy making these materials photoactive under sunlight, a natural and cheaper irradiation source. The obtained photocatalytic formulations are conformed by a mixture of crystalline phases that work as coupled semiconductors. The V_2O_5-20Ag catalyst was the most active for the MG degradation using simulated sunlight.

Competing Interests

The authors declare that they have no competing interests.

Acknowledgments

The authors thank CONACyT for the financial support through CB-168827 and CB-240998 projects and CB-239648 project and also thank the academic staff of CCIQS Alejandra Núñez, Lizbeth Triana, Citlalit Martínez, and Dr. Uvaldo Hernández Balderas. E. Rodriguez Castellon and A. Infantes thank the financial support of the CTQ2015-68951-C3-3-R project (Ministerio de Economía y Competitividad) and FEDER funds.

References

[1] A. Fujishima and K. Honda, "Electrochemical photolysis of water at a semiconductor electrode," *Nature*, vol. 238, no. 5358, pp. 37–38, 1972.

[2] L. Di, H. Hajime, O. Naoki, H. Shunichi, and Y. Yukio, "Synthesis of nanosized nitrogen-containing MO_x-ZnO (M = W, V, Fe) composite powders by spray pyrolysis and their visible-light-driven photocatalysis in gas-phase acetaldehyde decomposition," *Catalysis Today*, vol. 93–95, pp. 895–901, 2004.

[3] I. N. M. Posada Velázquez, "Los compuestos orgánicos persistentes y su toxicidad a la salud humana," 2009, http://es.scribd.com/doc/16026795/Los-compuestos-organicos-persistentes-y-su-toxicidad-a-la-salud-humana.

[4] D. Dhananjay, S. Bhatkhande, V. G. Pangarkar, and A. Beenackers, "Photocatalytic degradation for environmental applications—a review," *Journal of Chemical Technology and Biotechnology*, vol. 77, no. 1, pp. 102–116, 2002.

[5] A. I. Martínez, D. Acosta, and A. López, "Efecto del contenido de Sn sobre las propiedades físicas de películas delgadas de TiO_2," *Superficies y Vacío*, vol. 16, pp. 5–9, 2003.

[6] K. Melghit, A. K. Mohammed, and I. Al-Amri, "Chimie douce preparation, characterization and photocatalytic activity of nanocrystalline SnO_2," *Materials Science and Engineering B: Solid-State Materials for Advanced Technology*, vol. 117, no. 3, pp. 302–306, 2005.

[7] H. T. Tran, H. Kosslick, M. F. Ibad et al., "Photocatalytic performance of highly active brookite in the degradation of hazardous organic compounds compared to anatase and rutile," *Applied Catalysis B: Environmental*, vol. 200, pp. 647–658, 2017.

[8] S. Chen, D. Li, Y. Liu, and W. Huang, "Morphology-dependent defect structures and photocatalytic performance of hydrogenated anatase TiO_2 nanocrystals," *Journal of Catalysis*, vol. 341, pp. 126–135, 2016.

[9] K. Nakata and A. Fujishima, "TiO_2 photocatalysis: design and applications," *Journal of Photochemistry and Photobiology C: Photochemistry Reviews*, vol. 13, no. 3, pp. 169–189, 2012.

[10] C. McManamon, J. O'Connell, P. Delaney, S. Rasappa, J. D. Holmes, and M. A. Morris, "A facile route to synthesis of S-doped TiO_2 nanoparticles for photocatalytic activity," *Journal of Molecular Catalysis A: Chemical*, vol. 406, pp. 51–57, 2015.

[11] H. Li, G. Zhao, G. Han, and B. Song, "Hydrophilicity and photocatalysis of $Ti_{1-x}V_xO_2$ films prepared by sol-gel method," *Surface and Coatings Technology*, vol. 201, no. 18, pp. 7615–7618, 2007.

[12] T. Bak, J. Nowotny, M. Rekas, and C. C. Sorrell, "Photoelectrochemical hydrogen generation from water using solar energy. Materials-related aspects," *International Journal of Hydrogen Energy*, vol. 27, no. 10, pp. 991–1022, 2002.

[13] X. Yong and M. A. A. Schoonen, "The absolute energy positions of conduction and valence bands of selected semiconducting minerals," *American Mineralogist*, vol. 85, no. 3-4, pp. 543–556, 2000.

[14] F. Amano, M. Tanaka, and B. Ohtani, "Alkali metal ion-modified vanadium mononuclear complex for photocatalytic mineralization of organic compounds," *Catalysis Letters*, vol. 140, no. 1, pp. 27–31, 2010.

[15] M. A. Rauf, S. B. Bukallah, A. Hamadi, A. Sulaiman, and F. Hammadi, "The effect of operational parameters on the photoinduced decoloration of dyes using a hybrid catalyst V_2O_5/TiO_2," *Chemical Engineering Journal*, vol. 129, no. 1-3, pp. 167–172, 2007.

[16] K. Teramura, T. Tanaka, M. Kani, T. Hosokawa, and T. Funabiki, "Selective photo-oxidation of neat cyclohexane in the liquid phase over V_2O_5/Al_2O_3," *Journal of Molecular Catalysis A: Chemical*, vol. 208, no. 1-2, pp. 299–305, 2004.

[17] C. Karunakaran and P. Anilkumar, "Photooxidation of iodide ion on immobilized semiconductor powders," *Solar Energy Materials and Solar Cells*, vol. 92, no. 4, pp. 490–494, 2008.

[18] X. Rui, Y. Tang, O. I. Malyi et al., "Ambient dissolution-recrystallization towards large-scale preparation of V_2O_5 nanobelts for high-energy battery applications," *Nano Energy*, vol. 22, pp. 583–593, 2016.

[19] J. Mu, J. Wang, J. Hao et al., "Hydrothermal synthesis and electrochemical properties of V_2O_5 nanomaterials with different dimensions," *Ceramics International*, vol. 41, no. 10, pp. 12626–12632, 2015.

[20] H. He, L. Zan, and Y. Zhang, "Effects of amorphous V_2O_5 coating on the electrochemical properties of $Li[Li_{0.2}Mn_{0.54}Ni_{0.13}Co_{0.13}]O_2$ as cathode material for Li-ion batteries," *Journal of Alloys and Compounds*, vol. 680, pp. 95–104, 2016.

[21] S. Sel, O. Duygulu, U. Kadiroglu, and N. E. Machin, "Synthesis and characterization of nano-V_2O_5 by flame spray pyrolysis, and its cathodic performance in Li-ion rechargeable batteries," *Applied Surface Science*, vol. 318, pp. 150–156, 2014.

[22] A. A. Mane, V. V. Ganbavle, M. A. Gaikwad, S. S. Nikam, K. Y. Rajpure, and A. V. Moholkar, "Physicochemical properties of sprayed V_2O_5 thin films: effect of substrate temperature," *Journal of Analytical and Applied Pyrolysis*, vol. 115, pp. 57–65, 2015.

[23] C. Julien, E. Haro-Poniatowski, M. A. Camacho-López, L. Escobar-Alarcón, and J. Jímenez-Jarquín, "Growth of V_2O_5 thin films by pulsed laser deposition and their applications in lithium microbatteries," *Materials Science and Engineering B*, vol. 65, no. 3, pp. 170–176, 1999.

[24] F. Gonzalez-Zavala, L. Escobar-Alarcón, D. A. Solís-Casados, C. Rivera-Rodríguez, R. Basurto, and E. Haro-Poniatowski, "Preparation of vanadium oxide thin films modified with Ag using a hybrid deposition configuration," *Applied Physics A: Materials Science and Processing*, vol. 122, no. 4, article 461, 2016.

[25] N. Serpone and E. Pelizzetti, *Photocatalysis: Fundamentals and Applications*, John Wiley & Sons, New York, NY, USA, 1989.

[26] D. Bahnemann, "Photocatalytic water treatment: solar energy applications," *Solar Energy*, vol. 77, no. 5, pp. 445–459, 2004.

[27] Y. He, Y. Wu, T. Sheng, and X. Wu, "Photodegradation of acetone over V-Gd-O composite catalysts under visible light," *Journal of Hazardous Materials*, vol. 180, no. 1-3, pp. 675–682, 2010.

[28] T. Ishihara, N. S. Baik, N. Ono, H. Nishiguchi, and Y. Takita, "Effects of crystal structure on photolysis of H_2O on K–Ta mixed oxide," *Journal of Photochemistry and Photobiology A: Chemistry*, vol. 167, no. 2-3, pp. 149–157, 2004.

[29] T. Ohno, S. Izumi, K. Fujihara, and M. Matsumura, "Electron-hole recombination via reactive intermediates formed on PdO-doped $SrTiO_3$ electrodes. Estimation from comparison of photoluminescence and photocurrent," *Journal of Photochemistry and Photobiology A: Chemistry*, vol. 129, no. 3, pp. 143–146, 1999.

[30] F. Han, V. S. R. Kambala, M. Srinivasan, D. Rajarathnam, and R. Naidu, "Tailored titanium dioxide photocatalysts for the degradation of organic dyes in wastewater treatment: a review," *Applied Catalysis A: General*, vol. 359, no. 1-2, pp. 25–40, 2009.

[31] G. Lente, *Deterministic Kinetics in Chemistry and Systems Biology*, Springer, 2015.

[32] A. B. Murphy, "Band-gap determination from diffuse reflectance measurements of semiconductor films, and application to photoelectrochemical water-splitting," *Solar Energy Materials and Solar Cells*, vol. 91, no. 14, pp. 1326–1337, 2007.

[33] R. López and R. Gómez, "Band-gap energy estimation from diffuse reflectance measurements on sol-gel and commercial TiO_2: A Comparative Study," *Journal of Sol-Gel Science and Technology*, vol. 61, no. 1, pp. 1–7, 2012.

[34] P. Malathy, K. Vignesh, M. Rajarajan, and A. Suganthi, "Enhanced photocatalytic performance of transition metal doped Bi_2O_3 nanoparticles under visible light irradiation," *Ceramics International*, vol. 40, no. 1, pp. 101–107, 2014.

[35] D. Solís, E. Vigueras-Santiago, S. Hernández-López, A. Gómez-Cortés, M. Aguilar-Franco, and M. A. Camacho-López, "Textural, structural and electrical properties of TiO_2 nanoparticles using Brij 35 and P123 as surfactants," *Science and Technology of Advanced Materials*, vol. 9, no. 2, 2008.

[36] J. M. Lee, H.-S. Hwang, W.-I. Cho, B.-W. Cho, and K. Y. Kim, "Effect of silver co-sputtering on amorphous V_2O_5 thin-films for microbatteries," *Journal of Power Sources*, vol. 136, no. 2, pp. 122–131, 2004.

[37] T. Ono, Y. Tanaka, T. Takeuchi, and K. Yamamoto, "Characterization of K-mixed V_2O_5 catalyst and oxidative dehydrogenation of propane on it," *Journal of Molecular Catalysis A: Chemical*, vol. 159, no. 2, pp. 293–300, 2000.

[38] Libro, "Tablas para la elucidación estructural de compuestos orgánicos por métodos espectroscópicos".

[39] M. D. Soriano, A. Vidal-Moya, E. Rodríguez-Castellón, F. V. Melo, M. T. Blasco, and J. M. López-Nieto, "Partial oxidation of hydrogen sulfide to sulfur over vanadium oxides bronzes," *Catalysis Today*, vol. 259, pp. 237–244, 2016.

[40] J. C. Badot, D. G. F. Bourdeau, N. Baffier, and A. Tabuteau, "Electronic Properties of $Na_{0.33}V_2O_5$ bronze obtained by sol-gel process," *Journal of Solid State Chemistry*, vol. 92, pp. 8–17, 1991.

[41] M. E. Tousley, A. W. Wren, M. R. Towler, and N. P. Mellott, "Processing, characterization, and bactericidal activity of undoped and silver-doped vanadium oxides," *Materials Chemistry and Physics*, vol. 137, no. 2, pp. 596–603, 2012.

Relevance of the Physicochemical Properties of Calcined Quail Eggshell (CaO) as a Catalyst for Biodiesel Production

Leandro Marques Correia,[1] Juan Antonio Cecilia,[2] Enrique Rodríguez-Castellón,[2] Célio Loureiro Cavalcante Jr.,[1] and Rodrigo Silveira Vieira[1]

[1]*Grupo de Pesquisa em Separações por Adsorção (GPSA), Departamento de Engenharia Química, Universidade Federal do Ceará (UFC), Campus do Pici, Bl. 709, 60455-760 Fortaleza, CE, Brazil*
[2]*Departamento de Química Inorgánica, Cristalografía y Mineralogía, Facultad de Ciencias, Universidad de Málaga, Campus de Teatinos, 29071 Málaga, Spain*

Correspondence should be addressed to Juan Antonio Cecilia; jacecilia@uma.es

Academic Editor: Eri Yoshida

The CaO solid derived from natural quail eggshell was calcined and employed as catalyst to produce biodiesel via transesterification of sunflower oil. The natural quail eggshell was calcined at 900°C for 3 h, in order to modify the calcium carbonate present in its structure in CaO, the activity phase of the catalyst. Both precursor and catalyst were characterized using Hammett indicators method, X-ray fluorescence (XRF), X-ray diffraction (XRD), thermogravimetric analysis (TG/DTG), CO_2 temperature-programmed desorption (CO_2-TPD), X-ray photoelectronic spectroscopy (XPS), Fourier infrared spectroscopy (FTIR), scanning electron microscopy (SEM), N_2 adsorption-desorption at −196°C, and distribution particle size. The maximum biodiesel production was of 99.00 ± 0.02 wt.% obtained in the following transesterification reaction conditions: X_{MR} (sunflower oil/methanol molar ratio of 1 : 10.5 mol : mol), X_{CAT} (catalyst loading of 2 wt.%), X_{TIME} (reaction time of 2 h), stirring rate of 1000 rpm, and temperature of 60°C.

1. Introduction

The increase of the world population has led to a depletion of fossil resources and the generation of high residue content. In the last years, both society and governments have become aware of this issue, attempting these wastes without commercial interest in the other one with high added value. The eggshell is an example of these products without interest that could be treated to reuse in several applications. Eggshells are mainly composed of a network of protein fibers, associated with crystals of calcium carbonate ($CaCO_3$), magnesium carbonate ($MgCO_3$), and calcium phosphate $Ca_3(PO_4)_2$ as well as organic substances and water, calcite ($CaCO_3$) being the main constituent [1].

The major uses for this material are associated with the agriculture as fertilizer, pH correction of the acidic soils. The thermal treatment of the eggshell wastes as raw material can lead to the formation of an inexpensive and environment-friendly basic catalyst. This material is one alternative of the nonrenewable calcium carbonate mineral sources. The use of this new source can minimize the impact on the natural reserves of limestone and synthesis of $CaCO_3$ or dolomite $Ca_xMg_{(1-x)}CO_3$, which have been highly used for research purposes, obtaining excellent results for biodiesel production [2–4].

Biodiesel is obtained by the transesterification reaction of triglycerides, which are derived from vegetable oils or animal fats, with short chain alcohols (methanol, ethanol, and propanol) to obtain a mixture of alkyl esters and glycerol as by-product [5, 6]. Biodiesel is the most potential alternative energy to fossil fuels since it is biodegradable, renewable, and nontoxic [7, 8].

Traditionally, the synthesis of biodiesel has taken place using homogeneous catalysts, such as NaOH [9], KOH [10], and H_2SO_4 [11], obtaining high conversion values; however the use of these catalysts presents several disadvantages such as the inability to reuse the catalyst and the high generation of waste and effluents which can be hazardous and corrosives. For this reason, the use of solid heterogeneous catalysts as alternative to the traditional homogenous catalysts has been proposed. The use of heterogeneous catalysts minimizes environmental damage and improves the effectiveness of the process which reduces the biodiesel cost. In addition, solid catalysts can be easily separated from the reaction medium and regenerated for several reaction cycles [12]; besides that, the use of solid catalysts diminishes the volume of wash water and organic solvent required in the purification step of the biodiesel.

Heterogeneous basic catalysts commonly used in the transesterification reaction are Ca-Al hydrocalumite [13], MgO [14], CaO [14–16], BaO [17], Li-CaO [18], CaO-ZnO [19], CaO/SBA-15 [20], MgO-CaO [20], CaO/Al_2O_3 [21], $CaO-La_2O_3$ [22] calcium/chitosan spheres [23], and natural calcium [24]. Of these, CaO has been the active phase that has exhibited the highest activity under mild reaction conditions with a long lifetime and a low catalyst cost [25]. A natural source, such as the quail eggshell waste, could emerge as a potential catalytic precursor for biodiesel production due to its low cost and high efficiency of its active phase in the transesterification reaction after a thermal treatment to form the CaO phase, as is reported in

$$CaCO_3 \longrightarrow CaO + CO_2 \qquad (1)$$

In this sense, several researches have reported catalytic precursors based on calcium oxide, which were obtained from natural sources such as eggshell [25–28], bivalve clam shell [29], waste mussel shell [30, 31], waste of animal bone [31], snail shell [32], golden apple snail shell [32], crab shell [25, 33], oyster shell [34], waste fish (Labeo rohita) [35].

The objective of this research was the use of another natural calcium source not reported in the literature as quail eggshell wastes as raw material to obtain CaO by a thermal treatment (900°C for 3 h). The catalytic behavior of this catalyst was evaluated in the biodiesel production using methanol and sunflower refined oil as reagents. The material was characterized using Hammett indicators method, X-ray fluorescence (XRF), X-ray diffraction (XRD), thermogravimetric analysis (TG/DTG), CO_2-thermoprogrammed desorption (CO_2-TPD), X-ray photoelectronic spectroscopy (XPS), Fourier infrared spectroscopy (FTIR), scanning electron microscopy (SEM), N_2 adsorption-desorption at −196°C, and distribution particle size, to evaluate whether the calcium carbonate was converted into calcium oxide as well as verify its catalytic activity. The biodiesel production was performed using a factorial design studying the variables that influence the catalytic behavior, such as X_{MR} (sunflower oil/methanol to molar ratio), X_{CAT} (catalyst loading), and X_{TIME} (reaction time) in the biodiesel production.

2. Material and Methods

2.1. Materials and Reagents. Commercially edible grade sunflower oil was obtained from a supermarket (Liza, Brazil)

and methanol (99.95%) was obtained from Vetec (Rio de Janeiro, Brazil). The physicochemical properties and fatty acid composition of sunflower oil are summarized in Supplementary Material Table S1 (available online at https://doi.org/10.1155/2017/5679512). The gases used in the experiments were He (99.999%, Air Liquide), N_2 (99.999%, Air Liquide), and CO_2 (99.99%; Air Liquide). The quail eggshell was obtained from domestic residences (Fortaleza, Brazil). The fatty acid composition of sunflower oil was determined by promoting the esterification of the fatty acids and quantified by CG/FID. This esterification reaction was carried out according to the procedure described in the EN 14214:2008 [36].

2.2. Preparation and Activation of Basic Catalyst. Natural quail eggshell was thoroughly washed several times with warm water. The residue of quail eggshell was crushed in a household blender, sieved, and stored for characterization of the precursor catalytic. The catalysts were synthesized by a thermal treatment of the precursor in a muffle furnace at 900°C for 3 h with temperature ramp of 30°C min^{-1} (Figure 1). The purpose of the thermal treatment is the decomposition of the organic matter present in the natural quail eggshell and the decarbonation of the calcium carbonate to form the desired calcium oxide, which has catalytic activity for biodiesel production.

2.3. Material Characterization. Both natural quail eggshell and calcined quail eggshell were characterized by Hammett indicators method, XRD, XRF, TG/DTG, CO_2-TPD, XPS analysis, FTIR, SEM, N_2 adsorption-desorption at −196°C, and distribution particle size analysis. All of these characterization tests were performed to observe the differences after the thermal process modification.

Hammett Indicators Method. 0.1 g of calcined dried material was added to a test tube and stirred with 5 mL of anhydrous methanol. Then, one drop of 0.1% Hammett indicators (phenolphthalein, bromophenol blue, and phenol red) was added and left to equilibrate for 2 h. The changes of color were recorded.

The chemical composition was measured by fluorescence spectrometer X-ray Rigaku brand, model ZSX mini II, operating at a voltage of 40 kV and a current of 1.2 mA in the tube coupled to Pd.

The mineralogical identification was carried out using X-ray diffraction (XRD) on a diffractometer X'Pert Pro MPD in a tube CuKα operating at a voltage of 40 kV and a current 40 mA. To perform the analysis the powders were placed in the cavity of a support used as a sample holder. The spectra obtained swept the range 10–70°. The particle size of precursor and catalyst was estimated by Williamson-Hall method with a fitting of the diffraction profile. The crystallographic composition was obtained from Rietveld method by using the X'Pert HighScore Plus software.

The thermal decomposition of the quail eggshell was analyzed by thermogravimetric analysis/differential thermal analysis (TG/DTG), carried out on a Shimadzu TGA-50 analyzer operating under the following conditions: air flow

(a) Natural quail eggshell

(b) Edible part of quail eggshell

(c) Waste natural quail eggshell

(d) Waste powder natural quail eggshell

(e) Quail eggshell calcined at 900°C for 3 h

FIGURE 1: Method of obtaining the CaO from natural quail eggshell.

rate atmosphere ($50 \, cm^3 \, min^{-1}$); heating rate ($10°C \, min^{-1}$); and temperature ($25–900°C$).

The basicity of the catalyst was studied by temperature-programmed desorption of CO_2. Approximately 100 mg of sample was pretreated with a stream of helium at 800°C for 30 min ($10°C \, min^{-1}$ and $60 \, mL \, min^{-1}$). The reaction temperature was then decreased to 100°C, and a flow of pure CO_2 ($60 \, mL \, min^{-1}$) was subsequently introduced into the reactor for 30 min. The CO_2-TPD reaction was carried out between 100 and 800°C under a flow of helium ($10°C \, min^{-1}$, $30 \, mL \, min^{-1}$), and the amount of CO_2 evolved was analyzed using a quadrupole mass spectrometer (Balzer GSB 300 02) equipped with a Faraday detector (0–200 U), which monitors the mass of CO_2 (44 U) during the experiment.

X-ray photoelectron spectra were collected without exposure to the atmosphere using a physical electronics PHI 5700 spectrometer with nonmonochromatic Mg Kα radiation (300 W, 15 kV, and 1253.6 eV) with a multichannel detector. Spectra of each sample were recorded in the constant pass energy mode at 29.35 eV, using a $720 \, \mu m$ diameter analysis area. Charge referencing was measured against adventitious carbon (C 1s at 284.8 eV). A PHI ACCESS ESCA-V6.0 F software package was used for acquisition and data analysis. A Shirley-type background was subtracted from the signals. Recorded spectra were fitted using Gaussian–Lorentzian curves in order to determine the binding energies of the different element core levels more accurately. The calcined quail eggshell was treated at 900°C for 3 h and instantly was introduced into the analysis chamber and analyzed.

The FTIR analyses of catalyst were carried out using an FTIR spectrometer (model 8500; Shimadzu), which was operated in the range of 4000–400 wavenumber/cm^{-1}. A standard KBr technique was used for preparing the samples.

To observe the surface morphology of precursor and catalyst, both samples were initially covered with a thin layer of gold (10 nm) using a sputter coater (SCD 050; Baltec, Liechtenstein) and observed using a JEOL JXA-840A scanning electron microscope (20 kV) under a vacuum of 1.33 $\times 10^{-6}$ mbar (Jeol, Japan).

The textural parameters (S_{BET}, V_P, and d_P) were evaluated from nitrogen adsorption-desorption isotherms at –196°C as determined by an automatic ASAP 2020 system from Micromeritics. Prior to the measurements, samples were outgassed at 200°C and 10^{-4} mbar overnight. Surface areas were determined by using the Brunauer–Emmett–Teller (BET) equation and a nitrogen molecule cross section of 16.2 Å. The pore size distribution was calculated by applying the Barrett–Joyner–Halenda (BJH) method to the desorption branch of the N_2 isotherm. The total pore volume was calculated from the adsorption isotherm at $P/P_0 = 0.996$.

The particle size analyses were performed using the Mastersizer Particle Analyzer 2000, Hydro 2000 UM, Malvern (United Kingdom).

2.4. Transesterification Reaction. Transesterification reaction was conducted in a three-necked glass reactor with a condenser and magnetic stirrer, 60 mL of sunflower oil with different volumes of methanol, and varied amounts of catalyst

TABLE 1: Chemical composition (wt.%) of natural quail eggshell and calcined quail eggshell.

Material	Elements (wt.%)								
	Ca	P	S	K	Al	Cl	Si	Sr	Σ
Natural quail eggshell	97.71	1.30	0.31	0.19	0.16	0.15	0.13	0.06	100.00
Calcined quail eggshell	99.20	0.54	—	0.11	0.10	—	—	0.05	100.00

(over the weight of oil). The sunflower oil/methanol molar ratios (X_{MR}) used were 1 : 9, 1 : 10.5, and 1 : 12 and the catalyst loading (X_{CAT}) was 2, 2.5, and 3 wt.%. The temperature was fixed at 60°C and the reaction time (X_{TIME}) was 2, 2.5, and 3 h. After the reaction, the catalyst was separated by centrifuging the mixture at 2250 rpm for 30 min. The catalyst was separated by filtration, and the reaction mixture (biodiesel production and glycerin as coproduct) was placed in a funnel for phase separation. The residual methanol was evaporated using a rotary evaporator at 100°C for 15 min.

2.5. Quantification of the Biodiesel. A suitable experimental design was chosen to optimize the reaction condition based on the catalyst loading (X_{CAT}) and sunflower oil/methanol molar ratios (X_{MR}) as well as evaluate the optimal reaction conditions corresponding to maximum Y_{FAME} yield. Response surface methodology was used to analyze the results from the experiments. STATISTICA (version 7.0, StatSoft, Tulsa, OK) was used for statistical analysis of the experimental data. Analyses were carried out at a 95% confidence level.

The content of biodiesel was determined according to the procedure described in standard EN 14103 [37]. Analyses were done using a Varian CP-3800 gas chromatograph equipped with a CP-WAX 52CB (30 m × 0.25 mm × 0.05 μm) capillary column and a flame ionization detector (FID). For sample preparation, 250 mg of ester phase to be analyzed was added to 5 mL of a solution containing methyl heptadecanoate (at the final concentration of 10 mg L^{-1}, serving as internal standard) in heptane. 1 μL of the mixture was injected in the chromatography. The content of methyl esters of fatty acids was calculated as described in

$$Y_{FAME} \text{ (wt.\%)} = \left[\frac{\sum A_t - A_{pi}}{A_{pi}} \right] \frac{C_{pi} \cdot V_{pi} \cdot 100}{W}, \quad (2)$$

where $\sum A_t$ is the total area of the peaks of methyl esters; A_{pi} is the peak area of internal standard (methyl heptadecanoate). C_{pi} is the concentration of internal standard solution (10 mg L^{-1}), V_{pi} is the added volume in 5 mL solution of methyl heptadecanoate, and W is weight in 250 mg of sample (biodiesel).

2.6. Reuse of Catalyst. The calcined quail eggshell (900°C for 3 h) was reused in the transesterification reaction by three cycles. The transesterification reaction conditions were optimized: X_{MR} (1 : 10.5), X_{CAT} (2 wt.%), and X_{TIME} (2 h). Between each cycle the catalyst was separated from reaction medium by centrifuging the mixture at 2250 rpm for 30 min, washed twice with methanol, dried at 100°C during 30 minutes, and stored in a desiccator. Prior to the reuse, the

material was recalcinated to remove the recarbonation of the CaO species. This procedure was followed for each reaction cycle.

3. Results and Discussion

3.1. Material Characterization

3.1.1. Hammett Indicator Method. The Hammett indicator experiments can provide qualitative information about the strength of the basic sites obtained by thermal treatment of the quail eggshell. The results obtained, compiled in the Supplementary Material Table S2, indicated that the calcined quail eggshell has a basic character, in accordance with the colors observed in acidic and basic medium. From Hammett indicators the changes of color were observed for phenolphthalein from colorless to pink, for bromothymol blue from yellow to blue, and for phenol red from yellow to red. This data reveals that calcined quail eggshell displays basic sites with high strength, which can be employed for the transesterification reaction of sunflower oil with methanol.

3.1.2. X-Ray Fluorescence Analysis (XRF). Table 1 shows the chemical composition of natural quail eggshell and calcined quail eggshell, obtained by X-ray fluorescence analysis. The analysis of the natural quail eggshell reveals a high amount of calcium 97.71 wt.% together with phosphorous and sulfur possibly due to the presence of phospholipid in the membrane of the natural quail eggshell. After the thermal process, the percentage of calcium increases, reaching 99.20 wt.%. Moreover, the presence of lower amounts of other elements (Mg, Fe, P, S, K, Al, Cl, and Si) is noticeable, which will be considered as impurities of the CaO catalyst. The high calcium content of the calcined quail eggshell used in this work suggests that this material could be a promising material to biodiesel production [25].

3.1.3. X-Ray Diffraction Analysis (XRD). Figure 2(a) shows the XRD patterns for natural and calcined quail eggshell used in this research. Natural quail eggshell presents diffraction lines located at 2θ/degree = 23.3, 29.6, 31.6, 36.1, 39.6, 43.4, 47.3, 47.7, 48.7, 56.8, 57.6, 60.9, 61.6, 63.2, 64.8, 65.8, and 66.0, which have been ascribed to the presence of rhombohedra calcite ($CaCO_3$) (01-085-1108 reference code of card ICSD), discarding the presence of other crystallographic phases. After the activation process, the diffraction lines attributed to $CaCO_3$ disappear, with new diffraction lines arising about 2θ/degree = 32.3, 37.4, 53.7, 63.9, and 67.3 corresponding to lime (CaO) (00-044-1481 reference code of card ICSD). The particle sizes of precursor and catalyst were estimated by the Williamson-Hall method through the fitting of the diffraction

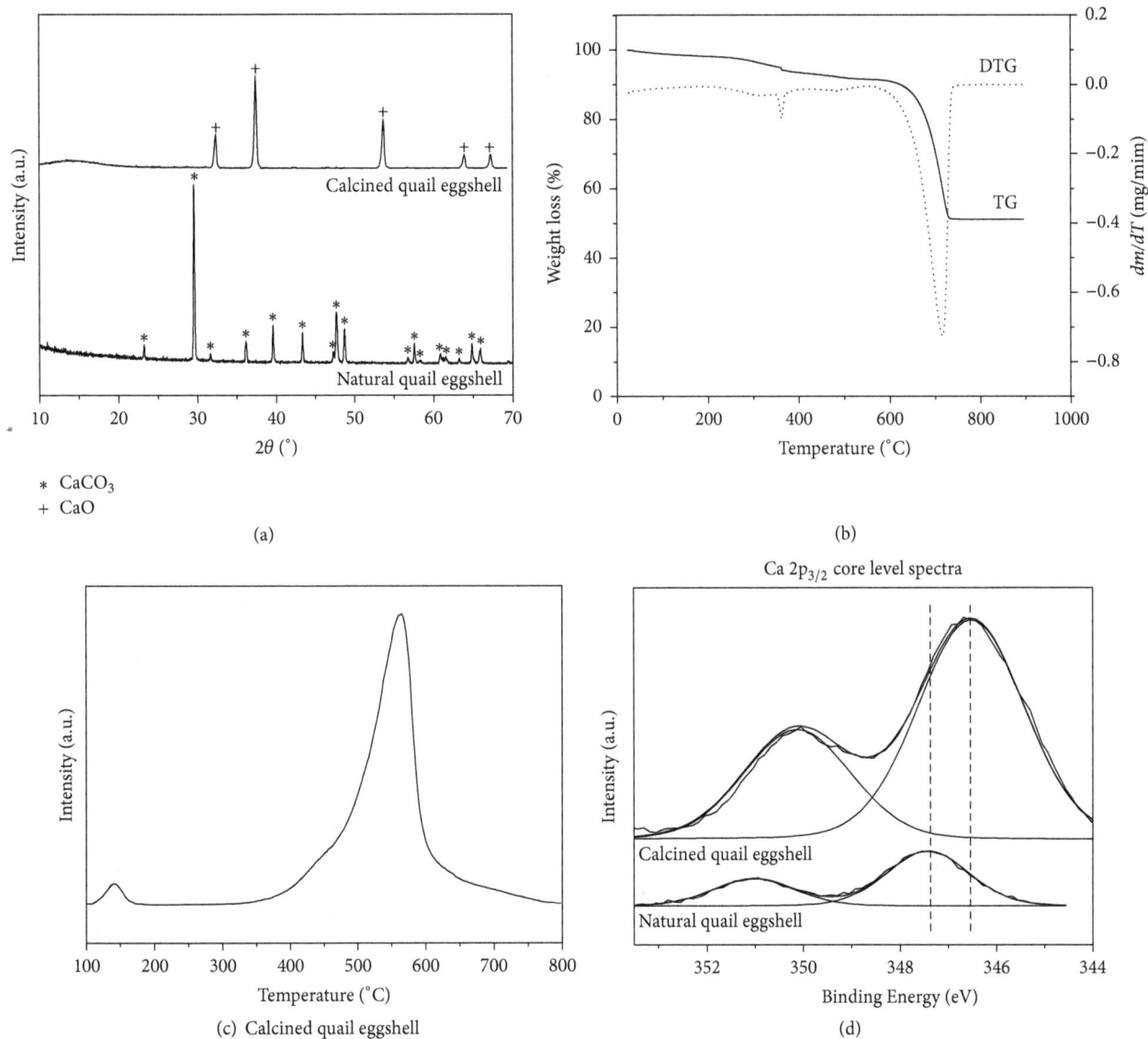

* CaCO$_3$
+ CaO

(a)

(b)

(c) Calcined quail eggshell

Ca 2p$_{3/2}$ core level spectra

(d)

FIGURE 2: X-ray diffraction of natural and calcined quail eggshell (a), TG/DTG of natural quail eggshell (b), CO$_2$-TPD profile of calcined quail eggshell (c), and Ca 2p core level spectra of natural and calcined quail eggshell (d).

profile. The quail eggshell shows a particle size of 315 nm (CaCO$_3$), while the calcined quail eggshell displays a particle size of 240 nm (CaO). The decrease of the crystal size can be attributed to the exothermic conditions of the decarbonation process. The lower intensity peaks for calcined quail eggshell could be related to the reduction in the crystallite size.

3.1.4. Thermogravimetric Analysis (TG/DTG).

Figure 2(b) shows the thermal analysis results (TG/DTG) of the natural quail eggshell. The TG/DTG curves present two events of mass losses. The first loss (7.8 wt.%), between 50–525°C, is attributed to the loss of the physisorbed water onto the surface of the material by hydrogen bonds and the decomposition of the organic matter because the membrane of quail eggshell natural is primarily composed of proteins with disulfide groups carbohydrates, phospholipids, and an outer organic layer, which is rich in cysteine. The second

mass loss (40.4 wt.%), between 550 and 875°C, is associated with the decomposition of calcium carbonate. In summary, about 51.8 wt.% of catalyst is obtained after the decarbonation process. The data of the natural quail eggshell residue is in the same range compared with X-ray fluorescence data and it is similar to those shown for limestone in previous research [25, 38].

3.1.5. CO$_2$-TPD Analysis.

CO$_2$-TPD profile was analyzed to evaluate the basicity strength of the calcined quail eggshell (Figure 2(c)). The CO$_2$-TPD presents two desorption peaks. The first one located between 100 and 190°C is attributed to the presence of small amount of weak basic sites. The second one, more intense, presents a maximum centered at 565°C. The calcined quail eggshell shows an amount of CO$_2$ adsorbed of 17.43 mmol·g^{-1}, where 2.15 mmol·g^{-1} of CO$_2$ comes from weak basic sites and 15.27 mmol·g^{-1} of CO$_2$

comes from strong basic sites. These values are close to the theoretical CO_2 adsorption which is equal to $17.82 \, mmol \cdot g^{-1}$, which is in accordance with what was shown in X-ray fluorescence (Table 3) and TG/TGA analysis (Figure 2(b)). The desorption of CO_2 at high temperature confirms the existence of strong basic sites on CaO surface, making it a potential catalyst in the transesterification reaction for the biodiesel production [39].

3.1.6. X-Ray Photoelectronic Spectroscopy (XPS).

In order to evaluate the surface composition of the natural quail eggshell and calcined quail eggshell, XPS measurements were carried out. Supplementary Material Table S3 compiles the binding energy values (in eV) and the atomic concentration (in wt.%) of each analyzed element. Both Ca $2p_{3/2}$ and Mg $2p_{3/2}$ core level spectra show a contribution located at 347.4 and 89.3 eV, respectively, ascribed to the presence of Ca^{2+} and Mg^{2+} in the form of carbonate. C 1s core level spectrum of the natural quail eggshell presents three contributions attributed to the adventitious carbon and organic matter (284.8 eV), C=O, C-N, and C-S about 287.3 eV, assigned to proteins residues such as amino acids and disulfide groups, and carbonate species at 290.0 eV, respectively. O 1s core level spectrum shows a unique contribution about 531.7 eV corresponds to the oxygen species in the form of carbonate. N 1s core level spectrum also exhibits a band about 399.2 eV attributed to amine groups, while S 2p core level spectrum presents two contributions located at 168.2 and 162.5 eV, ascribed to sulfate and sulfide species, respectively. Finally, P $2p_{3/2}$ core level spectrum exhibits a contribution about 133.8 and 133.6 eV in the form of phosphate species (Supplementary Material Table S3). The presence of the band located in the N 1s, S $2p_{3/2}$, and P $2p_{3/2}$ core level is attributed to the existence of proteins residues in the form of amino acids and phospholipid species on the surface of the natural quail eggshell.

After the thermal activation, C 1s core level spectrum shows how the signals attributed to the organic matter and carbonate species suffer a significant decrease. In the same way, the contribution of both N 1s and S $2p_{3/2}$ also disappears after calcination step. This fact provokes an increase of the other atomic concentrations. In addition, both Ca $2p_{3/2}$ in Figure 2(d) and Mg $2p_{3/2}$ suffer a shift at lower binding energy that suggests the formation of their respective oxide species.

3.1.7. Fourier Transform Infrared Spectroscopy Analysis (FTIR).

Figures 3(a) and 3(b) show the FTIR spectra for natural quail eggshell and calcined quail eggshell. The spectrum of natural quail eggshell in Figure 3(a) shows the presence of the out-of-plane bending, the asymmetric stretching, and the in-plane bending modes of the carbonate groups, characteristic of natural dolomite, located at $869 \, cm^{-1}$, $1426 \, cm^{-1}$, and $714 \, cm^{-1}$, respectively. Besides the internal modes, the combination of the previous bending modes has also been observed at $1801 \, cm^{-1}$ and $2522 \, cm^{-1}$. Finally, the band located at $3444 \, cm^{-1}$ has been attributed to H-bonded water of the humidity [25, 38, 40]. Moreover, a band about $2900 \, cm^{-1}$ appears assigned to alkyl C-H stretch due to the organic matter of the quail eggshell.

After the thermal treatment of quail eggshell in Figure 3(b) a new hydroxyl band appears about $3637 \, cm^{-1}$, which can be assigned to OH groups of calcium [25]. The bands located at $1647 \, cm^{-1}$ and $1454 \, cm^{-1}$ can be assigned to the symmetric and asymmetric stretching vibrations of O-C-O bonds of unidentate carbonate at the surface of the calcium oxide. The band at $1050 \, cm^{-1}$ also arises from these carbonates groups. The band centered at $3454 \, cm^{-1}$ is assigned to H-bonded water [25, 41].

3.1.8. Scanning Electron Microscopy Analysis (SEM).

Figures 4(a) and 4(b) show the SEM micrographs of natural quail eggshell and calcined quail eggshell. The SEM micrograph displays that natural quail eggshell has structure of irregular and heterogeneous distribution of particle sizes ($CaCO_3$ and organic matter), while the SEM micrograph of calcined quail eggshell exhibits particles with lower particle size and more homogeneous distribution probably due to exothermic process that takes place in the decarbonation step, as was indicated previously in the XRD data (Figure 1(a)).

3.1.9. N_2 Adsorption-Desorption at $-196°C$ and Distribution Particle Size Analysis.

The textural properties of both natural quail eggshell and calcined quail eggshell are summarized in Supplementary Material Table S4. The textural properties reveal that both precursor and catalyst lack porosity. Nonetheless, after the activation process, S_{BET} increases slightly due to the porosity generated between particles of a lower size as revealed by the XRD in Figure 2(a) and the SEM micrographs in Figures 4(a) and 4(b).

Figures 3(c) and 3(d) display the grain-size distribution of natural quail eggshell and calcined quail eggshell. The particle size was estimated as statistic average. The natural quail eggshell exhibits a wide range of grain-size distribution with a maximum of 242 μm. After the thermal treatment, the grain-size is narrower and diminishes, displaying a maximum of 9 μm due to the decomposition of the calcium carbonate by the removal of CO_2 is a exothermic process, leading to particles with a lower grain-size, confirming data shown in XRD, SEM micrographs, and the textural properties.

3.2. Experimental Design Analysis.

A set of experiments were carried out to determine the influence of the variables (X_{MR}, X_{CAT}, and X_{TIME}) on the transesterification reaction of sunflower oil with methanol for the production of biodiesel production. The factors were preestablished with the intention of obtaining an optimum catalytic activity for catalyst studied. The optimization experiments were performed according to a matrix of the experimental designs $3^{**}(k\text{-}p)$ and Box-Behnken 3^2 with a central point in triplicate (Table 2). The dependent variable, Y_{FAME}, corresponds to the conversion of triglycerides into methyl esters. The parameters evaluated were the sunflower oil/methanol molar ratio (X_{MR}), catalyst loading (X_{CAT}), and reaction time (X_{TIME}). The stirring rate was set at 1000 rpm and the temperature was 60°C for all the experiments. Negative values on the scale correspond to lower values of the variable, while positive values represent higher values of the same, according to the parameters studied.

TABLE 2: Matrix of the experimental designs 3^{**} $(k\text{-}p)$ and Box-Behnken. Data are presented as mean ± standard deviation, $n = 3$.

| Run | Coded variables | | | Real variables | | | Response |
	X_{MR}	X_{CAT}	X_{TIME}	X_{MR}	X_{CAT}	X_{TIME}	Y_{FAME}
1	−1	−1	0	1:9	2.0	2.5	87.88 ± 0.01
2	1	−1	0	1:12	2.0	2.5	88.16 ± 0.07
3	−1	1	0	1:9	3.0	2.5	91.18 ± 0.30
4	1	1	0	1:12	3.0	2.5	89.39 ± 0.02
5	−1	0	−1	1:9	2.5	2.0	88.61 ± 0.04
6	1	0	−1	1:12	2.5	2.0	86.72 ± 0.13
7	−1	0	1	1:9	2.5	3.0	91.67 ± 0.02
8	1	0	1	1:12	2.5	3.0	98.35 ± 0.02
9	0	−1	−1	1:10.5	2.0	2.0	99.00 ± 0.20
10	0	1	−1	1:10.5	3.0	2.0	72.53 ± 0.40
11	0	−1	1	1:10.5	2.0	3.0	87.17 ± 0.03
12	0	1	1	1:10.5	3.0	3.0	88.29 ± 0.01
13	0	0	0	1:10.5	2.5	2.5	76.74 ± 0.09
14	0	0	0	1:10.5	2.5	2.5	76.80 ± 0.14
15	0	0	0	1:10.5	2.5	2.5	75.56 ± 0.90

The investigation of the model was performed by analysis of variance (ANOVA; Table 3). The calculated F-value (35.02) was higher than that shown for F tabulated (4.77). Equations (3) represented the linear model and quadratic model for the studied variables (X_{MR}, X_{CAT}, and X_{TIME}), considering all the regression coefficients for 95% confidence.

$$Y_{FAME} = 91.62 + 0.41 * X_{MR} - 0.29 * X_{MR}^2 - 3.65$$

$$* X_{CAT} - 2.18 * X_{CAT}^2 - 0.74$$

$$= 91.62 + 0.41 * X_{MR} - 0.29 * X_{MR}^2 + 3.68 \qquad (3)$$

$$* X_{TIME}$$

$$= 91.62 - 3.65 * X_{CAT} - 2.18 * X_{CAT}^2 + 3.68$$

$$* X_{TIME}.$$

A response surface shown in Figures 5(a), 5(b), and 5(c) for a calcined quail eggshell model shows that, for low concentrations of catalyst, the conversion of biodiesel decreased. In addition, Figure 5 also indicates that the significant factor is X_{MR}, X_{CAT}, and X_{TIME}, which has a positive effect on transesterification reaction at a confidence level of 95%, as noted in the response surface plot. This fact is attributed to the fact that an increase of the catalyst concentration leads to higher amount of available basic sites, which are the active phase in the transesterification reaction, leading to higher conversions to biodiesel. Another key factor is the alcohol : acid ratio. Thus, an excess of methanol leads to the chemical equilibrium shifted toward the products favoring the formation of higher proportions of methyl esters and subsequently increasing the conversion values. Nonetheless the use of higher proportion of methanol also has an adverse effect in the separation process due to the existence of gravity between the phases formed and glycerin ester, increasing

their miscibility and promoting the displacement of the balance in the opposite direction toward the formation of mono-, di-, and triglycerides, thereby decreasing the production of methyl esters.

The maximum Y_{FAME} value was 99.99 ± 0.20 wt.%, which was obtained in the central point of the factorial design, namely, the following reaction conditions: X_{MR} (1:10.5), X_{CAT} (2 wt.%), temperature (60°C), and X_{TIME} (2 h). Considering the best experimental transesterification reaction conditions, it can be concluded that 1 kg of quail eggshell leads to the obtaining of 82.5 kg of biodiesel production.

The mechanism of interaction between heterogeneous catalysts based on calcium oxide was described by [42]. In the 1st stage of the transesterification reaction, methyl alcohol reacts with calcium oxide to form the methoxide anion, and then the 2nd stage is the anion methoxide attacking the carbon of the carbonyl group present in triglycerides (refined vegetable oil sunflower) for formation of an intermediate carbonyl group.

Therefore, in the 3rd stage alkoxides groups present in the intermediate carbonyl formed by CaO-H$^+$ produced CaO and another carbonyl intermediate group and in the 4th step the intermediate group is reorganized and form methyl esters (organic phase/biodiesel) and diglycerides. Other authors have pointed out that calcium diglyceroxide Ca(C$_3$H$_7$O$_3$)$_2$ is the active site in the transesterification reaction to obtain biodiesel when CaO is used as solid basic catalyst. This species is formed by the interaction between Ca^{2+} cations and glycerol obtained as by-product. The oxygen anion of the diglyceroxide can abstract H atoms from the OH in methanol for hydrogen bond, yielding a surface methoxide anion. Later, nucleophilic MeO$^-$ species attack the carbonyl groups of the triglyceride to form the diglyceride molecule. This process takes place sequentially to form the monoglyceride and glycerol species [43].

The interesting catalytic behavior of the CaO catalyst obtained from a natural calcium source, such as quail eggshell, by a thermal treatment can be attributed to the decrease of both particle and crystal size in the calcination step which supposes an increase of the available basic sites, as indicated in the CO$_2$-TPD profile TPD (Figure 2(b)), which are necessary for the transesterification reaction. In this sense, previous researches have established that the choosing of the calcium source is a key parameter in the catalytic behavior in the transesterification reaction of triglycerides with short chain alcohols, such as methanol or ethanol [16]. Thus, the decomposition of carbonate precursors has shown the highest conversion values, while the use of calcium nitrate as precursor leads to lower dispersion of the basic sites and subsequent lower conversion values.

Table 4 compares the Y_{FAME} obtained in this study with the other ones from the literature. Data reported by other authors show that it was not necessary to use extreme transesterification reaction conditions (X_{MR} and X_{CAT}) to obtain Y_{FAME} reported in the literature.

3.3. Characterization of Biodiesel Sunflower. Biodiesel from sunflower oil was analyzed according to the requirements for the standard EN 14214 [35] (acid value, density at 20°C,

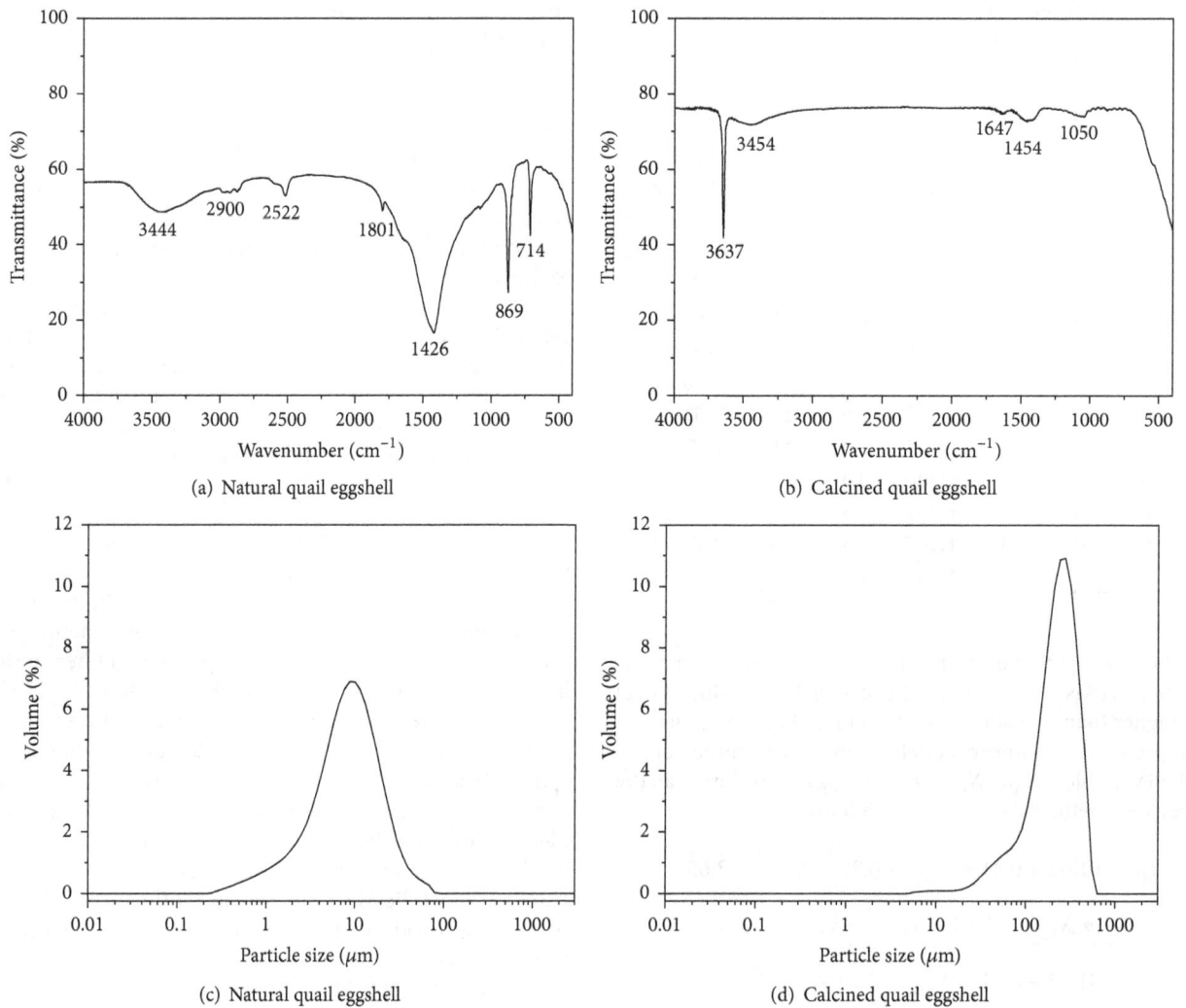

(a) Natural quail eggshell

(b) Calcined quail eggshell

(c) Natural quail eggshell

(d) Calcined quail eggshell

FIGURE 3: FTIR spectra ((a) and (b)) and distribution particle sizes analysis ((c) and (d)) of natural and calcined quail eggshell.

(a)

(b)

FIGURE 4: SEM images for natural quail eggshell ((a) and (b)) and calcined quail eggshell, with an increase of 200x.

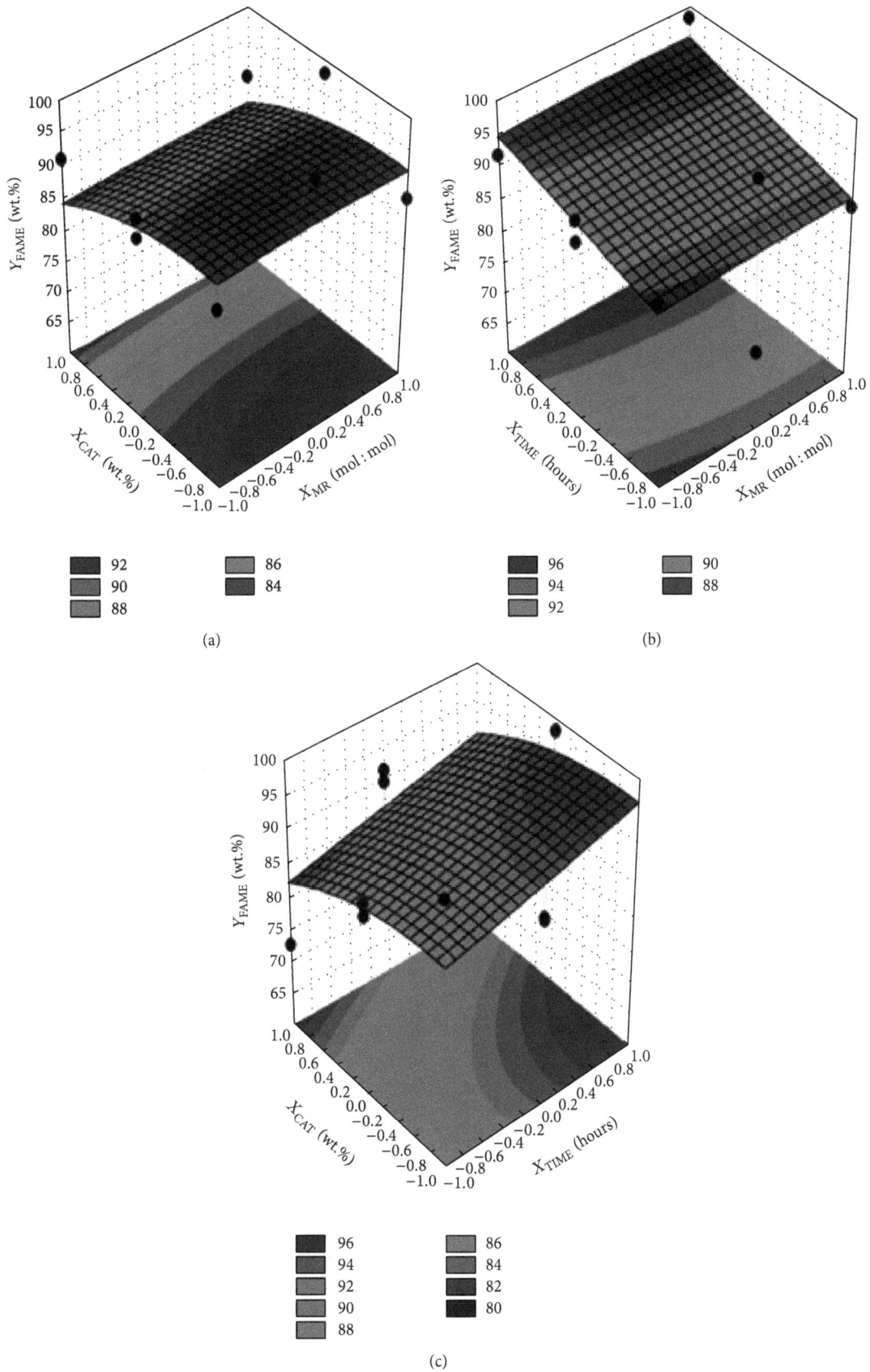

FIGURE 5: Response surface for model of quail eggshell calcined.

TABLE 3: ANOVA, analysis of variance, for the conversion of sunflower oil to biodiesel by empirical model of quail eggshell calcined at 900°C for 3 h.

Source of variation	Sum of squares	Degrees of freedom	Mean square	F-value
Regression	273.85223	4	68.46305	35.02
Residual	9.77317	5	1.954634	
Total	283.6254	9		$F_{9,5} = 4.77$

TABLE 4: Comparison of biodiesel production using different wastes eggshell catalysts in transesterification reaction.

Oil/alcohol	Material	X_{MR}	X_{CAT}	Y_{FAME}	Reference
Sunflower oil/methanol	Quail eggshell	1:10.5	2	99.00	This work
Sunflower oil/methanol	Eggshell	1:9	3	97.75	[25]
Soybean oil/methanol	Eggshell	1:9	3	95.00	[40]
Palm kernel oil/methanol	Eggshell	1:18	15	96.7	[26]
Palm olein oil/methanol	Eggshell	1:12	10	>90.00	[38]
Soybean oil/methanol	Fly ash and eggshell	1:6.9	1	96.97	[27]
Waste frying oil	Eggshell	1:12	5	94.52	[28]

TABLE 5: Properties of sunflower biodiesel. Data are presented as mean ± standard deviation, $n = 3$.

Property	Biodiesel from sunflower	EN 14214 standard
Acid value (mg KOH/g)	0.40 ± 0.09	0.5
Density (20°C, kg·m^{-3})	882 ± 1.30	860–900
Kinematic viscosity (40°C, mm^2/s)	4.49 ± 0.04	3.50–5.00
Water content (kg·m^{-3})	101 ± 8	500
Y_{FAME} (wt.%)	99.00 ± 0.20	96.50

kinematic viscosity at 40°C, water, magnesium and calcium content, and conversion of sunflower oils to biodiesel) which are shown in Table 5.

3.4. Reuse and Leaching of Catalyst. The reuse and stability of solid catalysts have an important role in determining the economic application for biodiesel production on a large industrial scale. The reuse of the catalyst was evaluated considering the optimized reaction conditions X_{MR} (1:10.5), X_{CAT} (2 wt.%), X_{TIME} (2 h), and temperature (60°C).

The decrease of Y_{FAME} with reuse of the catalyst in the transesterification reaction was demonstrated for 3 cycles. The calcined quail eggshell as heterogeneous catalyst showed only a slight decrease in the catalytic activity after each reuse which can be attributed to the washing and sample handling between each cycle. Thus, it is observed that the conversion of Y_{FAME} decreases to 99.00 ± 0.02 wt.% (1st run), 86.14 ± 0.02 wt.% (2nd run), and 78.26 ± 0.04 wt.% (3rd run) of Y_{FAME}. In this study the decrease in the biodiesel production was similar to the results described by [23] for calcined dolomite. As was previously indicated, CaO species can be solubilized in the glycerol obtained as by-product to form calcium diglyceroxide species, which is the active phase in the biodiesel production. This phase is more soluble than CaO under the reaction conditions used in the transesterification reaction with solid basic catalysts. In

any case, Ca^{2+} concentration is significantly lower than that observed for homogenous catalysis [44]. In addition, the solid CaO can be reutilized for several runs, so it seems clearly that CaO obtained from quail eggshell is a robust catalyst in the transesterification reaction.

4. Conclusion

The calcined quail eggshell used as a material promise biodiesel production by transesterification reaction using sunflower oil refined with methanol. The material process thermal treatment (900°C for 3 h) produces small CaO particles which show high basicity and they are responsible for the catalytic activity in the transesterification reaction. The optimization of the reaction parameters led to a maximum biodiesel production of 99.00 ± 0.02 wt.% obtained under the next conditions: X_{MR} (1:10.5), X_{CAT} (2 wt.%), X_{TIME} (2 h), magnetic stirring (1000 rpm), and temperature (60°C). Considering the best experimental conditions, it can be concluded that 1 kg of quail eggshell leads to obtaining of 82.5 kg of biodiesel. The reuse of the CaO catalyst under the best reaction parameters revealed a small, but progressive, loss of the biodiesel production from 99.00% to 78.26% after the third cycle by the loss of active sites by formation of carbonaceous deposits and/or the loss of sample in the washing or recalcination step between each cycle.

Competing Interests

The authors declare that there is no conflict of interests regarding the publication of this paper.

Acknowledgments

The authors wish to acknowledge the financial support provided by the Fundação Cearense de Apoio ao Desenvolvimento Científico e Tecnológico (FUNCAP), Conselho Nacional de Desenvolvimento Científico e Tecnológico (CNPq), Coordenação de Aperfeiçoamento de Pessoal de Nível Superior (CAPES), DGU Project 216/12, Bioinorganic Laboratory (UFC-Chemical), X-Rays Laboratory (UFC-Physical), Polymer Laboratory (LAPOL-UFC-Chemical), Núcleo de Tecnologia do Estado do Ceará (NUTEC), Universidad de Málaga (UMA), and Ministerio de Economia y Competitividad Project CTQ 2012-37925-C03-03 and FEDER funds.

References

[1] D. A. Oliveira, P. Benelli, and E. R. Amante, "A literature review on adding value to solid residues: egg shells," *Journal of Cleaner Production*, vol. 46, pp. 42–47, 2013.

[2] L. M. Correia, N. de Sousa Campelo, D. S. Novaes et al., "Characterization and application of dolomite as catalytic precursor for canola and sunflower oils for biodiesel production," *Chemical Engineering Journal*, vol. 269, pp. 35–43, 2015.

[3] O. Ilgen, "Dolomite as a heterogeneous catalyst for transesterification of canola oil," *Fuel Processing Technology*, vol. 92, no. 3, pp. 452–455, 2011.

[4] S. Jaiyen, T. Naree, and C. Ngamcharussrivichai, "Comparative study of natural dolomitic rock and waste mixed seashells as heterogeneous catalysts for the methanolysis of palm oil to biodiesel," *Renewable Energy*, vol. 74, pp. 433–440, 2015.

[5] F. Ma and M. A. Hanna, "Biodiesel production: a review," *Bioresource Technology*, vol. 70, no. 1, pp. 1–15, 1999.

[6] Z. Du, Z. Tang, H. Wang, J. Zeng, Y. Chen, and E. Min, "Research and development of a sub-critical methanol alcoholysis process for producing biodiesel using waste oils and fats," *Chinese Journal of Catalysis*, vol. 34, no. 1, pp. 101–115, 2013.

[7] W. N. R. W. Isahak, M. Ismail, J. M. Jahim, J. Salimon, and M. A. Yarmo, "Characterisation and performance of three promising heterogeneous catalysts in transesterification of palm oil," *Chemical Papers*, vol. 66, no. 3, pp. 178–187, 2012.

[8] E. M. Shahid and Y. Jamal, "A review of biodiesel as vehicular fuel," *Renewable and Sustainable Energy Reviews*, vol. 12, no. 9, pp. 2484–2494, 2008.

[9] D. Y. C. Leung and Y. Guo, "Transesterification of neat and used frying oil: optimization for biodiesel production," *Fuel Processing Technology*, vol. 87, no. 10, pp. 883–890, 2006.

[10] A. Demirbas, "Biodiesel from waste cooking oil via base-catalytic and supercritical methanol transesterification," *Energy Conversion and Management*, vol. 50, no. 4, pp. 923–927, 2009.

[11] S. Zheng, M. Kates, M. A. Dubé, and D. D. McLean, "Acid-catalyzed production of biodiesel from waste frying oil," *Biomass and Bioenergy*, vol. 30, no. 3, pp. 267–272, 2006.

[12] N. Chammoun, D. P. Geller, and K. C. Das, "Fuel properties, performance testing and economic feasibility of *Raphanus sativus* (oilseed radish) biodiesel," *Industrial Crops and Products*, vol. 45, pp. 155–159, 2013.

[13] M. J. Campos-Molina, J. Santamaría-González, J. Mérida-Robles et al., "Base catalysts derived from hydrocalumite for the transesterification of sunflower oil," *Energy & Fuels*, vol. 24, no. 2, pp. 979–984, 2010.

[14] N. Degirmenbasi, S. Coskun, N. Boz, and D. M. Kalyon, "Biodiesel synthesis from canola oil via heterogeneous catalysis using functionalized CaO nanoparticles," *Fuel*, vol. 153, pp. 620–627, 2015.

[15] M. L. Granados, M. D. Z. Poves, D. Martín-Alonso et al., "Biodiesel from sunflower oil by using activated calcium oxide," *Applied Catalysis B: Environmental*, vol. 73, no. 3, pp. 317–326, 2007.

[16] D. Martín-Alonso, F. Vila, R. Mariscal, M. Ojeda, M. López-Granados, and J. Santamaría-González, "Relevance of the physicochemical properties of CaO catalysts for the methanolysis of triglycerides to obtain biodiesel," *Catalysis Today*, vol. 158, no. 1-2, pp. 114–120, 2010.

[17] P. D. Patil and S. Deng, "Transesterification of camelina sativa oil using heterogeneous metal oxide catalysts," *Energy & Fuels*, vol. 23, no. 9, pp. 4619–4624, 2009.

[18] D. Martín-Alonso, R. Mariscal, M. López-Granados, and P. Maireles-Torres, "Biodiesel preparation using Li/CaO catalysts: activation process and homogeneous contribution," *Catalysis Today*, vol. 143, no. 1-2, pp. 167–171, 2009.

[19] J. M. Rubio-Caballero, J. Santamaría-González, J. Mérida-Robles et al., "Calcium zincate derived heterogeneous catalyst for biodiesel production by ethanolysis," *Fuel*, vol. 105, pp. 518–522, 2013.

[20] M. C. G. Albuquerque, I. Jiménez-Urbistondo, J. Santamaría-González et al., "CaO supported on mesoporous silicas as basic catalysts for transesterification reactions," *Applied Catalysis A: General*, vol. 334, no. 1-2, pp. 35–43, 2008.

[21] M. Zabeti, W. M. A. W. Daud, and M. K. Aroua, "Optimization of the activity of CaO/Al$_2$O$_3$ catalyst for biodiesel production using response surface methodology," *Applied Catalysis A: General*, vol. 366, no. 1, pp. 154–159, 2009.

[22] S. Yan, M. Kim, S. O. Salley, and K. Y. S. Ng, "Oil transesterification over calcium oxides modified with lanthanum," *Applied Catalysis A: General*, vol. 360, no. 2, pp. 163–170, 2009.

[23] L. M. Correia, N. D. S. Campelo, R. D. F. Albuquerque et al., "Calcium/chitosan spheres as catalyst for biodiesel production," *Polymer International*, vol. 64, no. 2, pp. 242–249, 2015.

[24] C. Ngamcharussrivichai, P. Nunthasanti, S. Tanachai, and K. Bunyakiat, "Biodiesel production through transesterification over natural calciums," *Fuel Processing Technology*, vol. 91, no. 11, pp. 1409–1415, 2010.

[25] L. M. Correia, R. M. A. Saboya, N. de Sousa Campelo et al., "Characterization of calcium oxide catalysts from natural sources and their application in the transesterification of sunflower oil," *Bioresource Technology*, vol. 151, pp. 207–213, 2014.

[26] P. Khemthong, C. Luadthong, W. Nualpaeng, P. Changsuwan, P. Tongprem, and N. Viriya-Empikul, "Industrial eggshell wastes as the heterogeneous catalysts for microwave-assisted biodiesel production," *Catalysis Today*, vol. 190, pp. 112–126, 2010.

[27] R. Chakraborty, S. Bepari, and A. Banerjee, "Transesterification of soybean oil catalyzed by fly ash and egg shell derived solid catalysts," *Chemical Engineering Journal*, vol. 165, no. 3, pp. 798–805, 2010.

[28] S. Niju, M. S. Begum, and N. Anantharaman, "Modification of egg shell and its application in biodiesel production," *Journal of Saudi Chemical Society*, vol. 18, no. 5, pp. 702–706, 2014.

[29] N. Girish, S. P. Niju, K. M. Meera Sheriffa Begum, and N. Anantharaman, "Utilization of a cost effective solid catalyst derived from natural white bivalve clam shell for transesterification of waste frying oil," *Fuel*, vol. 111, pp. 653–658, 2013.

[30] R. Rezaei, M. Mohadesi, and G. R. Moradi, "Optimization of biodiesel production using waste mussel shell catalyst," *Fuel*, vol. 109, pp. 534–541, 2013.

[31] A. Obadiah, G. A. Swaroopa, S. V. Kumar, K. R. Jeganathan, and A. Ramasubbu, "Biodiesel production from Palm oil using calcined waste animal bone as catalyst," *Bioresource Technology*, vol. 116, pp. 512–516, 2012.

[32] A. Birla, B. Singh, S. N. Upadhyay, and Y. C. Sharma, "Kinetics studies of synthesis of biodiesel from waste frying oil using a heterogeneous catalyst derived from snail shell," *Bioresource Technology*, vol. 106, pp. 95–100, 2012.

[33] P.-L. Boey, G. P. Maniam, and S. A. Hamid, "Biodiesel production via transesterification of palm olein using waste mud crab (*Scylla serrata*) shell as a heterogeneous catalyst," *Bioresource Technology*, vol. 100, no. 24, pp. 6362–6368, 2009.

[34] S. Jairam, P. Kolar, R. Sharma-Shivappa Ratna, J. A. Osborne, and J. P. Davis, "KI-impregnated oyster shell as a solid catalyst for soybean oil transesterification," *Bioresource Technology*, vol. 104, pp. 329–335, 2012.

[35] R. Chakraborty, S. Bepari, and A. Banerjee, "Application of calcined waste fish (Labeo rohita) scale as low-cost heterogeneous catalyst for biodiesel synthesis," *Bioresource Technology*, vol. 102, no. 3, pp. 3610–3618, 2011.

[36] European Committee for Standardization, *European Standard: Automotive Fuels-Fatty Acid Methyl Esters (FAME) for Diesel Engines-Requirements and Test Methods*, EN 14214:2008+A1:2009, European Committee for Standardization, Brussels, Belgium, 2008.

[37] European Committee for Standardization, "European standard: fat and oil derivatives—Fatty Acid Methyl Esters (FAME)—determination of ester and linolenic acid methyl ester contentes—requirements and test methods," EN 14103:2003, European Committee for Standardization, Brussels, Belgium, 2003.

[38] N. Viriya-empikul, P. Krasae, B. Puttaswat, B. Yoosuk, N. Chollacoop, and K. Faungnawakij, "Waste shells of mollusk and egg as biodiesel production catalysts," *Bioresource Technology*, vol. 101, no. 10, pp. 3765–3767, 2010.

[39] Y. B. Cho and G. Seo, "High activity of acid-treated quail eggshell catalysts in the transesterification of palm oil with methanol," *Bioresource Technology*, vol. 101, no. 22, pp. 8515–8519, 2010.

[40] Z. Wei, C. Xu, and B. Li, "Application of waste eggshell as low-cost solid catalyst for biodiesel production," *Bioresource Technology*, vol. 100, no. 11, pp. 2883–2885, 2009.

[41] J. G. Pereira, F. Okumura, L. A. Ramos, É. T. G. Cavalheiro, and J. A. Nóbrega, "Termogravimetria: um novo enfoque para a clássica determinação de cálcio em cascas de ovos," *Química Nova*, vol. 32, no. 6, pp. 1661–1666, 2009.

[42] J. F. Puna, M. J. N. Correia, A. P. S. Dias, J. Gomes, and J. Bordado, "Biodiesel production from waste frying oils over lime catalysts," *Reaction Kinetics, Mechanisms and Catalysis*, vol. 109, no. 2, pp. 405–415, 2013.

[43] L. León-Reina, A. Cabeza, J. Rius, P. Maireles-Torres, A. C. Alba-Rubio, and M. López Granados, "Structural and surface study of calcium glyceroxide, an active phase for biodiesel production under heterogeneous catalysis," *Journal of Catalysis*, vol. 300, pp. 30–36, 2013.

[44] M. López-Granados, D. Martín-Alonso, I. Sádaba, R. Mariscal, and P. Ocón, "Leaching and homogeneous contribution in liquid phase reaction catalysed by solids: the case of triglycerides methanolysis using CaO," *Applied Catalysis B: Environmental*, vol. 89, no. 1-2, pp. 265–272, 2009.

12

Assessment of a Physicochemical Indexing Method for Evaluation of Tropical River Water Quality

Siong Fong Sim ⓘ and Szewei Elaine Tai ⓘ

Faculty of Resource Science & Technology, Universiti Malaysia Sarawak, 94300 Kota Samarahan, Sarawak, Malaysia

Correspondence should be addressed to Siong Fong Sim; sfsim@unimas.my

Academic Editor: Samuel B. Dampare

This paper attempts to evaluate the Malaysian water quality indexing method that has been criticized for its ineffectiveness. The indexing method is referred to as the Water Quality Index of the Department of Environment, Malaysia (DOE-WQI). This index was assessed against two other indices (River Ganga Index and Minimal Water Quality Index (WQI_{min})) and a modified DOE-WQI was also proposed. DOE-WQI and WQI_{min} are physicochemical indices, whereas the River Ganga Index and modified DOE-WQI are indices with the inclusion of a microbiological parameter. The assessment was conducted based on the water quality of tropical rivers in Malaysia, with specific reference to Sarawak River and its tributaries. Water quality in terms of pH, dissolved oxygen, conductivity, turbidity, total suspended solids, biochemical oxygen demand, chemical oxygen demand, ammoniacal nitrogen, and fecal coliform count (FCC) was measured from 18 stations in December 2015, January 2016, and March 2016. Generally, the river was characterized with high FCC with the four indices significantly correlated. The results demonstrated the shortcomings of the existing DOE-WQI; the physicochemical index assigned water quality to a better class than its actual conditions without taking into consideration the FCC.

1. Introduction

Water quality monitoring is an area of utmost concern for many developing countries due to the rapid growth in population and urbanization. In Malaysia, many river catchments are at risk as a result of increasing land use. As reported by the Department of Environment (2010) [1], the percentage of clean river basins has consistently reduced with increasing numbers of rivers categorized as moderately polluted/ polluted. The water quality of Malaysian rivers is primarily monitored based on the water quality index (WQI) developed by the Department of Environment using six physicochemical parameters, namely, dissolved oxygen (DO), biochemical oxygen demand (BOD), chemical oxygen demand (COD), ammoniacal nitrogen (NH_3-N), total suspended solids (TSS), and pH. This approach, like other water quality indices, transforms the water quality data into a single numerical value to depict the overall water quality with a score between 0 and 100 [2, 3]. It has been widely used to facilitate the decision-making process and management of regulatory programs relating to water quality assessment in Malaysia.

Numerous water quality indexing methods were developed for different uses and water sources; the first water quality index was introduced in 1965 [4]. They differ primarily in the determinants used, the mathematical derivation strategies, and the classification criteria [2]. This has been comprehensively reviewed by Gitau et al. [5]; for example, the water quality index of the U.S. National Sanitation Foundation (NSF) uses nine variables, that is, DO, fecal coliform count (FCC), pH, biochemical oxygen demand (BOD), temperature (Temp), total phosphate (TP), nitrate (NO_3^-), turbidity (Turb), and total solids (TS). For each parameter, a score, Q_n, is determined from a rating curve and multiplied by the weightage assigned, W_n, yielding the weighted subindices. The sum of the weighted subindices describes the water quality status where a score of 90–100 denotes excellent water quality whereas a score of 70–90 indicates water in good condition [$\sum_{n=1}^{N} Q_n W_n$ (n is the determinant involved)]. A lower score of 50–70, 25–50, and 0–25 refers to medium,

TABLE 1: The GPS coordinates of the sampling stations.

River	Station	GPS coordinates	Description
Sarawak Kanan	1	1°23′51.8″N, 110°06′59.4″E	At the bridge near Ruiz Suba Buan Village
	2	1°24′54.5″N, 110°08′06.1″E	Wind Cave
	3	1°25′25.2″N, 110°09′00.5″E	At the bridge near Bau Water Treatment Plant
Sarawak Kanan	4	1°27′08.0″N, 110°11′02.1″E	At Buso Village
	5	1°26′50.1″N, 110°13′10.7″E	At Siniawan Village
Sarawak Kiri	6	1°27′07.5″N, 110°16′54.6″E	At Bau Bridge Road
	7	1°27′03.8″N, 110°16′52.8″E	Near Sk. Batu Kitang and village
Sarawak	8	1°30′01.0″N, 110°14′47.0″E	At the bridge of Batu Kawa Road
	9	1°32′22.4″N, 110°17′09.9″E	At Sejijak Village
Maong	10	1°32′34.0″N, 110°18′40.2″E	At Maong River Village with construction activities
	11	1°32′00.5″N, 110°19′27.3″E	Near housing areas & primary schools
	12	1°31′35.0″N, 110°20′12.0″E	Near commercial areas
Sarawak	13	1°33′17.0″N, 110°19′25.8″E	Near Satok Bridge
	14	1°33′36.6″N, 110°20′46.4″E	Kuching Waterfront Bazaar
	15	1°33′27.0″N, 110°22′50.0″E	Pending-Petra Jaya Toll Bridge, industrial area
Sarawak	16	1°34′33.2″N, 110°24′33.2″E	Downstream of Sarawak Barrage
	17	1°34′31.2″N, 110°24′36.9″E	Upstream of Sarawak Barrage, before discharge from any industries
	18	1°32′38.7″N, 110°23′21.9″E	At the bridge at Setia Raja Road

bad, and very bad water quality, respectively [2, 6]. The Oregon Water Quality Index (OWQI) on the other hand adopts eight parameters of NSF except Turb [2, 4]. There are also indices derived using less than a handful of parameters which are cost-effective and useful for developing countries with a limited budget for water quality assessment [7]. A simple index with three key parameters of DO, BOD, and NH_3-N was used to evaluate the level of organic pollution by the Environmental Protection Department, Hong Kong [8]. Simões et al. [9] employed total phosphorus, Turb, and DO to infer the effects of aquaculture activities on water quality.

The Malaysian Water Quality Index of six variables has been continuously reported with limitations; it is perceived as insufficiently inclusive as only physicochemical parameters are considered [10]. Al-Manun and Idris [11] suggested the addition of biological indicators to the existing index as microbial pollution is a widespread issue in rivers of Malaysia due to the warm and humid tropical climate, conducive for microbial growth [12–15]. Naubi et al. [16] examined the water quality of a Malaysian river, concluding that the existing WQI is not an effective measure as the index did not take into account heavy metals, nutrients, and FCC that are found to be problematic. Al-Manun and Idris [11] proposed a revisited version of DOE-WQI with turbidity and total phosphorus replacing pH and COD; the revised WQI was found to produce a score 10–20 points lower than the original index. Despite criticisms and recommendations, there are yet limited studies to objectively demonstrate the shortcomings

FIGURE 1: The sampling locations at the Sarawak River and its tributaries.

of this indexing method. Hence, this paper attempts to evaluate the effectiveness of Malaysian physicochemical WQI by comparisons with other indices and to propose a modified index for assessment of tropical rivers with specific reference to rivers in Sarawak, Malaysia.

2. Materials and Methods

2.1. Sampling. Water samples were collected from 18 stations located at the Sarawak River and its tributaries. The sampling stations are shown in Figure 1 with the corresponding GPS positions summarized in Table 1. Sampling was carried out during the wet season in December 2015, January 2016, and March 2016. The samples were collected from subsurface of 0–20 cm into 2 L polyethylene bottles for ex situ analyses including TSS, COD, and NH_3-N. No acid was added for preservation; the samples were analyzed within three days. For BOD analysis, the water samples were collected in a 2 L polyethylene bottle and shaken vigorously. The initial DO reading was recorded. The water sample was then filled into a 300 mL BOD glass bottle and wrapped with aluminum foil. For FCC analysis, water samples were collected in sterilized glass bottles and the analysis was conducted within 24 hours. All the samples were kept in the cooler box for transportation to the laboratory.

2.2. Water Quality Analyses.

The in situ parameters including pH, DO, specific conductivity (Cond), and turbidity (Turb) were recorded using Horiba U-52 Multiparameter in triplicate. The meters were calibrated according to the standard procedures prior to sampling. The samples were subjected to ex situ analysis for TSS, BOD, COD, NH_3-N, and FCC. Total suspended solids (TSS) were determined according to the standard method of gravimetric analysis [17]. A total of 500 mL of the water sample was filtered through a membrane filter of 0.45 μm. The membrane filter was preweighed to 0.1 mg. After filtration, the membrane filter was oven-dried at 105°C overnight and left to cool in a desiccator before its final weight was recorded. Biochemical oxygen demand (BOD) was determined according to the standard method of 5-day BOD test [17]. The BOD bottles were stored in the dark and the DO reading was recorded after 5 days. COD and NH_3-N were measured using DR900 according to the reactor digestion method and salicylate method, respectively. Fecal coliform count (FCC) was determined based on the membrane filtration method.

2.3. Evaluation of Water Quality Indices.

The water quality data obtained was evaluated with the water quality indexing method developed by the Department of Environment (DOE), Malaysia, referred to as DOE-WQI. The data was also subjected to evaluation with the River Ganga Index, Minimal Water Quality Index (WQI_{min}), and an index modified from the existing DOE-WQI (modified DOE-WQI), proposed by the authors.

2.3.1. Department of Environment Water Quality Index (DOE-WQI) [1].

DOE-WQI is the sum of weighted subindices for six variables (pH, DO, BOD, COD, NH_3-N, and TSS). The subindices are derived based on segmented nonlinear functions as follows.

DO (in % saturation):

$$SI_{DO} = 0, \quad x \le 8,$$
$$SI_{DO} = 100, \quad x \ge 92, \quad (1)$$
$$SI_{DO} = -0.395 + 0.030x^2 - 0.00020x^3, \quad 8 < x < 92.$$

BOD (mg/L):

$$SI_{BOD} = 100.4 - 4.23x, \quad x \le 5,$$
$$SI_{BOD} = 108e^{(-0.055x)} - 0.1x, \quad x > 5. \quad (2)$$

COD (mg/L):

$$SI_{COD} = -1.33x + 99.1, \quad x \le 20,$$
$$SI_{COD} = 103e^{(-0.0157x)} - 0.04x, \quad x > 20. \quad (3)$$

NH_3-N (mg/L):

$$SI_{NH3\text{-}N} = 100.5 - 105x, \quad x \le 0.3,$$
$$SI_{NH3\text{-}N} = 94e^{(-0.573x)} - 5\,|x - 2|, \quad 0.3 < x < 4, \quad (4)$$
$$SI_{NH3\text{-}N} = 0, \quad x \ge 4.$$

TSS (mg/L):

$$SI_{SS} = 97.5e^{(-0.00676x)} + 0.05x, \quad x \le 100,$$
$$SI_{SS} = 71e^{(-0.0061x)} - 0.015x, \quad 100 < x < 1000, \quad (5)$$
$$SI_{SS} = 0, \quad x \ge 1000.$$

pH:

$$SI_{pH} = 17.02 - 17.2x + 5.02x^2, \quad x < 5.5,$$
$$SI_{pH} = -242 + 95.5x - 6.67x^2, \quad 5.5 \le x < 7,$$
$$SI_{pH} = -181 + 82.4x - 6.05x^2, \quad 7 \le x < 8.75, \quad (6)$$
$$SI_{pH} = 536 - 77.0x + 2.76x^2, \quad x \ge 8.75.$$

The sum of the weighted subindices is a score between 0 and 100 that is categorized into five classes (I–V) to suggest the beneficial uses of water.

$$DOE\text{-}WQI = (0.22 \times SI_{DO}) + (0.19 \times SI_{BOD})$$
$$+ (0.16 \times SI_{COD}) + (0.15 \times SI_{NH3\text{-}N}) \quad (7)$$
$$+ (0.16 \times SI_{TSS}) + (0.12 \times SI_{pH}).$$

Class I (>92.7): conservation of natural environment; Water Supply 1, practically no treatment necessary; Fishery 1, very sensitive aquatic species.

Class II (76.5–92.7): Water Supply II, conventional treatment required; Fishery II, sensitive aquatic species.

Class III (51.9–76.5): Water Supply III, extensive treatment required; Fishery III, common and tolerant species; for livestock drinking.

TABLE 2: Values of parameters after normalization for WQI_{min} (Pesce and Wunderlin, 2000).

Class	Normalization factor (Ci)										
	I		II		III		IV			V	
	100	90	80	70	60	50	40	30	20	10	0
Cond (μS/cm)	<750	<1000	<1250	<1500	<2000	<2500	<3000	<5000	<8000	≤12,000	>12,000
DO (mg/L)	≥7.5	>7.0	>6.5	>6.0	>5.0	>4.0	>3.5	>3.0	>2.0	≥1.0	<1.0
Turb (NTU)	<5	<10	<15	<20	<25	<30	<40	<60	<80	≤100	>100

Class IV (31.0–51.9): irrigation.

Class V (<31.0): none of the above.

2.3.2. The River Ganga Index [2].

The River Ganga Index, like DOE-WQI, is a weighted arithmetic index. It is calculated based on four variables, namely, DO, FCC, pH, and BOD. The subindices are derived according to the following equations with the ultimate score representing four classes [63–100: Class I (good-excellent/nonpolluted); 50–63: Class II (medium-good/nonpolluted); 38–50: Class III (bad/polluted); and <38: Class IV (bad-very bad/heavily polluted)]. Bhutiani et al. [18] employed this index to assess the water quality of the River Ganga from 2000 to 2010.

The River Ganga Index

$$= (0.31 \times SI_{DO}) + (0.19 \times SI_{BOD}) + (0.22 \times SI_{pH}) \quad (8)$$

$$+ (0.28 \times SI_{FCC}).$$

DO (% saturation):

$$SI_{DO} = 0.18 + 0.66x, \quad 0\text{–}40\% \text{ saturation},$$

$$SI_{DO} = -13.5 + 1.17x, \quad 40\text{–}100\% \text{ saturation}, \quad (9)$$

$$SI_{DO} = 263.34 - 0.62x, \quad 100\text{–}140\% \text{ saturation}.$$

BOD_5 (mg/L):

$$SI_{BOD5} = 96.67 - 7x, \quad 0\text{–}10,$$

$$SI_{BOD5} = 38.9 - x, \quad 10\text{–}30, \quad (10)$$

$$SI_{BOD5} = 2, \quad >30.$$

pH:

$$SI_{pH} = 16.1 + 7.35x, \quad 2\text{–}5,$$

$$SI_{pH} = -142.67 + 33.5x, \quad 5\text{–}7.3,$$

$$SI_{pH} = 316.96 - 29.85x, \quad 7.3\text{–}10, \quad (11)$$

$$SI_{pH} = 96.17 - 8.0x, \quad 10\text{–}12,$$

$$SI_{pH} = 0, \quad <2, >12.$$

FCC (counts/100 mL):

$$SI_{FCC} = 97.2 - 26.60 \log_{10} x, \quad 1\text{–}10^3,$$

$$SI_{FCC} = 42.33 - 7.75 \log_{10} x, \quad 10^3\text{–}10^5, \quad (12)$$

$$SI_{FCC} = 2, \quad >10^5.$$

2.3.3. Minimal Water Quality Index [19].

WQI_{min} is an index of arithmetic mean for normalized DO, Cond, and Turb; the normalized values are determined based on Table 2 with the index value categorized into five classes (Class I: 80–100; Class II: 60–80; Class III: 40–60; Class IV: 20–40; Class V: 0–20).

$$WQI_{min} = \frac{C_{DO} + C_{cond} + C_{turb}}{3}, \quad (13)$$

where C is the normalized value

2.3.4. Modified DOE-WQI.

The modified DOE-WQI is calculated based on six variables, of which four are identical to that of DOE-WQI (DO, BOD, COD, and NH_3-N), whereas pH and TSS are replaced with FCC and Turb. This index was modified from DOE-WQI undergoing three important stages for selection of essential parameters, derivation of subindices, and assignment of suitable weightages as suggested by Sutadian et al. [20]. Considering the essential determinants, pH and TSS are replaced with FCC and Turb. The rationale of substituting TSS with turbidity is because the latter is able to measure the clarity of water in real time. On the other hand, pH is omitted for the reason that most rivers in Malaysia are almost consistent and near neutral and hence do not contribute significantly to the resulting index [11]. Fernandez et al. [21] compared 36 water quality indices revealing that DO, TS, pH, FCC, BOD, total phosphorus, and nitrate are the most common parameters followed by Turb, Temp, and ammonia with similar importance. Although pH is obviously a parameter most conveniently measured for the calculation of WQI, Benvenuti et al. [22] asserted that pH and Temp are insignificant discriminants for the assessment of water quality. For subindices derivation, multiple functions were employed. The subindices of DO, turbidity, and FCC were assigned based on segmented nonlinear functions; this implies that a designated equation is applied over a specific range of the variable [2]. A nonlinear function was used for subindices conversion of BOD and NH_3-N while COD was expressed as a linear function. The subindex functions of modified DOE-WQI are as follows.

TABLE 3: The guidelines of water quality according to the National Water Quality Standards (NWQS) for Malaysia.

Parameters	Classes					
	I	IIA	IIB	III	IV	V
pH	6.5–8.5	6–9	6–9	5–9	5–9	-
Conductivity (μS/cm)	1000	1000	-	-	6000	-
TSS (mg/L)	25	50	50	150	300	300
Turbidity (NTU)	5	50	50	-	-	-
DO (mg/L)	7	5–7	5–7	3–5	<3	<1
BOD (mg/L)	1	3	3	6	12	>12
COD (mg/L)	10	25	25	50	100	>100
NH_3-N (mg/L)	0.1	0.3	0.3	0.9	2.7	>2.7
FCC (counts/100 mL)	10	100	400	5000 (20,000*)	5000 (20,000*)	-

*Maximum not to be exceeded.

DO (% saturation):

$$SI_{DO} = 0, \quad x \leq 13, \ x \geq 150,$$

$$SI_{DO} = 0.000004x^2 + 1.264x - 16.44, \quad 13 < x \leq 92,$$

$$SI_{DO} = 100, \quad 92 < x \leq 120,$$

$$SI_{DO} = -1.538x + 284.6, \quad 120 < x < 150.$$

(14)

BOD_5 (mg/L):

$$SI_{BOD5} = 0.098x^2 - 6.96x + 100, \quad x \leq 20,$$

$$SI_{BOD5} = 0, \quad x > 20.$$

(15)

COD (mg/L):

$$SI_{COD} = -0.666x + 100, \quad x \leq 150,$$

$$SI_{COD} = 0, \quad x > 150.$$

(16)

NH_3-N (mg/L):

$$SI_{NH3-N} = 0.966x^2 - 24.83x + 100, \quad x \leq 5.0,$$

$$SI_{NH3-N} = 0, \quad x > 5.0.$$

(17)

Turbidity (NTU):

$$SI_{TURB} = -0.8x + 100, \quad x \leq 25,$$

$$SI_{TURB} = -32.3 \ln(x) + 185.4, \quad 25 < x \leq 300,$$

$$SI_{TURB} = 0, \quad x > 300.$$

(18)

Fecal Coliform (counts/100 mL):

$$SI_{FCC} = -2x + 100, \quad x \leq 10,$$

$$SI_{FCC} = -0.051x + 80.51, \quad 10 < x \leq 400,$$

$$SI_{FCC} = -15.3 \ln(x) + 153.3, \quad 400 < x \leq 20000,$$

$$SI_{FCC} = 0, \quad x > 20000.$$

(19)

The modified index is adapted from the original WQI according to the National Water Quality Standards in Malaysia with five classes of water quality (as shown in Table 3). The original DOE-WQI was established by a team of multidisciplinary experts from universities [16]. In the original WQI, the subindex of DO is a nonsegmented function where DO greater than 92% is consistently assigned a value of 100. Typically, DO over 90% saturation is an indication of good water quality, while below 50% it will cause negative effects on biological communities [23]. Under supersaturated water with 115–120% DO for a period of time, fish and invertebrates may develop gas bubble disease [24]. This condition is taken into account in the modified WQI. The subindices of BOD and NH_3-N are expressed as nonlinear functions where a subindex value of zero is assigned when BOD and NH_3-N are greater than 20 and 5 mg/L, respectively. For COD on the other hand, the subindex is a linear function, differing from the original WQI of a combination of linear and nonlinear functions.

For relative weightage determination, a weight between 1 and 4 is assigned to each parameter; a higher value suggests greater importance. The relative weightage is calculated by dividing the assigned weight by the total weights. The sum of the weighted subindices is categorized into five classes to describe the water quality status (Class I: >91; Class II: 75–91; Class III: 57–75; Class IV 25–57; Class V: <25).

$$Modified\ WQI = (0.20 \times SI_{DO}) + (0.18 \times SI_{BOD})$$
$$+ (0.18 \times SI_{COD})$$
$$+ (0.14 \times SI_{NH3-N})$$
$$+ (0.15 \times SI_{TURB})$$
$$+ (0.15 \times SI_{FCC}).$$

(20)

2.4. Statistical Analyses and Map Representation. The water quality indices derived with different approaches were compared statistically using the paired t-test at significance level of 0.05. Pearson's correlation was used to examine the relationships between different indices. The original and modified DOE-WQI were evaluated against a dataset simulated based on the National Water Quality Standards (NWQS) for Malaysia, serving as the benchmark. Note that the DOE-WQI is rooted on NWQS that categorizes water quality into five classes according to parameters as shown in Table 3. The dataset consists of 500 samples comprising 8 parameters (DO, BOD, COD, NH_3-N, Turb, FCC, TSS, and pH). The simulated samples are generated according to the stipulated water quality of five classes in NWQS. For example, a sample of Class I shall exhibit BOD < 1.0 mg/L according to NWQS. An algorithm is programmed to simulate a total of 100 samples with BOD < 1.0 mg/L according to uniform distribution; this is done in turn for the remaining classes and parameters, yielding a dataset with dimensions 500 × 8. The strategy of simulation is detailed in Sim et al. [25]. The river samples were classified based on Euclidean distance with the simulated data serving as the training set. The classification results were compared with the classes determined using the original and modified DOE-WQI. Principal Component Analysis (PCA) was used to illustrate the clustering pattern of the simulated samples according to classes. Prior to PCA, the data was standardized (mean centering and scaling by standard deviation) to ensure all variables are comparable. The statistical analyses were performed with Matlab 2013a. Quantum Geographic Information System (QGIS) was used to map the classes of water quality according to indexing methods for visualization.

3. Results and Discussion

3.1. Overall Water Quality.
Table 4 summarizes the average water quality of 18 stations over three sampling campaigns in Dec 2015, Jan 2016, and Mar 2016. The measurements obtained are compared against the National Water Quality Standards (NWQS) for Malaysia to characterize the water quality status. Overall, pH is found to be between 6.95 and 7.58. As concluded by Al-Manun and Idris [11], pH is rather consistent for river water in Malaysia. For conductivity, the measurements vary between 47.67 and 4468 μS/cm with elevated readings recorded at ST15 and ST18 near the industrial area and ST16 and ST17 before and after Sarawak Barrage, respectively. This suggests input from industrial discharge and influence by barrage operation. Law et al. [26] monitored the salinity at 1.5 and 6 km upstream of Sarawak Barrage during flooding-in and flushing operations over 33 hours. The conductivity was found to fluctuate considerably between 0 and 24000 μS/cm (14.5 ppt) at 1.5 km upstream but was negligibly low at the station of 6 km (0–1000 μS/cm or <0.5 ppt). During sampling in Dec 2015 and Jan 2016, the conductivity was seen to be elevated possibly due to the extended dry period and low flow condition. The turbidity varies between 21.14 and 208.71 NTU, with the highest measurement recorded at ST16, after barrage. Relatively turbid water is also recorded at ST6–ST8 near the confluence of Sarawak Kiri and Sarawak Kanan River. The DO level lies between 1.98 and 5.18 mg/L. At ST11–ST13, located at the highly populated Maong River catchment, the DO is particularly low, likely associated with the input of

untreated domestic waste. The DO at ST1–ST10 is noticeably higher than that at ST11–ST18 as the upper reaches of Sarawak River (situated at the steep terrain) are typically narrower and shallower and hence the water is relatively fast flowing facilitating higher DO [27]. Generally, DO is rated between Classes III and IV according to NWQS where extensive treatment is required. The TSS ranges between 20.71 and 227.96 mg/L (Class I to Class V) with ST16 recording the highest concentration, corresponding to the measurement of turbidity. The BOD level is typically between 1.42 and 2.53 mg/L while COD ranges between 0.79 and 17.73 mg/L. NH_3-N is generally present in traces across all stations (<0.2 mg/L) except in ST10–ST13 with a concentration above 1.4 mg/L suggesting input of sewage effluent. Fecal coliform count varies extensively from 6500 to 94,805 CFU/100 mL with various stations recording measurements above the maximum NWQS limit of 20,000 counts/100 mL, suggesting microbiological contamination in water. Povlsen [27] likewise observed poor FCC of more than 16,000 counts/100 mL in catchments of Sarawak River and its tributaries, postulating discharge of untreated sewage effluents from inefficient septic tanks.

3.2. Assessment of Water Quality Index.
The physicochemical DOE-WQI is compared with the River Ganga Index, WQI_{min}, and the modified WQI. Table 5 shows the water quality index of 18 stations over three sampling campaigns with the classes assigned according to the four indexing methods. The WQIs derived using different methods are significantly correlated ($p < 0.05$), suggesting that these indices exhibit corresponding behavior on the water quality (Table 6). Essentially, the water quality is categorized between Classes II and III using the original DOE-WQI although the index attained in Jan (62–82) and Mar 2016 (59–87) is significantly lower than that recorded in Dec 2015 (62–94) ($p < 0.05$). With the modified strategy, the water quality is rated between Classes II and IV with a similar observation of poorer water quality in Jan and Mar 2016 (Dec: 50–80; Jan: 52–75; Mar: 48–73). It is also found that the modified index consistently assigns samples to a lower class compared to the original DOE-WQI with a significant difference ($p < 0.05$).

Table 7 shows the cumulative sum of weighted subindices, calculated for the average samples of three samplings, using the original and modified DOE-WQI. As observed, the cumulative sums attained using both methods are comparable when 5 variables are considered (DO, BOD, COD, NH_3-N, and TSS/Turb); however, when FCC replaces pH, there is a marked difference in the aggregated values. The modified index only experiences an increment of 2 points upon addition of the weighted SI_{FCC} while the original DOE-WQI encounters an addition of 12 points with inclusion of $SIpH$. This implies considerable influence of FCC over the resultant index. Benvenuti et al. [22] evaluated the water quality of Sinos River Basin in South Brazil using three different indexing methods. It was concluded from the results that an index that comprises significant discriminants with adequate weightage is essential for a more representative evaluation of water quality.

TABLE 4: The average water quality of 18 stations over three samplings in Dec 2015, Jan 2016, and Mar 2016.

ST	pH	DO (mg/L)	Conductivity (µS/cm)	Turbidity (NTU)	TSS (mg/L)	BOD (mg/L)	COD (mg/L)	NH$_3$-N (mg/L)	FCC (CFU/100 mL)
1	6.95 ± 0.13	4.80 ± 0.63	47.67 ± 10.97	39.63 ± 13.68	52.96 ± 16.63	1.49 ± 0.77	1.61 ± 2.14	0.02 ± 0.02	6500 ± 4501
2	7.07 ± 0.19	4.67 ± 0.97	53.33 ± 6.95	37.18 ± 7.36	44.96 ± 11.10	1.57 ± 0.78	2.90 ± 2.48	0.03 ± 0.05	9611 ± 7209
3	7.04 ± 0.18	4.41 ± 0.78	49.67 ± 2.65	40.64 ± 5.16	50.63 ± 8.63	1.44 ± 0.69	0.79 ± 0.77	0.02 ± 0.03	7638 ± 5491
4	7.04 ± 0.12	5.18 ± 1.20	67.33 ± 40.10	67.19 ± 47.13	68.78 ± 41.48	1.42 ± 0.72	5.55 ± 5.33	0.12 ± 0.18	25694 ± 13385
5	7.17 ± 0.19	4.06 ± 1.81	99.11 ± 16.34	28.29 ± 13.64	40.26 ± 17.27	1.66 ± 0.68	4.76 ± 1.95	0.14 ± 0.14	33722 ± 35238
6	7.17 ± 0.48	4.44 ± 0.87	109.11 ± 63.90	101.41 ± 52.20	191.28 ± 247.95	1.90 ± 1.04	13.88 ± 19.64	0.05 ± 0.07	17166 ± 10655
7	7.04 ± 0.24	3.83 ± 1.00	65.83 ± 26.87	57.53 ± 34.53	67.81 ± 39.06	1.61 ± 0.86	2.88 ± 4.08	0.02 ± 0.03	18361 ± 17759
8	6.60 ± 0.53	4.27 ± 1.12	66.33 ± 17.33	103.17 ± 89.15	86.44 ± 64.33	2.05 ± 1.40	4.80 ± 6.51	0.06 ± 0.05	8055 ± 6660
9	6.75 ± 0.25	4.33 ± 1.84	76.33 ± 24.28	22.12 ± 11.27	20.71 ± 4.44	2.06 ± 1.36	17.73 ± 23.86	0.07 ± 0.02	7750 ± 4883
10	7.25 ± 0.22	3.46 ± 1.70	137.33 ± 63.31	27.70 ± 6.78	29.26 ± 14.16	1.61 ± 0.44	5.99 ± 8.41	2.08 ± 1.69	94805 ± 77343
11	7.46 ± 0.38	1.98 ± 0.61	192.67 ± 26.74	21.14 ± 10.66	22.52 ± 7.48	2.01 ± 0.37	16.82 ± 16.45	1.45 ± 1.00	73638 ± 32315
12	7.58 ± 0.41	2.43 ± 1.08	255.67 ± 29.87	38.55 ± 27.66	15.96 ± 10.31	2.06 ± 0.61	13.43 ± 12.10	2.23 ± 0.52	75500 ± 31716
13	7.19 ± 0.12	2.75 ± 0.27	150.56 ± 111.59	29.41 ± 13.85	26.00 ± 6.82	2.32 ± 0.67	5.97 ± 3.13	1.48 ± 1.83	60555 ± 50737
14	6.96 ± 0.07	3.58 ± 1.13	68.11 ± 24.31	38.37 ± 29.64	37.41 ± 23.84	2.23 ± 0.62	8.41 ± 4.84	0.10 ± 0.08	14722 ± 8511
15	6.97 ± 0.16	3.22 ± 0.50	632.56 ± 843.61	30.37 ± 13.45	31.37 ± 9.79	2.51 ± 1.03	13.87 ± 11.28	0.10 ± 0.09	37694 ± 9794
16	7.25 ± 0.15	3.38 ± 0.92	4100.67 ± 4739.69	208.71 ± 193.36	227.96 ± 249.41	2.53 ± 0.86	10.82 ± 7.99	0.08 ± 0.08	22138 ± 19138
17	7.27 ± 0.10	3.00 ± 0.20	2951.33 ± 2190.91	32.52 ± 11.13	33.48 ± 5.63	1.84 ± 0.62	9.95 ± 2.48	0.11 ± 0.09	32916 ± 29389
18	7.36 ± 0.20	2.65 ± 0.27	4468.33 ± 2592.71	58.54 ± 35.83	55.00 ± 27.62	1.66 ± 0.55	8.39 ± 3.60	0.70 ± 0.52	44416 ± 9555

TABLE 5: The index values and the classes of water quality assigned with DOE-WQI, modified DOE-WQI, River Ganga, and WQI_{min} according to sampling months.

ST	Dec 15				Jan 16				Mar 16			
	DOE-WQI	Modified WQI	River Ganga	WQI_{min}	DOE-WQI	Modified WQI	River Ganga	WQI_{min}	DOE-WQI	Modified WQI	River Ganga	WQI_{min}
1	91 (II)	77 (II)	61 (II)	73 (III)	80 (II)	71 (III)	55 (II)	59 (III)	87 (II)	73 (III)	59 (II)	62 (II)
2	91 (II)	75 (II)	62 (II)	68 (II)	82 (II)	73 (III)	58 (II)	60 (II)	83 (II)	70 (III)	53 (II)	60 (II)
3	89 (II)	74 (II)	60 (II)	63 (II)	82 (II)	75 (II)	58 (II)	60 (II)	82 (II)	69 (III)	53 (II)	60 (II)
4	88 (II)	77 (II)	66 (I)	81 (I)	82 (II)	68 (III)	55 (II)	60 (II)	81 (II)	64 (III)	56 (II)	50 (III)
5	94 (I)	80 (II)	65 (I)	82 (I)	73 (III)	66 (III)	48 (III)	62 (II)	76 (III)	62 (III)	46 (III)	50 (III)
6	82 (II)	67 (III)	59 (II)	67 (II)	73 (III)	64 (III)	53 (II)	50 (III)	82 (II)	63 (III)	50 (II)	47 (III)
7	89 (II)	75 (II)	60 (II)	73 (II)	78 (II)	62 (III)	51 (II)	47 (III)	74 (III)	64 (III)	46 (III)	50 (III)
8	87 (II)	76 (II)	58 (II)	77 (II)	72 (III)	62 (III)	47 (III)	43 (III)	80 (II)	65 (III)	54 (II)	53 (III)
9	85 (II)	73 (III)	64 (I)	76 (II)	78 (II)	70 (III)	44 (III)	70 (II)	83 (II)	70 (III)	54 (II)	60 (II)
10	91 (II)	77 (II)	61 (II)	73 (II)	62 (III)	52 (IV)	43 (III)	50 (III)	65 (III)	53 (IV)	49 (III)	67 (II)
11	71 (III)	60 (III)	45 (III)	60 (III)	66 (III)	55 (IV)	43 (III)	50 (III)	62 (III)	53 (IV)	37 (III)	67 (II)
12	70 (III)	60 (III)	49 (III)	79 (II)	67 (III)	53 (IV)	44 (III)	47 (III)	59 (III)	48 (IV)	37 (III)	48 (III)
13	64 (III)	50 (IV)	45 (III)	51 (III)	73 (III)	63 (III)	46 (III)	63 (II)	78 (II)	63 (III)	45 (III)	60 (II)
14	83 (II)	69 (III)	59 (II)	59 (III)	77 (I)	67 (III)	47 (III)	67 (II)	77 (I)	62. (III)	43 (III)	60 (II)
15	80 (II)	65 (III)	51 (II)	57 (III)	77 (I)	66 (III)	47 (III)	51 (III)	73 (III)	59 (III)	42 (III)	57 (III)
16	83 (II)	70 (III)	57 (II)	62 (II)	65 (III)	55 (IV)	47 (III)	10 (V)	71 (III)	54 (IV)	43 (III)	40 (III)
17	76 (II)	66 (III)	52 (II)	30 (IV)	75 (II)	65 (III)	47 (III)	40 (IV)	78 (II)	62 (III)	45 (III)	53 (III)
18	77 (II)	66 (III)	49 (III)	38 (III)	64 (III)	52 (IV)	43 (III)	16 (V)	69 (III)	58 (III)	45 (III)	50 (III)

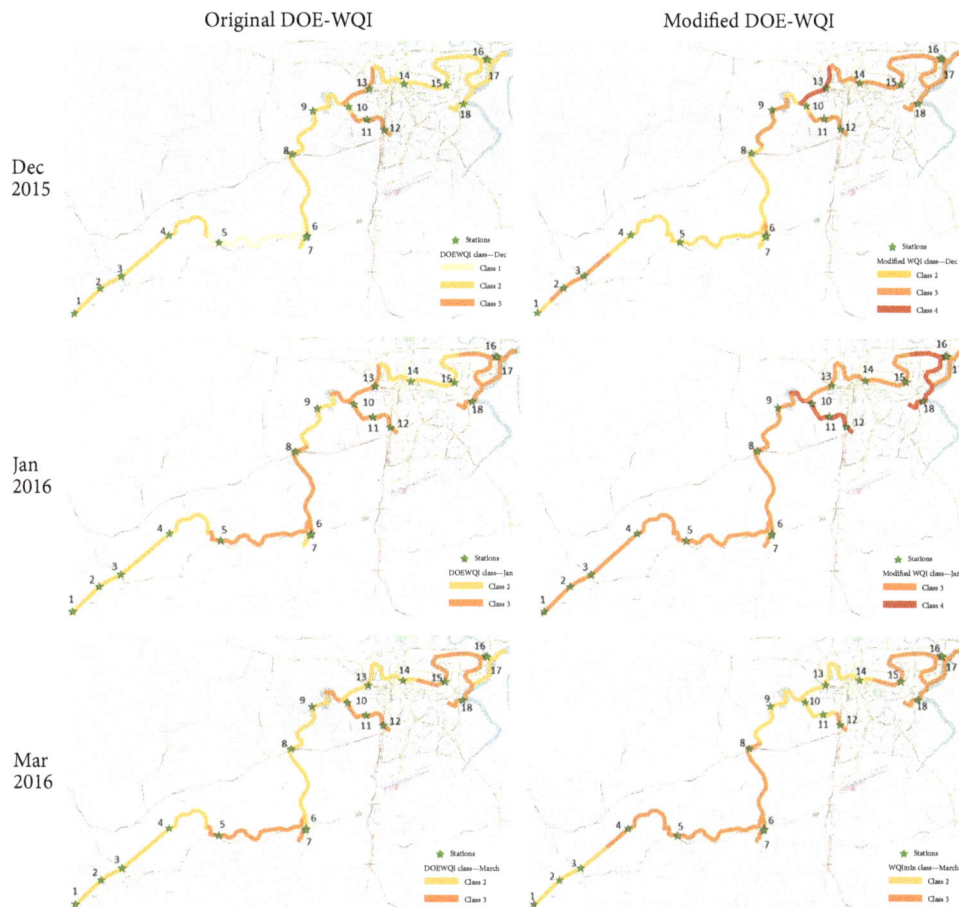

FIGURE 2: Map representations of the water quality of the Sarawak River and its tributaries.

TABLE 6: Correlation coefficients (R values) between the four indexing methods.

	DOE-WQI	Modified WQI	River Ganga	WQI$_{min}$
DOE-WQI	1			
Modified WQI	0.94	1		
River Ganga	0.87	0.86	1	
WQI$_{min}$	0.56	0.60	0.52	1

According to the River Ganga Index, the river is assigned to Classes II–IV (Dec: 45–66; Jan: 43–58; Mar: 37–59) with a similar observation of reducing water quality in Jan and Mar 2016. Many stations are rated as polluted to heavily polluted with an index value of <50. Like the modified WQI, the high FCC values contribute to small weighted subindices of 1–7 points to the overall aggregated value. Despite the reduced number of parameters used in the River Ganga Index, the index values attained remain responsive to the changes in water quality.

The three-parameter WQI$_{min}$ is an index of arithmetic mean for normalized DO, Turb, and Cond. This simple index was reported to be comparable to two other indices of 20 parameters for the assessment of water body under the influence of industrial discharge [28]; nonetheless, there is a tendency of overestimation as the parameters are assigned with similar weightage [29, 30]. As observed, water quality depicted with WQI$_{min}$ covers a wide spectrum from Class I to Class V with index values ranging between 10 and 82. The index similarly indicates poorer water quality conditions in Jan and Mar 2016, with Stations 16–18 (near industrial area and barrage) corresponding to lower index values due to high turbidity and conductivity. This suggests that WQI$_{min}$ is suitable for probing contamination from industrial discharge; nonetheless, some contradicting values are detected. For example, the water quality at ST4, collected in Dec 2015, was assigned a score of 81 indicative of good water quality, but it does not reflect the problem attributed to FCC with a measurement of 40,000 counts/100 mL. This misleading score is a typical consequence of a situation when the parameters chosen for WQI calculation do not represent the environmental stress.

Water quality of the 18 stations determined with the original and modified DOE-WQI is mapped with QGIS for better visualization and comparison (Figure 2). The water quality is represented with different colors according to classes (darker color indicates poorer water quality). As demonstrated, the

TABLE 7: The cumulative weighted subindices for the original and modified DOE-WQI.

Weighted subindices	Dec		Jan		Mar	
	Original	Modified	Original	Modified	Original	Modified
DO	14.9	12.9	7.6	7.3	7.6	7.3
DO + BOD	32.8	28.3	25.7	23.9	25.8	24.2
DO + BOD + COD	46.3	44.8	40.6	39.5	40.0	39.4
DO + BOD + COD + NH$_3$-N	58.5	58.3	52.0	52.6	52.5	52.8
DO + BOD + COD + NH$_3$-N + TSS (original)/Turb (modified)	71.4	68.9	61.8	61.5	63.6	62.1
DO + BOD + COD + NH$_3$-N + TSS (original)/Turb (modified) + pH (original)/FCC (modified)	83.2	70.1	73.5	63.7	75.3	62.5

rivers are characterized between Class I and Class III with the original DOE-WQI over three samplings; nonetheless, with the modified strategy, it is found that several stretches of the rivers are categorized in Class IV with most stations assigned to Class III. This is similarly observed with the River Ganga Index (maps are not shown).

3.3. Classification with Euclidean Distance. To assess the efficiency of DOE-WQI, training data simulated according to the guidelines of NWQS is used as the benchmark. The simulated data comprising eight variables (DO, BOD, COD, TSS, pH, NH$_3$-N, FCC, and Turb) was subjected to PCA to reveal the underlying clustering pattern. Figure 3 shows the scores plot of PC2 versus PC1; samples from different classes are noticeably distinguishable, indicative of their distinctive characteristics. The simulated data was used as the training samples for classification of the river water based on Euclidean distance. Table 8 summarizes water quality of the 18 stations classified based on Euclidean distance and the indexing methods. Most stations are categorized in Classes I and II according to Euclidean distance (based on 6 physicochemical parameters), matching the classes determined with DOE-WQI. Some minor discrepancies are detected; for example, a sample with a score of 81 is denoted as Class II using DOE-WQI but is appointed to Class III with Euclidean distance. This is possibly attributed to inconsistencies in the classification criteria. With the inclusion of FCC, the samples are mostly classified between Classes III and IV based on Euclidean distance; this corresponds better to the modified DOE-WQI. Evidently, DOE-WQI classifies water accordingly but did not pick up the environmental stress caused by the microbiological parameter. This signifies the pitfall of the indexing method when an important discriminant is not included.

4. Conclusions

Results show that the Sarawak River and its tributaries are generally rich in FCC with elevated conductivity identified at stations near industrial areas and barrage. The water quality is very much affected by the weather condition and the adjacent

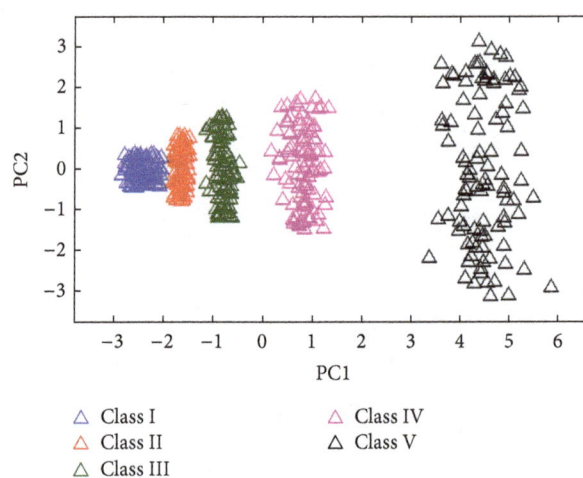

FIGURE 3: The scores plot of training data with five classes comprising eight parameters (DO, BOD, COD, TSS, pH, NH$_3$-N, FCC, and Turb).

land use activities. The water quality indices evaluated exhibit a corresponding behavior; nonetheless, it is found that the existing DOE-WQI tends to rate the river at better water quality than its actual condition as the deteriorating FCC is not included in the index. Assessment with Euclidean distance confirms that the existing DOE-WQI does not carry sufficient information to provide a representative assessment of water quality. The findings demonstrate that tropical rivers in this region are challenged with deteriorated water quality associated with domestic sewage; hence, the microbiological parameter is a crucial discriminant for the calculation of WQI. This paper demonstrates the shortcomings of a physicochemical indexing method when a significant discriminating variable is not taken into consideration.

Conflicts of Interest

The authors declare that there are no conflicts of interest regarding the publication of this article.

TABLE 8: The classes of water quality based on Euclidean distance and the indexing methods of DOE-WQI and modified DOE-WQI (the number refers to the score of WQI).

ST	DOE-WQI			Classification with Euclidean distance (based on 6 parameters of original DOE-WQI)			Modified DOE-WQI			Classification with Euclidean distance (based on 6 parameters of modified DOE-WQI)		
	Dec	Jan	Mar	Dec	Jan	Mar	Dec	Jan	Mar	Dec	Jan	Mar
1	91 (II)	80 (II)	87 (II)	II	II	II	77 (II)	71 (III)	73 (III)	III	II	IV
2	91 (II)	82 (II)	83 (II)	II	II	II	75 (II)	73 (III)	70 (III)	IV	II	IV
3	89 (II)	82 (II)	82 (II)	II	II	II	74 (II)	75 (II)	69 (III)	IV	II	IV
4	88 (II)	82 (II)	81 (II)	II	II	III	77 (II)	68 (III)	64 (III)	IV	IV	IV
5	94 (I)	73 (III)	76 (III)	I	II	II	80 (II)	66 (III)	62 (III)	IV	IV	IV
6	82 (II)	73 (III)	82 (II)	II	V	II	67 (III)	64 (III)	63 (III)	IV	III	IV
7	89 (II)	78 (II)	74 (III)	I	II	III	75 (II)	62 (III)	64 (III)	IV	III	IV
8	87 (II)	72 (III)	80 (II)	II	III	II	76 (II)	62 (III)	65 (III)	III	III	IV
9	85 (II)	78 (II)	83 (II)	II	I	I	73 (III)	70 (III)	70 (III)	III	III	V
10	91 (II)	62 (III)	65 (III)	I	II	I	77 (II)	52 (IV)	53 (IV)	IV	IV	IV
11	71 (III)	66 (III)	62 (III)	II	II	I	60 (III)	55 (IV)	53 (IV)	IV	IV	IV
12	70 (III)	67 (III)	59 (III)	I	II	I	60 (III)	53 (IV)	48 (IV)	IV	IV	IV
13	64 (III)	73 (III)	78 (II)	II	I	II	50 (IV)	63 (III)	63 (III)	IV	IV	IV
14	83 (II)	77 (II)	77 (II)	II	I	I	69 (III)	67 (III)	62 (III)	III	IV	IV
15	80 (II)	77 (II)	73 (III)	II	I	II	65 (III)	66 (III)	59 (III)	IV	IV	IV
16	83 (II)	65 (III)	71 (III)	II	V	III	70 (III)	55 (IV)	54 (IV)	IV	IV	IV
17	76 (II)	75 (II)	78 (II)	II	II	II	66 (III)	65 (III)	62 (III)	IV	IV	IV
18	77 (II)	64 (III)	69 (III)	I	II	II	66 (III)	52 (IV)	58 (III)	IV	IV	IV

Acknowledgments

The authors would like to thank the Ministry of Higher Education, Malaysia, for funding this project (ERGS/STWN (01)1015/2013(12)).

References

[1] DOE (Department of Environment) (2012), *Malaysia: Environmental quality report 2012*, Ministry of Science, Technology and the Environment, Putrajaya, Malaysia, 2012.

[2] T. Abbasi and S. A. Abbasi, *Water Quality Indices*, Elsevier, Oxford, UK, 2012.

[3] G. Singh and R. Kamal, "Application of water quality index for assessment of surface water quality status in Goa," *Current World Environment*, vol. 9, no. 3, pp. 994–1000, 2014.

[4] C. G. Cude, "Oregon water quality index: A tool for evaluating water quality management effectiveness," *Journal of the American Water Resources Association*, vol. 37, no. 1, pp. 125–137, 2001.

[5] M. W. Gitau, J. Chen, and Z. Ma, "Water quality indices as tools for decision making and management," *Water Resource Management*, vol. 30, no. 8, pp. 2591–2610, 2016.

[6] A. Said, D. K. Stevens, and G. Sehlke, "An innovative index for evaluating water quality in streams," *Journal of Environmental Management*, vol. 34, no. 3, pp. 406–414, 2004.

[7] E. D. Ongley, "Modernization of water quality programs in developing countries: Issues of relevancy and cost efficiency," *Water Quality International*, vol. 3, no. 4, pp. 37–42, 1998.

[8] Environment Protection Department of Hong Kong (2013), "Water: River water quality monitoring in Hong Kong," http://www.epd.gov.hk/epd/mobile/english/environmentinhk/water/river_quality/rwq_monitoring.html.

[9] F. D. S. Simões, A. B. Moreira, M. C. Bisinoti, S. M. N. Gimenez, and M. J. S. Yabe, "Water quality index as a simple indicator of aquaculture effects on aquatic bodies," *Ecological Indicators*, vol. 8, no. 5, pp. 476–484, 2008.

[10] A. A. Mamun and Z. Zainudin, "Sustainable river water quality management in Malaysia," *IIUM Engineering Journal*, vol. 14, no. 1, pp. 29–42, 2013.

[11] A. Al-Manun and A. Idris, "Revised water quality indices for the protection of rivers in Malaysia," in *Proceeding of the 12th International Water Technology Conference*, Alexandria, Egypt, 2008.

[12] F. Al-Badaii, M. Shuhaimi-Othman, and M. B. Gasim, "Water quality assessment of the Semenyih River, Selangor, Malaysia," *Journal of Chemistry*, vol. 2013, Article ID 871056, 10 pages, 2013.

[13] F. Othman, M. S. U. Chowdhury, and N. Sakai, "Assessment of microorganism pollution of Selangor River, Malaysia," *International Journal of Advances in Agricultural and Environmental Engineering*, vol. 1, no. 2, pp. 203–207, 2014.

[14] M. R. Nurul Ruhayu, J. A. Yii, and Y. Khairun, "Detection or river pollution using water quality index: A case study of tropical rivers in Penang Island," *Open Access Library Journal*, vol. 2, no. 3, Article no 68088, 8 pages, 2015.

[15] D. Kozaki, M. Hasbi bin Ab. Rahim, W. Mohd Faizal bin Wan Ishak et al., "Assessment of the river water pollution levels in Kuantan, Malaysia, using ion-exclusion chromatographic data, water quality indices, and land usage patterns," *Air, Soil and Water Research*, vol. 9, 11 pages, 2016.

[16] I. Naubi, N. H. Zardari, S. M. Shirazi, N. F. B. Ibrahim, and L. Baloo, "Effectiveness of water quality index for monitoring Malaysian river water quality," *Polish Journal of Environmental Studies*, vol. 25, no. 1, pp. 231–239, 2016.

[17] APHA (American Public Health Association) (2005), *Standard Methods for the Examination of Water and Wastewater*, American Public Health Association/American Water Works Association/Water Environment Federation, Wash, USA, 20th edition, 2005.

[18] R. Bhutiani, D. R. Khanna, D. B. Kulkarni, and M. Ruhela, "Assessment of Ganga river ecosystem at Haridwar, Uttarakhand, India with reference to water quality indices," *Applied Water Science*, vol. 6, no. 2, pp. 107–113, 2016.

[19] S. F. Pesce and D. A. Wunderlin, "Use of water quality indices to verify the impact of Cordoba City (Argentina) on Suquia River," *Water Research*, vol. 34, no. 11, pp. 2915–2926, 2000.

[20] A. D. H. Sutadian, N. Muttil, A. G. O. Yilmaz, and B. J. C. Perera, "Development of river water quality indices-a review," *Environmental Modeling & Assessment*, vol. 188, no. 1, p. 58, 2016.

[21] N. Fernandez, A. Ramirez, and F. Solano, "Physico-chemical water quality indices a comparative review," *Revisa Bistua*, pp. 19–30, 2004.

[22] T. Benvenuti, M. A. Kieling-Rubio, C. R. Klauck, and M. A. S. Rodrigues, "Evaluation of water quality at the source of streams of the Sinos River Basin, Southern Brazil," *Brazilian Journal of Biology*, vol. 75, no. 2, pp. S98–S104, 2015.

[23] G. Srivastava and P. Kumar, "Water quality index with missing parameters," *International Journal of Research in Engineering and Technology*, vol. 2, no. 1, pp. 609–614, 2013.

[24] D. E. Weitkamp and M. Katz, "A review of dissolved gas supersaturation literature," *Transactions of the American Fisheries Society*, vol. 109, no. 6, pp. 659–702, 1980.

[25] S. F. Sim, T. Y. Ling, S. Lau, and M. Z. Jaafar, "A novel computer-aided multivariate water quality index," *Environmental Modeling & Assessment*, vol. 187, no. 4, pp. 1–11, 2015.

[26] I. N. Law, Y. W. Oon, P. L. Law, and F. W. L. Kho, "Impacts of barrage flushing and flooding-in operations on saline intrusion upstream," *UNIMAS e-Journal of Civil Engineering*, vol. 2, no. 1, pp. 18–24, 2011.

[27] E. Povlsen, "Environment of Sungai Sarawak: relationships between city and river. river quality baseline study," Tech Rep. No. SUD-02-25, UM Colour Printing Company, Sarawak, Malaysia, 2001.

[28] R. Abrahão, M. Carvalho, W. R. Da Silva Jr., T. T. V. Machado, C. L. M. Gadelha, and M. I. M. Hernandez, "Use of index analysis to evaluate the water quality of a stream receiving industrial effluents," *Water SA*, vol. 33, no. 4, pp. 459–465, 2007.

[29] P. R. Kannel, S. Lee, Y.-S. Lee, S. R. Kanel, and S. P. Khan, "Application of water quality indices and dissolved oxygen as indicators for river water classification and urban impact assessment," *Environmental Modeling & Assessment*, vol. 132, no. 1-3, pp. 93–110, 2007.

[30] I. M. M. Rahman, M. M. Islam, M. M. Hossain et al., "Stagnant surface water bodies (SSWBs) as an alternative water resource for the Chittagong metropolitan area of Bangladesh: physicochemical characterization in terms of water quality indices," *Environmental Modeling & Assessment*, vol. 173, no. 1-4, pp. 669–684, 2011.

13

Synthesis, Spectroscopic, and Thermal Investigations of Metal Complexes with Mefenamic Acid

Karolina Kafarska, Michał Gacki, and Wojciech M. Wolf

Institute of General and Ecological Chemistry, Faculty of Chemistry, Lodz University of Technology, 116 Zeromskiego Street, 90-924 Lodz, Poland

Correspondence should be addressed to Karolina Kafarska; karolina.kafarska@p.lodz.pl

Academic Editor: Maria F. Carvalho

The novel metal complexes with empirical formulae $M(mef)_2 \cdot nH_2O$ (where M = Mn(II), Co(II), Ni(II), Cu(II), Zn(II), and Cd(II); mef is the mefenamic ligand) were synthesized and characterized by elemental analysis, molar conductance, FTIR-spectroscopy, and thermal decomposition techniques. All IR spectra revealed absorption bands related to the asymmetric (ν_{as}) and symmetric (ν_s) vibrations of carboxylate group. The Nakamoto criteria clearly indicate that this group is bonded in a bidentate chelate mode. The thermal behavior of complexes was studied by TGA methods under non-isothermal condition in air. Upon heating, all compounds decompose progressively to metal oxides, which are the final products of pyrolysis. Cu(II), Zn(II), and Cd(II) complexes were also characterized by the coupled TG-FTIR technique, which finally proved the path and gaseous products of thermal decomposition. Additionally, the coupled TG-MS system was used to determine the principal volatile products of thermolysis and fragmentation processes of $Mn(mef)_2 \cdot 3H_2O$ and $Co(mef)_2 \cdot 2H_2O$.

1. Introduction

Fenemates (N-arylated derivatives of anthranilic acid) are pharmaceutical compounds with distinct anti-inflammatory and analgesic, antipyretic activity. Their mode of biological action is based on inhibiting prostaglandin synthetase [1, 2]. In particular, 2-(2,3-dimethyl-phenyl) aminobenzoic acid (mefenamic acid: $(CH_3)_2C_5H_3NHC_5H_4COOH$ (Figure 1)) is an effective nonsteroidal agent widely used for the treatment of mild to moderate pains, inflammation, ache, and fever [3–5]. Moreover mefenamic acid, as other anti-inflammatory drugs, is emerging as novel chemopreventive agents against cancer [6, 7]. It is an active ingredient of numerous drugs which are present on the pharmaceutical market [8, 9]. However, NSAID-induced side effects, particularly in the gastrointestinal tract and kidney, often limit their applications. For this reason, considerable efforts have been made to increase their activity while minimizing side effects [10].

It is well known that several transition metal complexes with nonsteroidal drugs are more effective and show significantly lower toxicity than that of their parent drugs [11, 12]. In particular, divalent metal complexes with several NSAID are better anti-inflammatory candidates then NSAIDs, because they have unique structures that could interact with the target enzymes more specifically. In addition, metal ions introduce extra antioxidant activity [13] and antiproliferative activity against cancer [14–16]. Furthermore gastrointestinal toxicities associated with the administration of complexes are much lower than those of the parent drug and improved safety of these drugs [17].

Our studies were stimulated by the fact that majority of anti-inflammatory drugs are carboxylic acids with their carboxylate group prone to metal binding [11, 18–20]. Some of mefenamate complexes have been described in the literature. Brzyska and Ożga characterized complexes of rare earth metals with mefenamic ligand [21]. Tapacli and Ide described mefenamate compounds with Ca and Na ions [22]. Kovala-Demertzi et al. investigated the compounds with the formulae: $SnPh_3(mef)$ and $SnBu_2(mef)_2$ [23].

The aim of the present work was to obtain the mefenamate complexes of Mn(II), Co(II), Ni(II), Cu(II), Zn(II), and

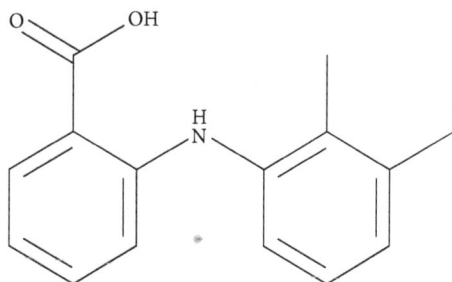

FIGURE 1: Chemical structure of mefenamic acid.

Cd(II), determine their chemical properties, and study their thermal decomposition patterns.

2. Materials and Methods

2.1. Materials. Pure mefenamic acid was obtained as a gift from Polfa Pabianice; metal chlorides $MCl_2 \cdot nH_2O$ (where M = Mn, Co, Ni, Cu, Zn, and Cd), DMSO, DMF, and EtOH p.a. were purchased from Aldrich and MeOH from Lab-Scan; other chemicals were from POCh-Gliwice.

2.2. Synthesis. All complexes were obtained according to similar procedures. The first step of synthesis was preparation sodium salt of ligand by dissolution of mefenamic (1 mmol) acid in 50 mL fresh precipitated aqueous-ethanol solution (1:1) of NaOH ($0.02 \, mol \cdot L^{-1}$). The mixture was heated up to 60°C and added to aqueous solution of metal chlorides (0.5 mol in 25 mL). The reaction mixture was kept in 60°C for 2 hours. After several days the solid precipitates were isolated by filtration, washed with hot water, and dried on air.

Complex $Mn(mef)_2 \cdot 3H_2O$: color: pale pink; IR (KBr, ν): 3358 (OH), 3067 (NH), 2859 (CH), 1652 (NH), 1578 (OCO^-), 1495 (NH), 1459 (CH_3), 1394 (OCO^-), 1283 (CH_3), 1183 (CH_3), 1159 (CH), 1093 (CH), 1043 (CH_3), 852 (CH), 749 (CH_3), 679 (MO) cm^{-1}; Anal. Calc. $C_{30}H_{34}MnN_2O_7$ (%): C, 61.12; H, 5.81; N, 4.75; Mn, 9.32; Found (%): C, 61.00; H, 5.50; N, 4.76; Mn, 9.36.

Complex $Co(mef)_2 \cdot 2H_2O$: color: pink; IR (KBr, ν): 3315 (OH), 3069 (NH), 1651 (NH), 1578 (OCO^-), 1504 (NH), 1454 (CH_3), 1393 (OCO^-), 1283 (CH_3), 1188 (CH_3), 1159 (CH), 1097 (CH), 1043 (CH_3), 854 (CH), 748 (CH_3), 675 (MO) cm^{-1}; Anal. Calc. $C_{30}H_{32}CoN_2O_6$ (%): C, 62.61; H, 5.60; N, 4.87; Co, 10.24; Found (%): C, 62.67; H, 5.60; N, 4.88; Co, 10.22.

Complex $Ni(mef)_2 \cdot 2H_2O$: color: pale green; IR (KBr, ν): 3346 (OH), 3069 (NH), 2860 (CH), 1653 (NH), 1578 (OCO^-), 1499 (NH), 1455 (CH_3), 1391 (OCO^-), 1285 (CH_3), 1190 (CH_3), 1159 (CH), 1097 (CH), 1043 (CH_3), 851 (CH), 748 (CH_3), 678 (MO) cm^{-1}; Anal. Calc. $C_{30}H_{32}NiN_2O_6$ (%): C, 62.64; H, 5.60; N, 4.87; Ni, 10.20; Found (%): C, 62.70; H, 5.49; N, 4.98; Ni, 10.21.

Complex $Cu(mef)_2 \cdot 2H_2O$: color: green; IR (KBr, ν): 3321 (OH), 3077 (NH), 2910 (CH), 1647 (NH), 1578 (OCO^-), 1506 (NH), 1458 (CH_3), 1393 (OCO^-), 1285 (CH_3), 1188 (CH_3), 1153 (CH), 1067 (CH), 1034 (CH_3), 854 (CH), 746 (CH_3), 680

(MO) cm^{-1}; Anal. Calc. $C_{30}H_{32}CuN_2O_6$ (%): C, 62.11; H, 5.56; N, 4.83; Cu, 10.95; Found (%): C, 62.08; H, 5.59; N, 4.85; Cu, 10.44.

Complex $Zn(mef)_2 \cdot 2H_2O$: color: white; IR (KBr, ν): 3338 (OH), 3066 (NH), 2858 (CH), 1651 (NH), 1576 (OCO^-), 1506 (NH), 1466 (CH_3), 1388 (OCO^-), 1283 (CH_3), 1183 (CH_3), 1155 (CH), 1069 (CH), 1043 (CH_3), 856 (CH), 748 (CH_3), 677 (MO) cm^{-1}; Anal. Calc. $C_{30}H_{32}ZnN_2O_6$ (%): C, 61.92; H, 5.54; N, 4.81; Zn, 11.23; Found (%): C, 61.88; H, 5.50; N, 4.88; Zn, 11.25.

Complex $Cd(mef)_2 \cdot 2H_2O$: color: white; IR (KBr, ν): 3312 (OH), 3067 (NH), 2858 (CH), 1651 (NH), 1576 (OCO^-), 1499 (NH), 1452 (CH_3), 1396 (OCO^-), 1283 (CH_3), 1190 (CH_3), 1159 (CH), 1043 (CH_3), 862 (CH), 750 (CH_3), 679 (MO) cm^{-1}; Anal. Calc. $C_{30}H_{32}CdN_2O_6$ (%): C, 57.29; H, 5.13; N, 4.45; Cd, 17.86; Found (%): C, 57.30; H, 5.15; N, 4.53; Cd, 17.90.

2.3. Measurements. The chemical compositions of all complexes were defined by the elemental analysis followed by the atomic absorption spectrometry. Hydrogen, carbon, and nitrogen contents were measured with the Vario EL III Elemental Analyzer. The metal content was determined in samples mineralized using the Anton Paar Multiwave 3000 closed system instrument. The mixture of concentrated HNO_3 (6 mL) and HCl (2 mL) was applied. Metal concentrations were measured by the FAAS with the GBC Scientific Equipment 932 plus spectrometer.

IR spectra were recorded on FTIR-8501 Shimadzu spectrophotometer over 4000–400 cm^{-1} range using KBr pellets. The thermal stabilities of complexes were studied by means of TGA techniques. The measurements were made with the Netzsch, TG 209 apparatus, and Q-1500 Derivatograph. Samples ($1 \cdot 10^{-2}$ g) were heated (in ceramic crucibles) up to 1000°C, at a heating rate 10°C min^{-1} in air atmosphere. The analysis of solid decomposition products was performed using TG and DTG curves and supported by the X-ray diffractograms (Siemens D-5000 diffractometer, graphite monochromatized CuK_α radiation) of sinters, obtained by heating the complex samples up to temperatures defined from TG curves. A coupled TG-MS system was applied for analysis of volatile products of thermal decomposition and fragmentation processes. Data were processed using online connected computer system with commercial software (Derivatograph TG/DTA-SETSYS-16/18, coupled to a Mass Spectrometer QMS-422 model ThermoStart from Balzers); platinum crucible, mass sample: 4–6 mg. Dynamic measurements were carried out in argon atmosphere (at a flow rate 20 mL·min^{-1}) with a heating rate 10°C·min^{-1} and an ion source temperature of *ca.* 150°C using 70 eV electron impact ionization. The TG-FTIR measurements were carried out in ceramic crucibles at flowing argon atmosphere (20 mL·min^{-1}) using the Netzsch TG 209 apparatus coupled with Bruker FTIR spectrophotometer. The samples were heated up to 1000°C at a heating rate 10°C·min^{-1}. Molar conductivity (Λ_M) of all synthesized compounds was measured in $1 \cdot 10^{-3}$ mol·L^{-1} solutions of MeOH, DMSO, and DMF, according to procedure as described in [24].

TABLE 1: Molar conductivity of complexes Λ_M/Ω^{-1} cm^2·mol^{-1} for 0,001 mol·L^{-1} solutions in MeOH, DMF, and DMSO at 25°C.

Complex	Λ_M [Ω^{-1} cm^2 mol^{-1}]		
	MeOH	DMF	DMSO
Mn(mef)$_2$·3H$_2$O	45.20	35.70	16.90
Co(mef)$_2$·2H$_2$O	11.23	13.03	3.07
Ni(mef)$_2$·2H$_2$O	17.07	20.05	7.09
Cu(mef)$_2$·2H$_2$O	12.05	13.15	4.27
Zn(mef)$_2$·2H$_2$O	21.0	28.11	3.17
Cd(mef)$_2$·2H$_2$O	18.18	24.34	4.55

TABLE 2: Principal IR bands (cm^{-1}) for carboxylate group in investigated complexes.

Compound	ν_{asym}	ν_{sym}	$\Delta\nu = \nu_{asym} - \nu_{sym}$
Na(mef)	1580,0	1380,0	200,0
Mn(mef)$_2$·3H$_2$O	1577,7	1393,6	184,1
Co(mef)$_2$·2H$_2$O	1577,7	1392,5	185,2
Ni(mef)$_2$·2H$_2$O	1577,7	1390,6	187,1
Cu(mef)$_2$·2H$_2$O	1577,7	1392,5	185,2
Zn(mef)$_2$·2H$_2$O	1575,7	1386,7	189,0
Cd(mef)$_2$·2H$_2$O	1575,7	1396,4	178,7

3. Results and Discussion

The empirical formulae of complexes showing number of water molecules and molar conductivities are presented in Table 1. All synthesized solid complexes are stable in air. They are practically insoluble in water, but on the contrary quite soluble in polar organic solvents (e.g., EtOH, MeOH). Analysis of X-ray powder diffraction data reveals high level of crystallinity and proves that neither pair of investigated compounds is isostructural. The molar conductivity data clearly indicate that all complexes in MeOH and DMSO as well as Co(II), Ni(II), and Cu(II) compounds in DMF are non-electrolytes. On the other hand, molar conductivities for Mn(mef)$_2$·3H$_2$O in MeOH and DMF and Zn(mef)$_2$·2H$_2$O and Cd(mef)$_2$·2H$_2$O point out that solutions of these compounds according to the Geary criterion [25] are intermediates between those of nonelectrolytes and 1:1 electrolytes.

3.1. FTIR Spectra. The IR spectra of all complexes exhibit a broad absorption band in the water stretching region (3300–3600 cm^{-1}). Additionally, bands related to the water bending vibrations are observed (1615–1620 cm^{-1}). In all spectra the valence vibrations of monodissociated carboxylic group are not observed. On the contrary, asymmetric (1575–1578 cm^{-1}) and symmetric (1386–1394 cm^{-1}) vibration of dissociated OCO$^-$ group are clearly observed (Table 2). These bands are affected by the coordination of mefenamic ligand to metal ions. The separation $\Delta\nu = \nu_{as}(OCO) - \nu_s(OCO)$ and the direction of the shifts of these bands in comparison to those values of sodium salt characterized the nature of the metal–carboxylate bonds. The bathochromic shifts of asymmetric (ν_{as}) and hypsochromic shifts of symmetric (ν_s) frequencies are also observed. The magnitude of separation and the bands direction show that carboxylate group of mefenamic ligand coordinated in bidentate chelate mode [26].

Apart from carboxylate group the mefenamic anion has amine group, available for coordination. The NH deformation vibrations in the IR spectra of mefenamic acid are very close to those observed in the IR spectra of its complexes. That indicates that the NH group does not participate directly in coordination.

3.2. Thermal Analysis. The data obtained from TG, DTG, and DTA curves supported by chemical and X-ray diffraction

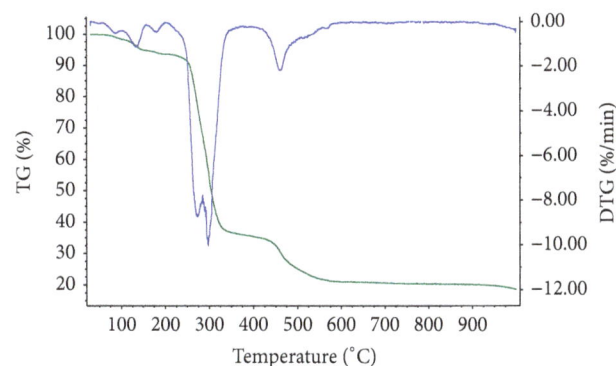

FIGURE 2: Thermoanalytical curves of Cd(mef)$_2$·2H$_2$O in air, sample mass 7,407 mg.

pattern investigations are collected in Table 3. The thermal decomposition curves of Cd(mef)$_2$·2H$_2$O are shown on Figure 2. The thermal decomposition of the complexes is a multistages process. All compounds started to decompose by dehydration accompanied by the endothermic effect. Majority of complexes lose their water molecules in one step. Only for Co(II) and Zn(II) complexes water elimination is two-stage transformation. Thermolyses of these complexes are quite similar. The two-stage dehydration (60–110°C, 130–240°C and 100–160°C, 160–190°C for Co(II) and Zn(II) compound, respectively) is followed by organic ligand thermodestruction and metal carbonate formation. Further heating leads to metal oxides: ZnO and CoO. The latter is obtained via Co$_3$O$_4$ intermediate. In the terminal step of pyrolysis the DTA curve exhibits exoeffects. Formation of oxides was confirmed by the powder X-ray diffraction technique (Figure 3).

Mn(mef)$_2$·3H$_2$O and Cu(mef)$_2$·2H$_2$O decompose in a similar way. One-step dehydration takes place at 60–240°C and 100–180°C for Mn(II) and Cu(II) complex, respectively. Further increase of temperature results in high mass loss (ca. 80%) caused by mefenamic ligand decomposition and Mn$_3$O$_4$ (via Mn$_2$O$_3$) and CuO formation. The lowest thermal stability was determined for Ni(II) complex. Its decomposition starts at 50°C with 6,5% mass loss related to water molecules elimination. In the following step, the anhydrous compound decomposes (190–460°C) to NiCO$_3$. This step is represented as two distinct exothermic peaks at 340°C and 430°C on the DTA curve. Further heating leads to formation of intermediate equimolar mixture of NiO and Ni, which

TABLE 3: Thermal decomposition data of complexes in air.

| Compounds | Ranges of decomp., °C | DTA peaks, °C | Mass loss, % | | Intermediate and final solid products |
			Found	Calc.	
$Mn(mef)_2 \cdot 3H_2O$	60–240	120 endo	9.0	9.17	$Mn(mef)_2$
	240–740	320,600 exo	78.0	77.68	Mn_2O_3
	740–900	860 exo	1.0	1.80	Mn_3O_4
$Co(mef)_2 \cdot 2H_2O$	60–110	100 endo	3.0	3.13	$Co(mef)_2 \cdot H_2O$
	130–240	-	3.0	3.13	$Co(mef)_2$
	250–520	340 endo 470 exo	73.0	73.07	$CoCO_3$
	530–660	570 exo	7.0	6.72	Co_3O_4
	>900	920 exo	1.0	0.93	CoO
$Ni(mef)_2 \cdot 2H_2O$	50–180	80 endo	6.5	6.30	$Ni(mef)_2$
	190–460	340, 430 exo	73.0	73.10	$NiCO_3$
	460–550	490 exo	8.5	9.0	$NiO+Ni$
	550–750	570 exo	1.5[a]	1.39[a]	NiO
$Cu(mef)_2 \cdot 2H_2O$	100–180	150 endo	6.5	6.66	$Cu(mef)_2$
	190–840	550, 640 exo	80.0	80.08	CuO
$Zn(mef)_2 \cdot 2H_2O$	100–160	130 endo	3.0	3.10	$Zn(mef)_2 \cdot H_2O$
	160–190	180 endo	3.0	3.10	$Zn(mef)_2$
	200–490	420 exo	72,0	72.26	$ZnCO_3$
	490–640	510 endo, 580 exo	8.0	7.56	ZnO
$Cd(mef)_2 \cdot 2H_2O$	110–160	150 endo	6.0	5.73	$Cd(mef)_2$
	210–380	300 endo, 340 exo	67.0	66.86	$CdCO_3$
	470–760	550 exo	7.0	6.59	CdO

[a] Mass increase on TG curve.

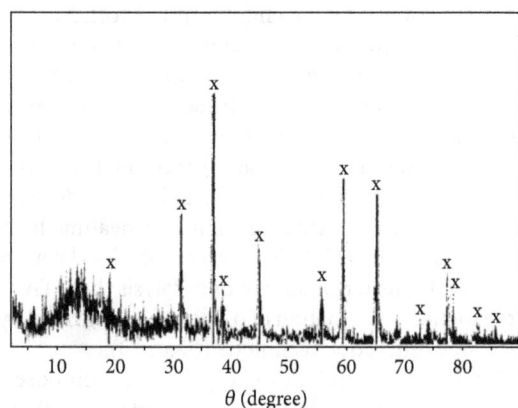

x: CO_3O_4

FIGURE 3: X-ray powder diffraction patterns of decomposition products of $Co(mef)_2 \cdot 2H_2O$ heated up to 660°C.

subsequently is fully oxided to nickel oxide (mass increase on the TG curve). Presence of NiO was confirmed by powder X-ray diffraction of sinters prepared by heating compound to 750°C. The highest thermal stability is shown by Cd(II) complex. Its pyrolysis pattern closely resembles that of Ni(II)

complex. Dehydration is single-step process and started at 110°C. The next mass loss (210–380°C) as observed on TG curve corresponds to destruction of organic ligand and leads to formation of $CdCO_3$. The latter is associated with endo- and exothermic effects as represented by two distinct DTA peaks at 300 and 340°C, respectively. Further weight loss is observed at 470–760°C and results from the CdO formation (maximum on DTA curves at 550°C).

3.3. *TG-MS Measurement.* Conventional thermoanalytic studies such as TG/DTA often do not allow for unequivocal identification of gaseous products. For this reason, the TG/MS techniques were employed to characterize products of dynamic decomposition and fragmentation of $Mn(mef)_2 \cdot 3H_2O$ and $Co(mef)_2 \cdot 2H_2O$. The determination was carried out under an argon atmosphere. The m/z values are given for ^1H, ^{12}C, ^{14}N, and ^{16}O. Figure 4 presents some profiles of ion current detected by mass spectrometer as a function of time for Co(II) complex. MS peaks of ion fragments corresponding to H_2^+, C^+, CH_3^+ OH^+, H_2O^+, $C_2H_2^+$, HCN^+, N_2^+, $C_2H_5^+$, NO^+, CO_2^+, $^{13}C^{16}O_2^+$, $^{13}C^{16}O^{18}O^+$ (m/z = 1, 12, 15, 17, 18, 26, 27, 28, 29, 30, 44, 45, 46) are monitored. Recorded data clearly indicated that all investigated complexes decompose progressively. For Mn(II) complex maxima

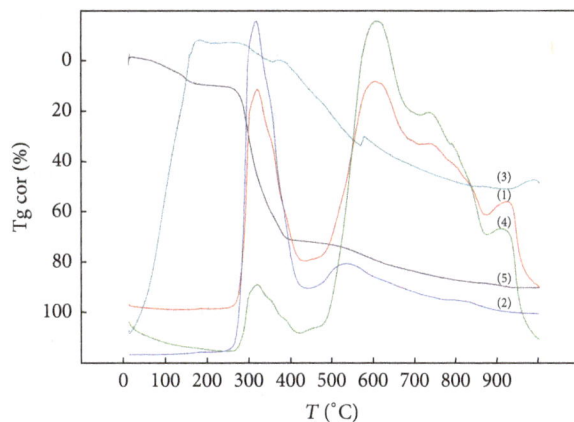

FIGURE 4: TG and corresponding MS analysis of $Co(mef)_2 \cdot 2H_2O$ (where: (1) $m/z = 12$; (2) $m/z = 15$; (3) $m/z = 18$; (4) $m/z = 44$; (5) TG).

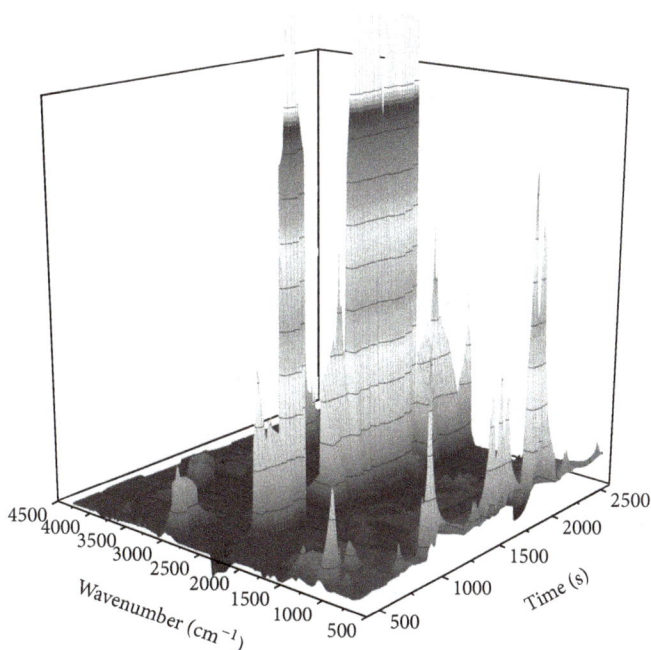

FIGURE 5: The stacked plot TG-FTIR spectra of the evolved gases for $Cd(mef)_2 \cdot 2H_2O$.

of ion current intensities were observed for temperature ranges 120–280°C, 270–450°C, and 600–900°C.

The peaks from H_2O^+ and H_2^+ observed at 110 and 120°C are related to dehydration of complex. Subsequently, fragmentation of organic ligand occurred within the 120–180°C temperature range. The following molecular ions were detected (respective temperature maxima are given in parentheses): $C_3H_7^+$ (160°C), CH_3^+ (150°C), $C_2H_2^+$ (170°C), $C_2H_5^+$ (165°C), C^+ (210°C), CO_2^+ (215°C), $^{13}C^{16}O_2^+$ (180°), (HCN^+, N_2^+). Three major temperature ranges of gaseous products emissions 120–230°C, 230–400°C, and 500–900°C were recorded for $Co(mef)_2 \cdot 2H_2O$. In the first area, CH_3^+, $C_2H_5^+$, $C_2H_2^+$ ions were detected (all maxima at 160°C). The second is related to C^+, CO_2^+ including combination of isotopes (ca. 300°C) while the third range corresponds to C^+, CH_3^+, $C_2H_5^+$, and CO_2^+ ions.

3.4. TG-FTIR Measurement. The coupled TG-FTIR technique was applied to $Cd(mef)_2 \cdot 2H_2O$, $Cu(mef)_2 \cdot 2H_2O$, and $Zn(mef)_2 \cdot 2H_2O$ complexes. All experiments were performed in argon atmosphere. The IR spectra were recorded using the Gramm-Schmidt curves. The respective stacked plot as registered for $Cd(mef)_2 \cdot 2H_2O$ is given in Figure 5.

Analysis of IR spectra of gases evolved during the thermal decomposition indicates that pyrolysis schemes of all investigated complexes are similar and closely related to those recorded in air atmosphere. All spectra clearly confirm that the thermal decomposition begins from the dehydration process. On the TG curve in range 100–160°C for $Cd(mef)_2 \cdot 2H_2O$ and $Cu(mef)_2 \cdot 2H_2O$ and 100–190°C for $Zn(mef)_2 \cdot 2H_2O$ ca. 6% mass losses connected with water losing were observed. The IR spectra which recorded up to 120°C for Cd(II) and Cu(II) complexes and up to 150°C for Zn(II) complex show bands in the wavenumbers 3750–3500

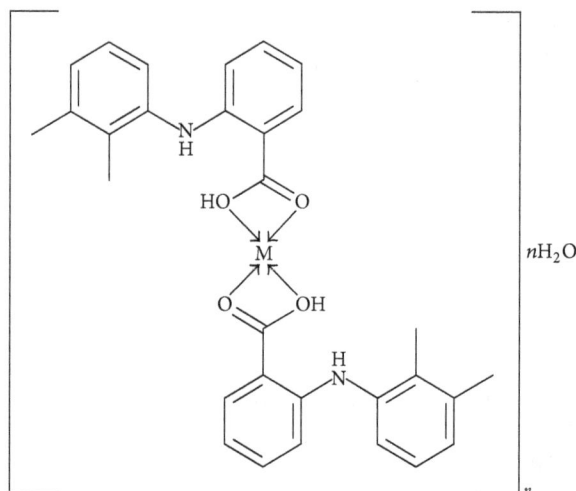

FIGURE 6: Proposed formula of the mefenamic-metal complexes.

and 1750–1400 cm^{-1} corresponding to stretching and deformation vibrations of liberating water molecules. Anhydrous complexes are stable up to about 170°C which is confirmed by the lack of IR spectra of gaseous products. Further heating leads to organic ligand degradation indicated by a significant mass loss observed on the TG curve. The spectra recorded at 260°C contain bands in frequencies ranging 3000–2600 cm^{-1} (assigned to the stretching vibrations of the CH bond from the CH_3 and CH_2 groups); 1450–1350 cm^{-1} (corresponding to bending vibrations of the CH in the CH_3 and CH_2 groups); 1225–1000 cm^{-1} and 900–720 cm^{-1} (from bending vibration of CH in aromatic ring). Additionally, this spectrum shows the trace of water molecules in split of organic ligand destruction (tied in with NH vibration). Spectra which recorded up to 290°C contain the same patterns as described above augmented by additional bands (in the range 2500–2250 cm^{-1} and 750–650 cm^{-1}) corresponding to CO_2 vibrations and trace CO oscillations (2250–2050 cm^{-1}). Heating the sample above 450°C resulted in rising intensity of bands related to carbon oxides while in the same time decreasing those of water and complete absence of aromatic ring vibrations.

4. Conclusions

All investigated complexes have been obtained as crystalline hydrates. Majority of synthesized compounds in MeOH, DMSO, and DMF solutions do not dissociate or dissociate only in a very limited degree. IR spectra firmly confirmed that mefenamic ligands are directly coordinated to metal ions only through carboxylate group, in a bidentate chelate mode. According to the chemical, spectroscopic, and thermal data, we have proposed the formula of obtained complexes (Figure 6).

All compounds decompose progressively starting with dehydration at the temperature range 60–150°C. Anhydrous compounds are stable up to almost 170°C. The most temperature persistence is Cd(II) complex, while the least one is the Ni(II) compound. Results of the TG-MS and TG-FTIR investigations correlate closely with those obtained by the

TG-DTG system. The differences are related to final products and paths of dehydration.

Conflicts of Interest

The authors declare that they have no conflicts of interest.

Supplementary Materials

The supplementary material includes the TG/DTG/DTA, TG-MS, and TG-FTIR plots for synthesized complexes. Figure 1A. TG/DTG/DTA curves of Mn(mef)2·3H2O. Figure 2A. TG/DTG/DTA curves of Co(mef)2·2H2O. Figure 3A. TG/DTG/DTA curves of Ni(mef)2·2H2O. Figure 4A. TG/DTG/DTA curves of Cu(mef)2·2H2O. Figure 5A. TG/DTG/DTA curves of Zn(mef)2·2H2O. Figure 6A. TG/DTG/DTA curves of Cd(mef)2·2H2O. Figure 7A. TG and corresponding MS analysis of Mn(mef)2·3H2O (where: (1) $m/z = 12$; (2) $m/z = 15$; (3) $m/z = 18$; (4) $m/z = 44$; (5) TG). Figure 8A. The stacked plot TG-FTIR spectra of the evolved gases for Cu(mef)2·2H2O. Figure 9A. The stacked plot TG-FTIR spectra of the evolved gases for Zn(mef)2·2H2O.

References

[1] J. R. Vane and R. M. Botting, "Mechanism of action of nonsteroidal anti-inflammatory drugs," *American Journal of Medicine*, vol. 104, no. 3 A, 1998.

[2] J. R. Vane, "Inhibition of prostaglandin synthesis as a mechanism of action for aspirin-like drugs," *Nature: New biology*, vol. 231, no. 25, pp. 232–235, 1971.

[3] F. A. Aly, S. A. Al-Tamimi, and A. A. Alwarthan, "Determination of flufenamic acid and mefenamic acid in pharmaceutical preparations and biological fluids using flow injection analysis with tris(2,2'-bipyridyl)ruthenium(II) chemiluminescence detection," *Analytica Chimica Acta*, vol. 416, no. 1, pp. 87–96, 2000.

[4] N. Cimolai, "The potential and promise of mefenamic acid," *Expert Review of Clinical Pharmacology*, vol. 6, no. 3, pp. 289–305, 2013.

[5] S. Cesur and S. Gokbel, "Crystallization of mefenamic acid and polymorphs," *Crystal Research and Technology*, vol. 43, no. 7, pp. 720–728, 2008.

[6] I. H. Sahin, M. M. Hassan, and C. R. Garrett, "Impact of nonsteroidal anti-inflammatory drugs on gastrointestinal cancers: Current state-of-the science," *Cancer Letters*, vol. 345, no. 2, pp. 249–257, 2014.

[7] P. Ghanghas, S. Jain, C. Rana, and S. N. Sanyal, "Chemopreventive action of non-steroidal anti-inflammatory drugs on the inflammatory pathways in colon cancer," *Biomedicine & Pharmacotherapy*, vol. 78, pp. 239–247, 2016.

[8] I. Jasya and K. Hideo, Jpn. Patent No. 59-175220, 1986.

[9] K. Hirojuki, J. Mitihiro, and S. Enuhisa, Jpn. Patent No. 69-134541, 1989.

[10] F. Dimiza, S. Fountoulaki, A. N. Papadopoulos et al., "Nonsteroidal antiinflammatory drug-copper(II) complexes: Structure and biological perspectives," *Dalton Transactions*, vol. 40, no. 34, pp. 8555–8568, 2011.

[11] S. B. Etcheverry, D. A. Barrio, A. M. Cortizo, and P. A. M. Williams, "Three new vanadyl(IV) complexes with non-steroidal anti-inflammatory drugs (Ibuprofen, Naproxen and Tolmetin). Bioactivity on osteoblast-like cells in culture," *Journal of Inorganic Biochemistry*, vol. 88, no. 1, pp. 94–100, 2002.

[12] C. Núñez, A. Fernández-Lodeiro, J. Fernández-Lodeiro, J. Carballo, J. L. Capelo, and C. Lodeiro, "Synthesis, spectroscopic studies and in vitro antibacterial activity of Ibuprofen and its derived metal complexes," *Inorganic Chemistry Communications*, vol. 45, pp. 61–65, 2014.

[13] J. Feng, X. Du, H. Liu et al., "Manganese-mefenamic acid complexes exhibit high lipoxygenase inhibitory activity," *Dalton Transactions*, vol. 43, no. 28, pp. 10930–10939, 2014.

[14] X. Totta, A. A. Papadopoulou, A. G. Hatzidimitriou, A. Papadopoulos, and G. Psomas, "Synthesis, structure and biological activity of nickel(II) complexes with mefenamato and nitrogen-donor ligands," *Journal of Inorganic Biochemistry*, vol. 145, pp. 79–93, 2015.

[15] A. Ashraf, W. A. Siddiqui, J. Akbar et al., "Metal complexes of benzimidazole derived sulfonamide: Synthesis, molecular structures and antimicrobial activity," *Inorganica Chimica Acta*, vol. 443, pp. 179–185, 2016.

[16] S. Ramzan, S. Saleem, B. Mirza, S. Ali, F. Ahmed, and S. Shahzadi, "Synthesis, characterization, and biological activity of transition metals complexes with mefenamic acid (NSAIDs)," *Russian Journal of General Chemistry*, vol. 85, no. 7, pp. 1745–1751, 2015.

[17] C. T. Dillon, T. W. Hambley, B. J. Kennedy et al., "Gastrointestinal toxicity, antiinflammatory activity, and superoxide dismutase activity of copper and zinc complexes of the antiinflammatory drug indomethacin," *Chemical Research in Toxicology*, vol. 16, no. 1, pp. 28–37, 2003.

[18] K. Kafarska, D. Czakis-Sulikowska, and W. M. Wolf, "Novel Co(II) and Cd(II) complexes with non-steroidal anti-inflammatory drugs: Synthesis, properties and thermal investigation," *Journal of Thermal Analysis and Calorimetry*, vol. 96, no. 2, pp. 617–621, 2009.

[19] R. P. Sharma, S. Singh, A. Singh, and V. Ferretti, "Spectra-structure relationship: Synthesis, characterization of copper(II) complexes with ibuprofenate, o-methoxybenzoate, p-ethoxybenzoate and single crystal X-ray structure determination of [trans-Cu(en)2(H2O)2](L)2 where en = ethylenediammine, L = o-methoxybenzoate/p-ethoxybenzoate," *Journal of Molecular Structure*, vol. 918, no. 1-3, pp. 188–193, 2009.

[20] C. Dendrinou-Samara, D. P. Kessissoglou, G. E. Manoussakis, D. Mentzafos, and A. Terzis, "Copper(II) complexes with anti-inflammatory drugs as ligands. Molecular and crystal structures of bis(dimethyl sulphoxide)tetrakis(6-methoxy-α-methyl-2-naphthaleneacetato)dicopper(II) and bis(dimethyl sulphoxide)tetrakis[1-methyl-5-(p-toluoyl)-1H-pyrrole-2-acetato]dicopper(II)," *Journal of the Chemical Society, Dalton Transactions*, no. 3, pp. 959–965, 1990.

[21] W. Brzyska and W. Ożga, "Preparation and properties of rare earth element complexes with mephenamic acid," *Polish Journal of Chemistry*, vol. 67, p. 619, 1993.

[22] A. Topacli and S. Ide, "Molecular structures of metal complexes with mefenamic acid," *Journal of Pharmaceutical and Biomedical Analysis*, vol. 21, no. 5, pp. 975–982, 1999.

[23] D. Kovala-Demertzi, V. Dokorou, Z. Ciunik, N. Kourkoumelis, and M. A. Demertzis, "Organotin mefenamic complexes-preparations, spectroscopic studies and crystal structure of a triphenyltin ester of mefenamic acid: Novel anti-tuberculosis agents,"

Applied Organometallic Chemistry, vol. 16, no. 7, pp. 360–368, 2002.

[24] D. Czakis-Sulikowska, A. Malinowska, and K. Kafarska, "Complexes of Mn(II), Cu(II) and Cd(II) with bipyridine isomers and lactates," *Polish Journal of Chemistry*, vol. 80, no. 12, pp. 1945–1958, 2006.

[25] W. J. Geary, "The use of conductivity measurements in organic solvents for the characterisation of coordination compounds," *Coordination Chemistry Reviews*, vol. 7, no. 1, pp. 81–122, 1971.

[26] K. Nakamoto, *Infrared and Raman Spectra of Inorganic and Coordination Compounds*, John Wiley & Sons, Inc., New York, NY, USA, 2009.

14

The Effect of Menisci on Kinetic Analysis of Evaporation for Molten Alkali Metal Salts (CsNO$_3$, CsCl, LiCl, and NaCl) in Small Cylindrical Containers

In-Hwan Yang ⓘ, Hee-Chul Yang, and Hyung-Ju Kim

Decontamination & Decommissioning Research Division, Korea Atomic Energy Research Institute, Daejeon 34057, Republic of Korea

Correspondence should be addressed to In-Hwan Yang; ihyang@unm.edu

Academic Editor: Jae Ryang Hahn

Using isothermal thermogravimetric data of alkali metal salts (CsNO$_3$, CsCl, LiCl, and NaCl), we conducted kinetic analysis on atmospheric evaporation to investigate the effect of meniscus on determining the condensation coefficient. In the process of evaporation into an atmospheric gas, molten salt decomposed at the interface between molten salt and an atmospheric gas reacts with chemical compositions of the atmospheric gas to be an equilibrium state. In this atmospheric evaporation, the interface shape of molten salts is affected by the container diameter and the contact angle at the container wall. In the analysis results, the formed concave/convex meniscus led to underestimating the condensation coefficient of molten salts. However, whether the values of the condensation coefficient of molten salts were affected by menisci, the range of the predicted values was still low from 10^{-3} to 10^{-5}. This result means that the presence of the foreign gas (air and Ar) is a dominant parameter in determining the condensation coefficient of atmospheric evaporation.

1. Introduction

There have been various researches on using molten salts as reaction and energy transfer medium for macro- and microscales of potential applications, because of the fascinating thermophysical and electrochemical properties of molten salts. Molten salts have relatively higher thermal stability and lower vapor pressure, compared to common solvents. These attractive properties have provided a wide and high operating temperature range, suppressing relevant mass loss with evaporation or decomposition at operating temperature. Examples include pyrochemical reprocessing of spent nuclear fuel [1, 2], molten salt electrolysis [3], heat transfer fluid in nuclear plant [4, 5], cooling medium for quenching process of alloys [6], alternative electrolytes in batteries [7], and thermal storage in solar thermal power plants [8, 9]. However, in spite of low vapor pressure of molten salts, there is inevitable loss with evaporation of metal species at higher operating temperature, which consequently affects the overall efficiency of the process.

To accurately predict this evaporation of molten salts, the kinetic theory of gas [10] has been widely used. In the kinetic theory of gas, the evaporation rate under the nonequilibrium condition is determined by the number of the gas molecules striking the surface from the equilibrium vapor and the actual gas pressure acting on the surface of the liquid phase. However, classic experimental studies [11–13] measuring the evaporation rate at the surface reported that the measured evaporation rate was lower than the value theoretically predicted by the kinetic theory of gas. To address this discrepancy between the theoretical predictions and experimental measurements with a correction coefficient, the Hertz-Knudsen equation [14–17] was suggested. As expressed in (1), the evaporation rate in the nonequilibrium condition is calculated by the pressure difference between the equilibrium vapor pressure, P^*, at temperature, T, and the pressure acting on the surface of the liquid phase, P, with the condensation coefficient, α.

$$\frac{dN}{dt} = \alpha \frac{(P^* - P)}{\sqrt{2\pi MRT}} A. \tag{1}$$

dN/dt is the evaporated molecular moles per unit of time, A is the cross-sectional area of the container, M

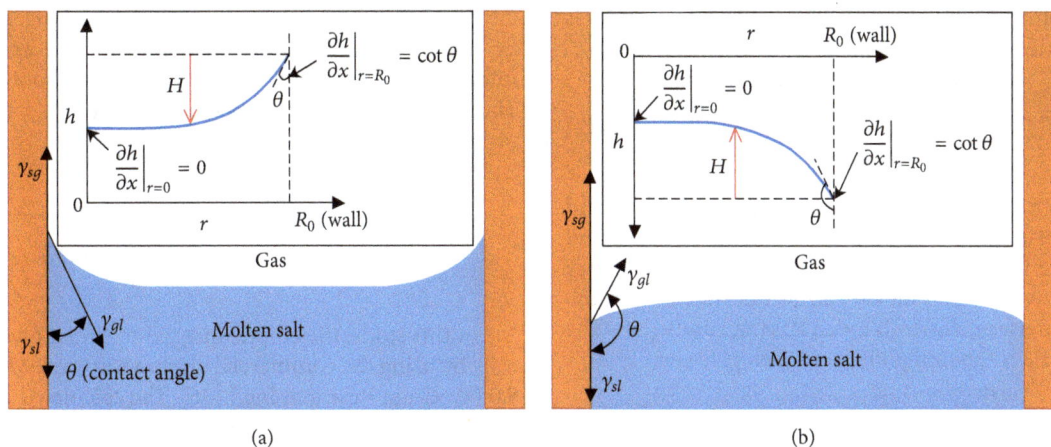

FIGURE 1: Schematic of the formed concave (a) and convex (b) meniscus of molten salts.

is molar mass, and R is the universal gas constant. The condensation coefficient used in the equation is effective for empirically correcting the predicted rate of evaporation into vacuum [16, 18, 19] or a gas atmosphere [1, 20–23]. However, determination of the coefficient values in numerous experimental studies on the evaporation process has produced inconsistency. In the case of water evaporation, for an example, the reported coefficient values varied from an order of 10^{-3} to 1 [24].

From the theoretical investigation on the inconsistency of the condensation coefficient [7, 25–30], it was suggested that the presence of a very thin layer above the surface of the condensed phase, known as the Knudsen layer, resulted from evaporation of the condensed phase. This very thin layer may work like diffusional resistance, reflecting gas molecules from the thin layer into the bulk gas phase. An equilibrium, therefore, cannot occur at the surface of the condensed phase but is achieved within the thin layer. This nonequilibrium within the Knudsen layer may cause the thermophysical properties of the layer, including density, pressure, and temperature, to diverge from those of the bulk gas phase, and consequently disturb determining the accurate condensation coefficient.

Based on this hypothesis that the condensation coefficient is the result of nonequilibrium within the Knudsen layer, several research groups have developed complicated models by modifying the Maxwell-Boltzmann distribution [25, 27] or by applying transition state theory [26]. These theoretical approaches for dealing with the effect of the Knudsen layer on the evaporation have managed to reduce the inconsistency of the condensation coefficient in the reported experimental results. However, the inconsistency of the condensation coefficient has not been fully resolved yet, because the analysis may leave behind unevaluated parameters, such as the meniscus.

This paper therefore examines experimental measurements of vaporization of alkali metal salts into an atmospheric gas in an attempt to quantify the effect of a meniscus on determining the evaporation flux and the condensation coefficient.

The actual evaporation area associated with morphological changes of the meniscus in various diameters of cylindrical containers was predicted and characterized by using an analytical equation derived from the Young-Laplace law [31, 32]. The equilibrium vapor pressure at the surface used in the kinetic analysis of evaporation was also predicted by using the commercial thermodynamic equilibrium software, HSC chemistry 7.1 (Outotec) [33]. The predicted values of the condensation coefficient in various sizes of a cylindrical container were compared. Inconsistent values in the determined condensation coefficient revealed the effect of the meniscus on evaporation of the molten salts.

2. Materials and Methods

2.1. Evaporation Area of Molten Salts. The interface separating a molten salt and an atmospheric gas in a small container is not a flat interface but should be either concave or convex meniscus (Figure 1) due to the force balance along the interface. In the Young-Laplace law, the interfacial tension force, the normal force associated with the local curvature of the interface, is equated to the difference in the pressure exerted onto the interface by the molten salt and the atmospheric gas. The force balance is thus given as

$$\gamma \nabla \cdot n = \Delta P. \qquad (2)$$

From the principle of the hydrostatic equilibrium, the term for the pressure jump across the interface in (2) can be replaced by the hydrostatic pressure of the molten salt and the atmospheric pressure. In the hydrostatic equilibrium state of the molten salt illustrated in Figures 1(a) and 1(b), the vertical gradient of the interface associated with the interfacial tension is equated to the hydrostatic pressure of the molten salt. Therefore, the curvature of the interface in (2) is rewritten as a function of the interface height from the horizontal reference plane where the hydrostatic pressure equals the atmospheric pressure. The interface profile is finally expressed by the next equation.

$$\nabla \cdot n = \frac{d^2h/dr^2}{\left(1 + (dh/dr)^2\right)^{3/2}} = \frac{\rho g \left(h - h_{ref}\right)}{\gamma}. \tag{3}$$

Here γ is the surface tension between the molten salt and the atmospheric gas, ρ is the density of the molten salt, and h and h_{ref} are the height of the interface and the reference plane, respectively. Equation (3) can be linearized when the slope of the interface, dh/dr, is much less than unity [34]. The interface profile is thus simplified and then given as

$$\frac{d^2h}{dr^2} = \frac{\rho g \left(h - h_{ref}\right)}{\gamma}. \tag{4}$$

This equation is solved subject to boundary conditions: that is, the slope of the interface is zero at the center of the axisymmetric interface ($r = 0$), the interface forms a contact angle θ with the container wall ($r = R_0$), and the reference plane locates at the intersection between the interface and the container wall ($r = R_0$).

$$\left.\frac{dh}{dr}\right|_{r=0} = 0,$$

$$\left.\frac{dh}{dr}\right|_{r=R_0} = \cot(\theta), \tag{5}$$

$$h_{ref}\big|_{r=R_0} = 0.$$

In these conditions, the contact angle, θ, is the angle formed by the intersection between the interface of the solid-condensed phase and the interface of the gas-condensed phase at the container wall, as shown in Figure 1. This contact angle is an intrinsic property of the three-phase contact line between the molten salt, the atmospheric gas, and the container wall. Solving (3) in conjunction with the boundary conditions, the concave or convex meniscus can be expressed by the changes in the local elevation along the container radius. The local elevation of the interface, H, is expressed in terms of the radial position, r, contact angle, θ, as in the following equation:

$$H = \frac{2L_c}{e^{R_0/L_c} - e^{-R_0/L_c}} \cot(\theta)$$
$$\cdot \left(\cosh\left(\frac{r}{L_c}\right) - \cosh\left(\frac{R_0}{L_c}\right)\right). \tag{6}$$

In the equation, the capillary length [35], L_c, is a characteristic length scale for an interface which is subject to the hydrostatic force ($\rho g L_c$) and the surface tension force (γ/L_c). When the capillary length is much larger than the diameter of the container, the surface tension force dominates the hydrostatic force at the interface of the molten salt. This capillary length is defined by the next equation,

$$L_c = \sqrt{\frac{\gamma}{\rho_l g}}. \tag{7}$$

The interface area is obtained by rotating its concave or convex meniscus, from the center of the axisymmetric interface to the cylindrical container wall, about the h-axis. The meniscus area, A_{meni}, becomes

$$A_{meni}$$
$$= \int_0^{R_0} 2\pi r \sqrt{1 + \cot(\theta)^2 \operatorname{csch}\left(\frac{R_0}{L_c}\right)^2 \sinh\left(\frac{r}{L_c}\right)^2} \, dr. \tag{8}$$

In this study, the elliptic integral of (8) is solved numerically by using the commercial algebra software, Mathematica 9.0 (Wolfram Research Inc.) [36]. The calculated area is used as the actual evaporation area in the analysis for evaporation of the molten salts in small cylindrical containers.

2.2. Thermogravimetric Experiment. The vaporization rates of the molten cesium salts were measured by using a custom-made TGA (thermogravimetric analyzer). The TGA system used in this study mainly consists of a furnace equipped with a maximum weight capacity of 10 grams, a control and data acquisition system, and a gas feeding system. Gaseous atmospheres in the furnace are maintained by flowing compressed air at a rate of 1 L/min during conducting experiment. The exhaust gas from the furnace is scrubbed with multiple metal-vapor scrubbing impingers, as shown in Figure 2.

In the experiment for measuring the evaporation rate at constant temperatures, five grams of the sample salt loaded in a cylindrical crucible with an inner diameter of 21 mm was placed inside the TGA furnace and was heated with a constant rate of 20 K/min from room temperature to the investigated isothermal temperature. The dynamics of sample weight loss with time at an isothermal temperature was measured and recorded until the change of the sample mass became constant.

The experimental evaporation rates of CsCl and $CsNO_3$ were determined by

$$\frac{dN}{dt} = \frac{1}{M}\frac{dW}{dt}. \tag{9}$$

In the equation, M is the molecular mass of the sample cesium species and dW is the mass change during the differential time, dt. The evaporation rates were consistent during the experiment, except the start-up and shut-down process of TGA. Thus, the evaporation rates were determined by averaging their values over time, weight losses from 20% to 80% of the initial mass. The results of the isothermal TGAs for CsCl and $CsNO_3$ at five different isothermal temperatures and the parameters used in the experiment are summarized in Table 1.

In this paper, in addition to the conducted experimental measurement for evaporation rates of CsCl and $CsNO_3$, the reported data of alkali chloride species (NaCl and LiCl) [1, 22, 23] were also used to extend the present kinetic analysis of evaporation. Table 1 also includes the evaporation rates and the parameters used in the reported thermogravimetric experiments.

TABLE 1: Measured evaporation rates and experimental conditions used in isothermal thermogravimetric experiments for the molten alkali metal salts.

Molten salt	Temperature (°C)	Evaporation rate (mol/min)	Atmospheric Gas	Contact angle [37] (°)	Diameter of container (cm)
CsCl	850	1.39×10^{-5}			
	890	2.45×10^{-5}			
	930	4.28×10^{-5}	Air	31	2.1
	970	7.22×10^{-5}			
	1000	1.78×10^{-4}			
CsNO₃	500	3.29×10^{-5}			
	540	6.27×10^{-5}			
	580	1.12×10^{-4}	Air	22	2.1
	620	1.90×10^{-4}			
	660	3.08×10^{-4}			
LiCl [1, 22]	850	9.68×10^{-5}			3.0
	900	4.19×10^{-4}	Ar	136	
	950	4.19×10^{-5}			0.4
NaCl [23]	935	2.00×10^{-4}			
	950	2.77×10^{-4}	Ar	113	3.81
	985	4.03×10^{-4}			

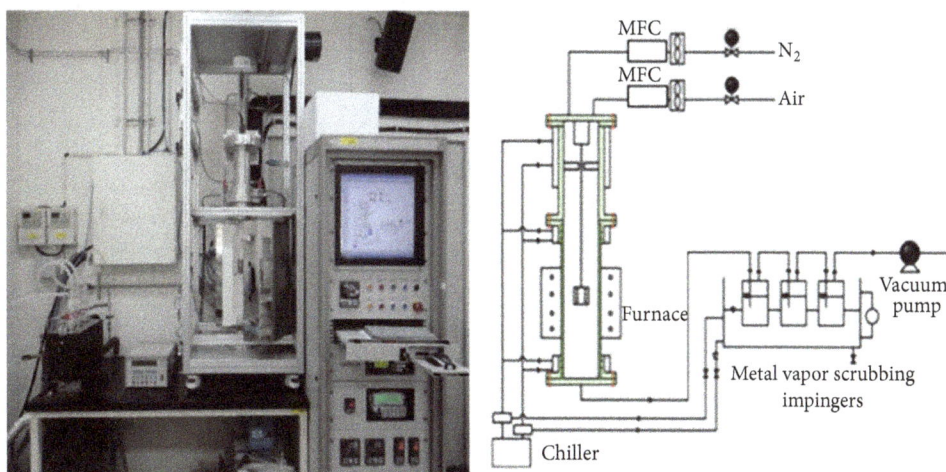

FIGURE 2: Image and schematic diagram of the TGA system used in this study.

3. Results and Discussion

3.1. Characterization of Meniscus. As indicated earlier, the profile of the interface between molten salt and atmospheric gas in a cylindrical container is directly determined by solving (6). The temperature dependent properties of the molten salts, such as the density and the interfacial tension coefficient, used the values found on the molten salts handbook [3]. The reported values from Baumli and Kaptay's measurement [37] were used for contact angles of the molten slats. The calculations for predicting the meniscus in an open cylindrical container with wide ranges of diameters (0.1 m–0.0005 m) were performed at the identical temperature conditions used in the thermogravimetric experiments. Figures 3(a) and 3(b) plot the local elevation of the interface, H, versus the

dimensionless radial location, r/R_0, for NaCl at 950°C and CsCl at 1000°C, respectively.

The comparison of the interface profiles in these figures shows that decreasing the diameter of the container causes changes in the shape of the interface between the molten salt and the atmospheric gas. For all of the alkali metal salts used in the present calculations, as the diameter of the container decreases, the curvature of the interface, which is initially located near the container wall, starts to extend toward the center of the interface profile, reducing its flat portion. This change of the curvature finally leads the meniscus to be parabolic and then reduces the maximum elevation.

Based on the analysis results of the interface profile, the meniscus of the investigated molten salts can be classified into three basic regimes, namely, fully developed interface,

(a) NaCl

(b) CsCl

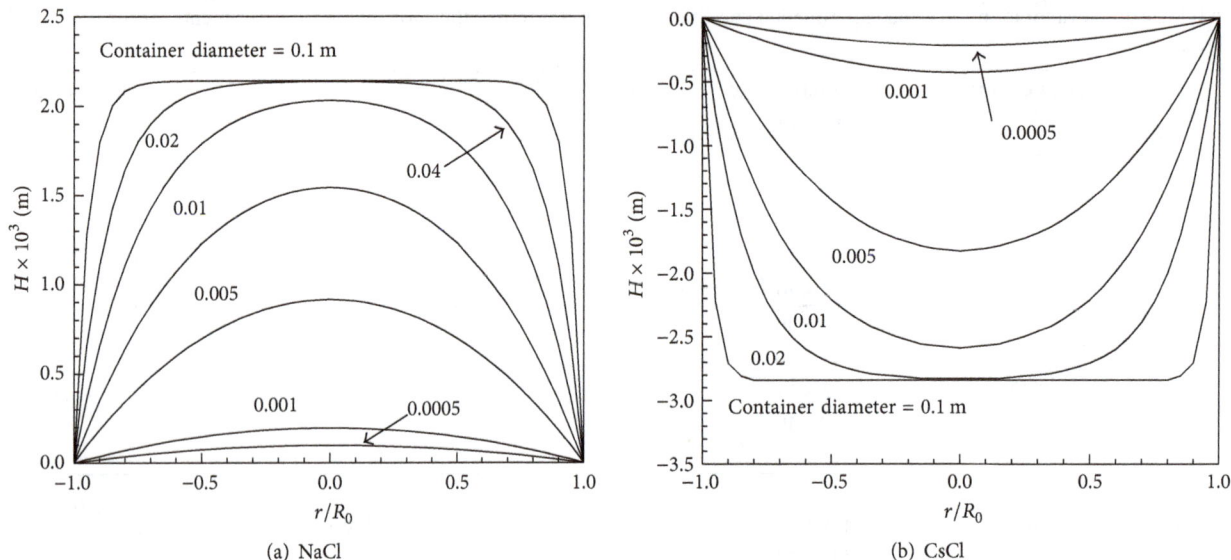

FIGURE 3: The meniscus profiles of molten NaCl at 950°C (a) and CsCl at 1000°C (b) in various diameters of cylindrical containers.

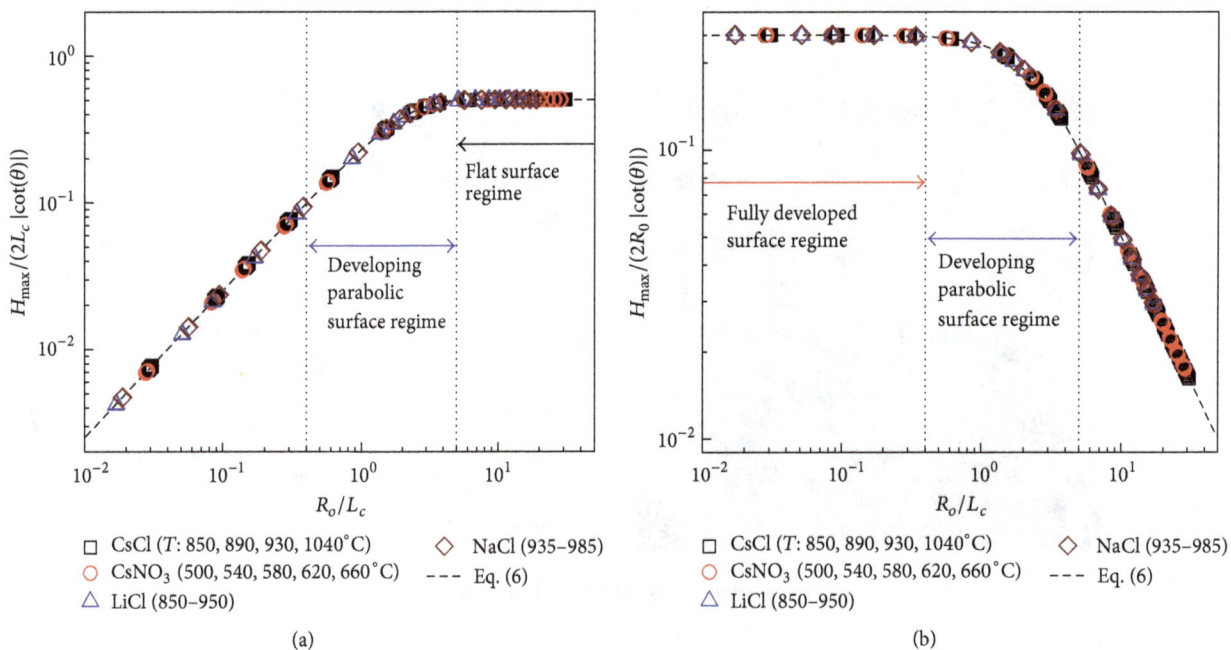

(a)

□ CsCl (T: 850, 890, 930, 1040°C)	◇ NaCl (935–985)
○ CsNO$_3$ (500, 540, 580, 620, 660°C)	--- Eq. (6)
△ LiCl (850–950)	

(b)

□ CsCl (T: 850, 890, 930, 1040°C)	◇ NaCl (935–985)
○ CsNO$_3$ (500, 540, 580, 620, 660°C)	--- Eq. (6)
△ LiCl (850–950)	

FIGURE 4: Maximum elevation ratio of molten salts to the capillary length (a) and to the container diameter (b) depending on the dimensionless radius of containers.

transition (or developing parabolic interface), and flat interface regime. Each interface regime is characterized by the dimensionless radius, which is the ratio of the container radius to the capillary length. In the flat interface regime, the gravitational force acting on the interface is dominant force over the interface. Thus, decreasing the container diameter decreases only the flat portion of the meniscus. In the maximum elevation, $H_{max}/(2L_c|\cot(\theta)|)$, presented in Figure 4(a), the values of each molten alkali metal salt are consistent when the ratio of the container radius to the capillary length is larger than 5.

Decreasing the values of the dimensionless radius (R_0/L_c) below 5 by decrease of the container diameter, the meniscus profile of the molten salts starts to evolve into a parabolic shape due to the extension of the interfacial tension force near the container wall. Figure 4(b) presents the results plotted in Figure 4(a) but plots the dimensionless elevation with respect to the container diameter; that is, $H_{max}/(2R_0|\cot(\theta)|)$. In the results, the boundary between the transition regime and the fully developed interface regime is identified by the changes in the gradient of the dimensionless elevation. The dimensionless elevation of the molten slats in the flat interface

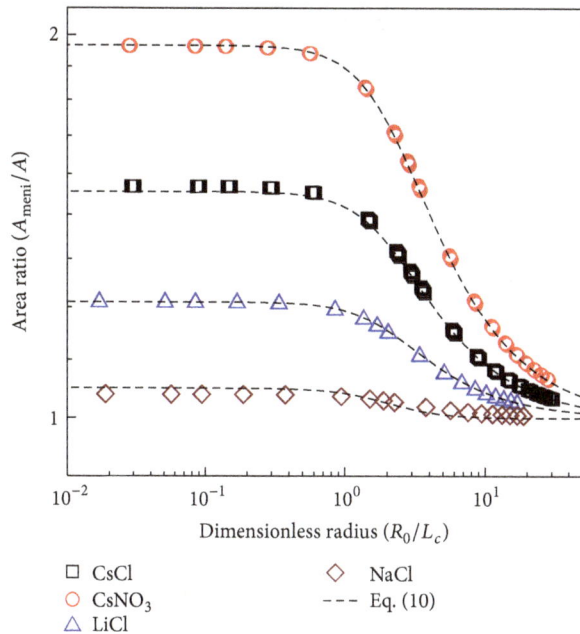

FIGURE 5: Comparison of semiempirical correlation (see (10)) with calculated values for the area ratio of the meniscus to the cross-section of container.

regime increases in exact proportion to the decrease of the container diameter, because the maximum elevation is not affected by the container dimension. However, after passing their flat interface regime, the gradients of the maximum elevation for all molten alkali metal salts start to decrease and then approach their asymptotic value of 0.25. Since the interface between the molten salt and atmospheric gas is completely governed by the interfacial tension force from the diameter corresponding to that value, further decrease in the radius of the container (R_o/L_c < 0.4) insignificantly affects the interface shape. It means that the changes of the maximum height are exactly proportional to that of the container radius in the fully developed interface regime.

3.2. Actual Evaporation Area in Containers.
Figure 5 plots the area ratio of the meniscus to the horizontal cross-section of the container (A_{mani}/A) versus the dimensionless radius (R_0/L_c). Similar to the earlier results of the interface profile, the area ratio also decreases from its asymptotic value in the fully developed interface regime, as the dimensionless radius increases. For the meniscus in the flat interface regime, the curvature near the wall makes the interface area be also larger than the cross-sectional area of the container. However, this effect of the curvature decreases with more increase of the dimensionless radius, due to the increasing flat portion of the interface.

The comparison results in Figure 5 also show that the contact angle of the molten salt affects the values of the meniscus area where evaporation of molten salt occurs. The contact angle of molten salts, irrelevant in determining the interface shape regime, increases the value of the maximum elevation in proportion to the absolute value of $\cot(\theta)$.

Using these compiled results of the analysis, a semiempirical correlation can be developed to account for the changes in the actual evaporation area of the molten salts in small cylindrical containers. The correlation is formulated in terms of the dimensionless radius of the container and the contact angle of molten salts at the container wall as

$$\frac{A_{meni}}{A} = 1 + \frac{0.222\,|\cot(\theta)|^{1.62}}{\left(1 + (0.5R/L_c)^{2.21}\right)^{0.445}}. \tag{10}$$

The empirical expression for the area ratio given by (10) is in excellent agreement with the predicted results.

3.3. Equilibrium Vapor Pressure and Maximum Evaporation Flux.
Along with the experimental evaporation rate, the equilibrium vapor pressures of molten salts should be known to determine the condensation coefficient. In evaporation of molten alkali metal salts into a foreign gas, the condensed species could be decomposed at its surface and react with chemical compositions of an atmospheric gas in the evaporation process. If this chemical reaction occurs according to minimization of the total Gibbs free energy, that reaction represents an equilibrium state. The commercial thermodynamic calculations code, HSC chemistry 7.1, was therefore used for predicting the equilibrium composition of $CsCl$, $CsNO_3$, $NaCl$, and $LiCl$ in an atmospheric gas (compressed air or argon gas). The air used in the calculations was composed of N_2 (78%), O_2 (20%), Ar (0.9%), and H_2O (1.1%). In the calculations, it was also assumed that the state of chemical equilibrium was maintained at the meniscus of molten salts with an atmospheric pressure. The thermodynamic calculation results of the equilibrium composition at the meniscus of each vaporizing species are summarized in Table 2.

The equilibrium compositions of the investigated alkali metal salts depending on the temperature are also plotted in Figure 6. In the predicted results of the equilibrium composition of the molten alkali metal salts, metal chloride species mainly evaporates into monomer ($CsCl$, $LiCl$, and $NaCl$), dimer (Cs_2Cl_2, Li_2Cl_2, and Na_2Cl_2), and trimer (Li_3Cl_3) and then become dominant species of the equilibrium gas phase. Other chemical species in each vaporization of molten salt also exists as an equilibrium species, but their concentrations are negligibly small in comparison with their dominant gas species as shown in the concentration distribution of equilibrium species (Figure 6).

In the calculated equilibrium composition of $CsNO_3$ (Figure 6(b)), however, the condensed phases of the cesium species such as $CsNO_3$, $CsNO_2$, and $CsOH$ exist as equilibrium species together with each of their gaseous species as dominant species. These cesium species resulted from the reaction of $CsNO_3$ with chemical compositions of air, that is, expressed as the following reaction equations ((R1) and (R2)):

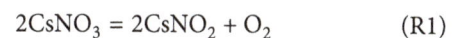

$$2CsNO_3 = 2CsNO_2 + O_2 \tag{R1}$$

$$4CsNO_3 + 2H_2O = 4CsOH + 2N_2 + 5O_2. \tag{R2}$$

The condensed phases of cesium species ($CsNO_3$, $CsNO_2$, and $CsOH$) are present in equilibrium with each of their

TABLE 2: Chemical species in the calculation of the equilibrium compositions of the molten alkali metal salts.

Vaporizing species	Condensed species	Gaseous species	Dominant gaseous species
CsCl	Cs, CsCl	Cs, Cs_2, CsCl, Cs_2Cl_2	CsCl, Cs_2Cl_2
$CsNO_3$	$CsNO_3$, $CsNO_2$, CsOH, CsO_2	$CsNO_3$, $CsNO_2$, CsOH, $Cs_2(OH)_2$, Cs, Cs_2O_2, CsO, Cs_2O, CsH, Cs_2	$CsNO_3$, $CsNO_2$, CsOH, $Cs_2(OH)_2$
NaCl	NaCl	NaCl, Na_2Cl_2, Na_3Cl_3, Na, Cl, Cl_2, Na_2, Cl_4, Cl_3	NaCl, Na_2Cl_2
LiCl	LiCl, Li_2Cl_2	LiCl, Li_2Cl_2, Li_3Cl_3, Li, Cl, Cl_2, Cl_4, Cl_3, Li_2	LiCl, Li_2Cl_2, Li_3Cl_3

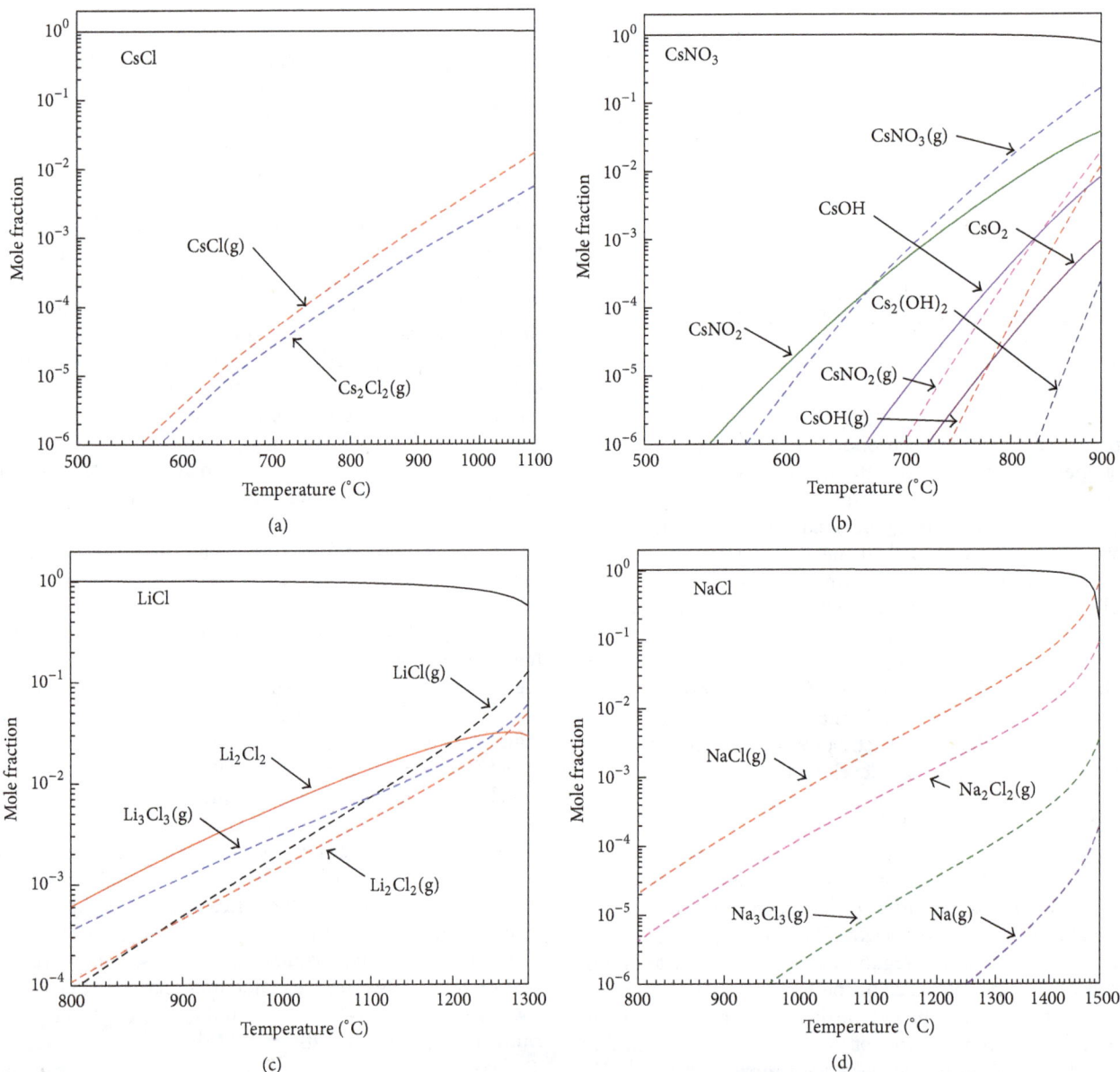

FIGURE 6: Calculated equilibrium compositions at the surfaces of molten CsCl (a), $CsNO_3$ (b), LiCl (c), and NaCl (d).

gaseous species at temperatures ranging from 550 to 900°C. $CsNO_3$ and $CsNO_2$ are in equilibrium with each of their gaseous species, and CsOH is in equilibrium with both the monomer and dimer forms of the gaseous cesium hydroxide in this temperature range.

From (1), the maximum mass flux from a molten salt to an atmospheric gas occurs when the condensation coefficient and the pressure acting on the surface of the liquid phase is unity and zero, respectively. The term for the vapor pressure of the condensed phase in (1) is equated to sum of

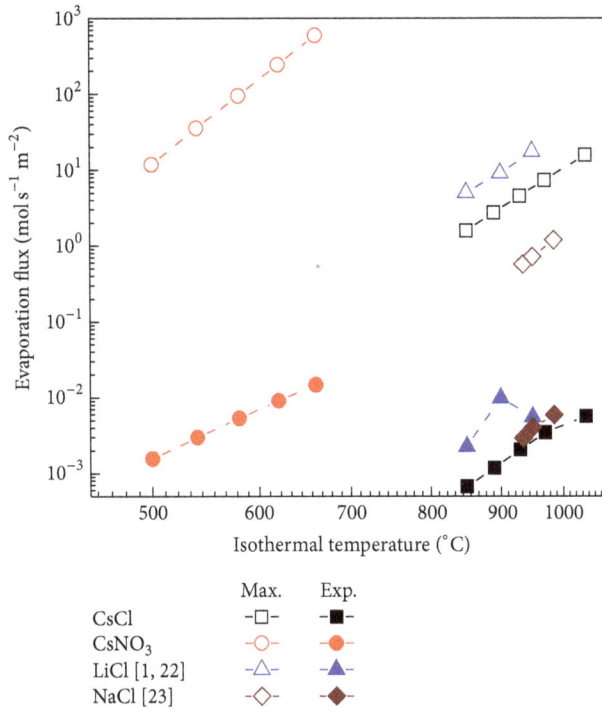

FIGURE 7: Comparison of the maximum theoretical vaporization fluxes and the experimental vaporization fluxes of the molten alkali metal salts (CsCl, CsNO$_3$, LiCl [1, 22], and NaCl [23]).

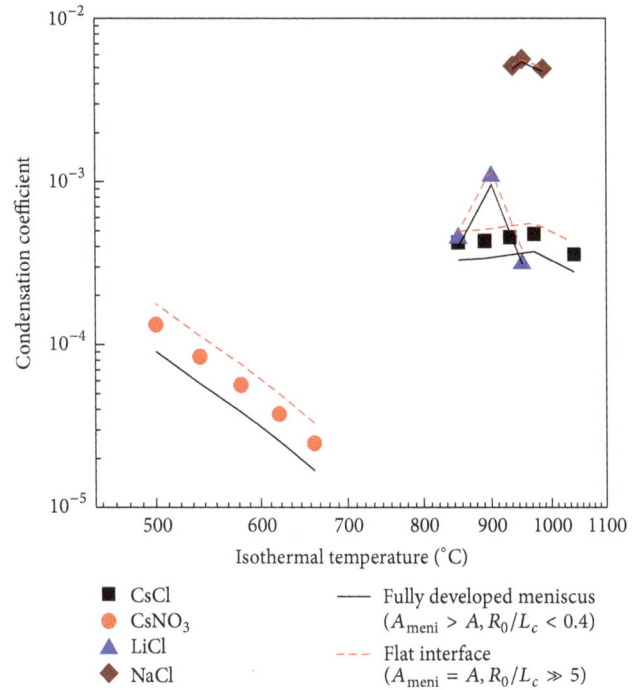

FIGURE 8: Comparison of the condensation coefficient of the molten alkali metal salts (CsCl, CsNO$_3$, LiCl, and NaCl).

equilibrium vapor pressure of monomer, dimer, and trimer species. Equation (1), therefore, can be expressed in terms of monomer species as

$$\frac{1}{A}\frac{dN}{dt}$$

$$= \left(\frac{P_m^*}{\sqrt{2\pi M_m RT}} + \frac{2P_d^*}{\sqrt{2\pi M_d RT}} + \frac{3P_{tr}^*}{\sqrt{2\pi M_{tr}RT}} \right) \quad (11)$$

$$\cdot \left(1 + \frac{0.222\,|\cot(\theta)|^{1.62}}{\left(1 + \left(0.5R/L_c\right)^{2.21}\right)^{0.445}} \right).$$

M is molar mass of monomer species, and subscripts m, d, and tr denote the monomer, dimer, and trimer species, respectively. The second term on the right hand side of (11) represents the area ratio of the molten salt surface to the cross-section of the container which accounts for the meniscus effect on the actual evaporation area.

The maximum theoretical vaporization flux of the molten salts obtained from (11) is compared with the experimental values determined from dividing the measured vaporization rate by the cross-sectional area of the container. In the thermogravimetric experiment, evaporation of the molten salts could be maximized by the purge gas flow which continuously removes vaporized gas species from the molten salt surface ($P = 0$). Therefore, the condensation coefficient of the molten salts can be determined by comparing the experimental flux to the maximum flux. Figure 7 shows the results of the molten salts comparing the values of the

calculated maximum flux with the values obtained from the experiment and the reported data of Hur et al. [1], Kim et al. [22], and Baughman et al. [23]. The predicted values of the maximum theoretical flux are very large compared to those of the experimental flux.

3.4. Condensation Coefficient. Figure 8 presents the results of the condensation coefficient determined from the experimental flux data and the predicted maximum theoretical evaporation flux plotted in Figure 7. The values of the condensation coefficient at isothermal temperatures are the ratio of the maximum theoretical flux to the experimental flux. The experimental results show underestimation of the condensation coefficient for CsNO$_3$, CsCl, and LiCl. The maximum theoretical evaporation flux calculated from (11) underpredicts the values of the condensation coefficient, due to the meniscus effect on the evaporation area.

The results are also compared with the condensation coefficient curves predicted by using various sizes of containers. Decrease of the container radius makes the value of the condensation coefficient decrease at a constant temperature, due to the changes of the interface shape of the molten salts. However, this meniscus effect could not decrease the values of the condensation coefficient any more as the container radius is less than the critical radius for the fully developed interface ($R_0/L_c = 0.4$), such that the profile of the interface is fully developed (Figure 4).

The effect of the contact angle on determining the condensation coefficient of fully developed meniscus was presented by comparing the relative condensation coefficient of the molten slats. In the comparison results of Figure 9,

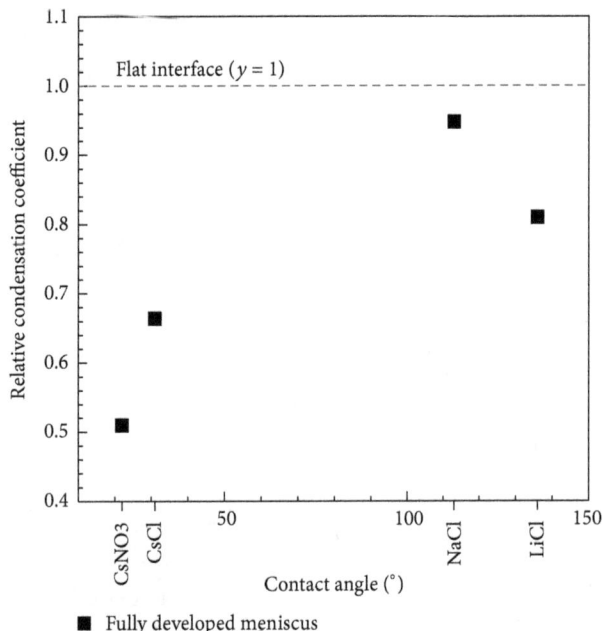

FIGURE 9: The relative condensation coefficient depending on the contact angle of fully developed meniscus of CsCl, CsNO$_3$, LiCl, and NaCl.

there is only 5% deviation in the condensation coefficient of NaCl relative to the value obtained from flat surface. However, for LiCl, CsCl, and CsNO$_3$, the deviation of the condensation coefficient increases by about 19%, 34%, and 49%, respectively, as the formed angle between the slope of the interface at the container wall and the horizontal plane increases.

The deviation of the condensation coefficient due to the morphological changes of the interface decreases as the container radius increases, or the contact angle is close to 90 degrees. This result suggests that assuming the cross-sectional area as an actual evaporation area would be invalid for the fully developed meniscus of the molten salt having higher or smaller contact angle at the container wall than 90 degrees. However, whether the condensation coefficient of all molten salts is affected by the interface shape, the range of their values is still from 10^{-3} to 10^{-5}. This result means that the presence of the foreign gas (air and Ar) would be a dominant parameter in determining the evaporation rate of alkali metal species. As the reported analysis on the mass transfer resistance of evaporation into atmosphere [20], the condensation coefficient of evaporation is the overall mass transfer resistance consisting of the atmospheric gas and the Knudsen layer, but the resistance values of the Knudsen layer are a hundred times lower than that of the atmospheric gas. It means that the contribution of the Knudsen layer to the values of the condensation coefficient is negligibly small even though the Knudsen layer is formed above the fully developed meniscus.

4. Summary and Conclusions

This paper has theoretically predicted the formed interface shape between the molten alkali metal salts and an atmospheric gas in the various diameters of small cylindrical containers and analyzed the meniscus effect on their evaporation. The deviation of the actual evaporation area from the cross-sectional area of small containers due to the meniscus decreases the predicted condensation coefficient. Results show that the meniscus of the molten salts is developing into a parabolic shape until the dimensionless radius is less than 0.4, increasing the area ratio of the meniscus with decreasing the container radius. However, after the meniscus is fully developed, decreasing the container radius negligibly affected the value of the area ratio of the meniscus to the cross-section of the container which was consistent but larger than unity.

The meniscus effect is important for accurately determining the condensation coefficient of molten salts. The values of the condensation coefficient of molten alkali metal salts were obtained from the experimental measurement and the reported analysis of the evaporation flux by Hur et al. [1], Kim et al. [22], and Baughman et al. [23], in conjunction with the analysis of the equilibrium vapor pressure. The meniscus effect on the evaporation of the molten salts could be quantified by comparing the condensation coefficients determined by using various diameters of cylindrical containers. The comparison results of the alkali metal slats show that the maximum decrement of the condensation coefficient due to the meniscus was CsNO$_3$ > CsCl > LiCl > NaCl, according to the angle formed at the intersection between the interface and the horizontal reference plane.

The meniscus effect of molten salts depending on the container radius and their contact angle at the container wall could significantly decrease the condensation coefficient by the changes in the actual evaporation area. However, considering the magnitude of the predicted condensation coefficient, it is suggested that the presence of a foreign gas would be a dominant resistance of mass transfer between a condensed phase and a gaseous phase in evaporation process of molten salts into a foreign gas.

Conflicts of Interest

The authors declare that they have no conflicts of interest.

Acknowledgments

This work was supported by the National Research Foundation of Korea (NRF) Grant funded by Korea government (MSIP) (no. NRF-2017M2A8A5015147).

References

[1] J.-M. Hur, S.-M. Jeong, and H. Lee, "Molten salt vaporization during electrolytic reduction," *Nuclear Engineering and Technology*, vol. 42, no. 1, pp. 73–78, 2010.

[2] Y. Sakamura, M. Kurata, and T. Inoue, "Electrochemical reduction of U O2 in molten Ca Cl2 or LiCl," *Journal of The Electrochemical Society*, vol. 153, no. 3, pp. D31–D39, 2006.

[3] G. J. Janz, *Molten Salts Handbook*, Academic Press, New York, NY, USA, 1967.

[4] C. W. Forsberg, P. F. Peterson, and P. S. Pickard, "Molten-salt-cooled advanced high-temperature reactor for production of

hydrogen and electricity," *Nuclear Technology*, vol. 144, no. 3, pp. 289–302, 2003.

[5] T. Abram and S. Ion, "Generation-IV nuclear power: a review of the state of the science," *Energy Policy*, vol. 36, no. 12, pp. 4323–4330, 2008.

[6] J. G. Speer, F. C. Rizzo Assunção, D. K. Matlock, and D. V. Edmonds, "The "quenching and partitioning" process: Background and recent progress," *Materials Research*, vol. 8, no. 4, pp. 417–423, 2005.

[7] H. Kim, D. A. Boysen, J. M. Newhouse et al., "Liquid metal batteries: Past, present, and future," *Chemical Reviews*, vol. 113, no. 3, pp. 2075–2099, 2013.

[8] U. Herrmann, B. Kelly, and H. Price, "Two-tank molten salt storage for parabolic trough solar power plants," *Energy*, vol. 29, no. 5-6, pp. 883–893, 2004.

[9] Z. Yang and S. V. Garimella, "Thermal analysis of solar thermal energy storage in a molten-salt thermocline," *Solar Energy*, vol. 84, no. 6, pp. 974–985, 2010.

[10] X.-T. Yan and Y. Xu, *Chemical Vapor Deposition: An integrated Engineering Design for Advanced Materials*, Springer, London, UK, 2010.

[11] E. H. Kennard, *Kinetic Theory of Gases with an Introduction to Statistical Mechanics*, McGraw-Hill, New York, NY, USA, 1938.

[12] M. Knudsen, "Die maximale Verdampfungsgeschwindigkeit des Quecksilbers," *Annalen der Physik*, vol. 352, no. 13, pp. 697–708, 1915.

[13] H. Hertz, "On the Evaporation of Liquids, Especially Mercury, in Vacuo," *Annals of Physics*, vol. 17, no. 177, 1882.

[14] A. H. Persad and C. A. Ward, "Expressions for the Evaporation and Condensation Coefficients in the Hertz-Knudsen Relation," *Chemical Reviews*, vol. 116, no. 14, pp. 7727–7767, 2016.

[15] P. Rahimi and C. A. Ward, "Kinetics of evaporation: Statistical rate theory approach," *Journal of Thermodynamics*, vol. 8, no. 1, pp. 1–14, 2005.

[16] C. T. Ewing and K. H. Stern, "Vaporization kinetics of solid and liquid silver, sodium chloride, potassium bromide, cesium iodide, and lithium fluoride," *The Journal of Physical Chemistry C*, vol. 79, no. 19, pp. 2007–2017, 1975.

[17] R. Hołyst, M. Litniewski, and D. Jakubczyk, "A molecular dynamics test of the Hertz-Knudsen equation for evaporating liquids," *Soft Matter*, vol. 11, no. 36, pp. 7201–7206, 2015.

[18] C. T. Ewing and K. H. Stern, "Equilibrium vaporization rates and vapor pressures of solid and liquid sodium chloride, potassium chloride, potassium bromide, cesium iodide, and lithium fluoride," *The Journal of Physical Chemistry C*, vol. 78, no. 20, pp. 1998–2005, 1974.

[19] J. Safarian and T. A. Engh, "Vacuum evaporation of pure metals," *Metallurgical and Materials Transactions A: Physical Metallurgy and Materials Science*, vol. 44, no. 2, pp. 747–753, 2013.

[20] H.-C. Yang, I.-H. Yang, M.-W. Lee, D.-Y. Chung, and J.-W. Choi, "Gas-surface interfacial and gas-phase resistances to vaporizing cesium hydroxide and chloride in air," *Vacuum*, vol. 123, pp. 86–90, 2016.

[21] K. Chatterjee, D. Dollimore, and K. S. Alexander, "Calculation of vapor pressure curves for hydroxy benzoic acid derivatives using thermogravimetry," *Thermochimica Acta*, vol. 392, no. 393, pp. 107–117, 2002.

[22] I. S. Kim, D. Y. Chung, M. S. Park, J. M. Hur, and J. K. Moon, "Evaporation of CsCl, BaCl2, and SrCl2 from the LiCl-Li2O molten salt of the electrolytic reduction process," *Journal of Radioanalytical and Nuclear Chemistry*, vol. 303, pp. 223–227, 2015.

[23] R. J. Baughman, R. A. Lefever, and W. R. Wilcox, "Evaporation of sodium chloride melts," *Journal of Crystal Growth*, vol. 8, no. 4, pp. 317–323, 1971.

[24] R. Marek and J. Straub, "Analysis of the evaporation coefficient and the condensation coefficient of water," *International Journal of Heat and Mass Transfer*, vol. 44, no. 1, pp. 39–53, 2001.

[25] T. Ishiyama, T. Yano, and S. Fujikawa, "Molecular dynamics study of kinetic boundary condition at an interface between argon vapor and its condensed phase," *Physics of Fluids*, vol. 16, no. 8, pp. 2899–2906, 2004.

[26] G. Nagayama and T. Tsuruta, "A general expression for the condensation coefficient based on transition state theory and molecular dynamics simulation," *The Journal of Chemical Physics*, vol. 118, no. 3, pp. 1392–1399, 2003.

[27] R. Meland and T. Ytrehus, "Evaporation and condensation Knudsen layers for nonunity condensation coefficient," *Physics of Fluids*, vol. 15, no. 5, pp. 1348–1350, 2003.

[28] A. V. Gusarov and I. Smurov, "Gas-dynamic boundary conditions of evaporation and condensation: numerical analysis of the Knudsen layer," *Physics of Fluids*, vol. 14, no. 12, pp. 4242–4255, 2002.

[29] G. Fang and C. A. Ward, "Temperature measured close to the interface of an evaporating liquid," *Physical Review E: Statistical, Nonlinear, and Soft Matter Physics*, vol. 59, no. 1, pp. 417–428, 1999.

[30] G. Fang and C. A. Ward, "Examination of the statistical rate theory expression for liquid evaporation rates," *Physical Review E: Statistical, Nonlinear, and Soft Matter Physics*, vol. 59, no. 1, pp. 441–453, 1999.

[31] E. Chibowski and R. Perea-Carpio, "Problems of contact angle and solid surface free energy determination," *Advances in Colloid and Interface Science*, vol. 98, no. 2, pp. 245–264, 2002.

[32] D. Y. Kwok and A. W. Neumann, "Contact angle interpretation in terms of solid surface tension," *Colloids and Surfaces A: Physicochemical and Engineering Aspects*, vol. 161, no. 1, pp. 31–48, 2000.

[33] *Proceedings of the International Symposium on Computer Software in Chemical and Extractive Metallurgy*, Elsevier, 1989.

[34] H. T. Davis, *Statistical Mechanics of Phases, Interfaces, and Thin Films*, VCH, New York, NY, USA, 1995.

[35] P.-G. Gennes, F. Brochard-Wyart, and D. Quere, *Capillarity and Wetting Phenomena: Drops, Bubbles, Pearls, Waves*, Springer, New York, NY, USA, 2002.

[36] Wolfram research Inc. Wolfram Mathmatica 9.0. 2013.

[37] P. Baumli and G. Kaptay, "Wettability of carbon surfaces by pure molten alkali chlorides and their penetration into a porous graphite substrate," *Materials Science and Engineering: A Structural Materials: Properties, Microstructure and Processing*, vol. 495, no. 1-2, pp. 192–196, 2008.

Study on Shale Adsorption Equation Based on Monolayer Adsorption, Multilayer Adsorption, and Capillary Condensation

Qing Chen,[1,2] Yuanyuan Tian,[1,2] Peng Li,[2] Changhui Yan,[1,2] Yu Pang,[3] Li Zheng,[2] Hucheng Deng,[2] Wen Zhou,[1] and Xianghao Meng[1]

[1]*State Key Laboratory of Oil and Gas Reservoir Geology and Exploration, Chengdu University of Technology, Chengdu 610059, China*
[2]*College of Energy Resource, Chengdu University of Technology, Chengdu 610059, China*
[3]*Petroleum Engineering, Texas Tech University, Lubbock, TX, USA*

Correspondence should be addressed to Yuanyuan Tian; yuanyuan-tian@outlook.com

Academic Editor: Davide Vione

Shale gas is an effective gas resource all over the world. The evaluation of pore structure plays a critical role in exploring shale gas efficiently. Nitrogen adsorption experiment is one of the significant approaches to analyze pore size structure of shale. Shale is extremely heterogeneous due to component diversity and structure complexity. Therefore, adsorption isotherms for homogeneous adsorbents and empirical isotherms may not apply to shale. The shape of adsorption-desorption curve indicates that nitrogen adsorption on shale includes monolayer adsorption, multilayer adsorption, and capillary condensation. Usually, Langmuir isotherm is a monolayer adsorption model for ideal interfaces; BET (Brunauer, Emmett, Teller) adsorption isotherm is a multilayer adsorption model based on specific assumptions; Freundlich isotherm is an empirical equation widely applied in liquid phase adsorption. In this study, a new nitrogen adsorption isotherm is applied to simultaneously depict monolayer adsorption, multilayer adsorption, and capillary condensation, which provides more real and accurate representation of nitrogen adsorption on shale. In addition, parameters are discussed in relation to heat of adsorption which is relevant to the shape of the adsorption isotherm curve. The curve fitting results indicate that our new nitrogen adsorption isotherm can appropriately describe the whole process of nitrogen adsorption on shale.

1. Introduction

Shale gas has attracted much attention in United States, China, Canada, and so forth, because of the gas storage mechanism and recovery potential of shale gas reservoirs [1, 2]. To investigate gas adsorption capacity and pore size distribution of shale rocks, high-pressure methane adsorption and low-pressure nitrogen or carbon dioxide adsorption experiments are conducted, respectively. Many researches have been done to find and modify adsorption equations suitable for describing methane adsorption. Considering methane adsorption as monolayer adsorption, Langmuir equation, L-F (Langmuir-Freundlich) equation, and M-L (modified Langmuir) equation are successfully applied to evaluate methane adsorption [3–5]. Furthermore, D-R (Dubinin-Radushkevich) equation, D-A (Dubinin-Astakhov) equation, and S-D-R (supercritical Dubinin-Radushkevich) equation are also used with consideration of methane adsorption as micropore filling [6–8]. For carbon dioxide adsorption, to take into account the monolayer adsorption property, both Langmuir equation and L-F equation are applied to depict variations of the adsorption capacity with pressure [9–11]. On the contrary, it is hard to find an equation to depict low-pressure nitrogen adsorption because of the complicated adsorption mechanism. On the basis of BDDT (Brunauer-Deming-Deming-Teller) adsorption isotherm classification, nitrogen adsorption belongs to type IV, which indicates that it includes three processes: monolayer adsorption, multilayer adsorption, and capillary condensation. Unfortunately, the majority of adsorption equations are developed based on only one kind of adsorption mechanism, and they can be categorized into three aspects: monolayer adsorption, multilayer adsorption, and micropore filling.

In terms of monolayer adsorption, a widely accepted one is Langmuir adsorption equation which assumed only one type of adsorption sites on the surface of adsorbent [12, 13].

When extending the Langmuir equation for gas-liquid-phase adsorption studies, two types of sites are considered and the relationship between equilibrium concentration and amount of adsorbate is obtained [14–19]. Because the Langmuir equation describes adsorption on homogeneous surface, Gaussian energy distribution is used to adjust monolayer adsorption theory to heterogeneous surface [20–22]. To study multicomponent, monolayer adsorption of multicomponent gas, the assumption that the saturated amount of adsorption for each component is equal based on Langmuir equation was derived [23, 24].

In the aspect of multilayer adsorption, BET (Brunauer, Emmett, and Teller) equation is the most popular one, and it proposes a multilayer adsorption model which assumes that the interaction on adsorbent surface is much larger than that between neighboring adsorbate molecules [25–27]. The theory is appropriate for adsorption on solid surfaces with homogeneous chemical properties, which is frequently applied to calculate specific surface area for porous media. To extend BET equation to multicomponent adsorption, three kinds of n-component BET equations were proposed considering that adsorbed layers have evaporation-condensation characters for liquid mixture, supposing that the adsorbed layer of gas mixture is an ideal solution according to statistic thermodynamics and assuming gas mixture is immiscible liquid [28–31].

Micropore filling is also a common adsorption mechanism, which is introduced on the basis of Polanyi adsorption potential theory [32]. According to thermodynamics, adsorption potential (ε) is transferring unit mass of adsorbate from gas phase to adsorbent surface. On account of thermodynamics, D-R and D-A equations were generated [33–36]. Changing micropore filling to surface coverage and keeping feature of Gaussian distribution of energy, D-R-K (Dubinin-Radushkevich-Kaganer) equation was built [37, 38]. For micropore filling on nonregular porous media, D-R equation was modified by fractal dimension function [39]. Furthermore, for supercritical fluid adsorption, S-D-R equation was built [7].

In fact, most adsorbents are heterogeneous porous media. Combination of adsorption equations is a solution to build equation for heterogeneous adsorbent. In studies of methane adsorption, empirical Freundlich equation was combined with Langmuir equation to obtain the L-F equation which is widely used in depicting CBM (coal bed methane) adsorption successfully [40, 41]. Moreover, heterogeneity of adsorbent surface has been taken into account, and an adsorption equation was built to express the relationship between equilibrium concentration and mass of adsorbate by combining Freundlich adsorption isotherm with Langmuir adsorption isotherm [42, 43].

It is clear that these adsorption equations only focus on one adsorption mechanism and cannot be applied to interpret monolayer adsorption, multilayer adsorption, and capillary condensation simultaneously. In addition, it is well known that shale consists of clay minerals (kaolinite, illite, chlorite, etc.), detrital minerals (quartz, feldspar, etc.), and some characteristic minerals (such as pyrite) [44–48], each with its specific adsorption property. On the other hand,

pore size distribution in shale is irregular [49], which results in an uneven distribution of adsorption potential. Compared with homogeneous materials used in other interfacial phenomenon studies, shale is an extremely heterogeneous adsorbent. However, most of current adsorption models assume that adsorbent is homogenous. Hence, our research is aiming at building a new adsorption equation for shale which enables us to depict complex adsorption including monolayer adsorption, multilayer adsorption, and capillary condensation.

2. Experiment

Adsorption and desorption data of shale measured by nitrogen at low temperature (77 K) is a fundamental method to analyze pore structure of shale. Samples were collected from Yanchang formation (Triassic, Ordos), Pingliang formation (Ordovician, Ordos), Wulalik formation (Ordovician, Ordos), Xujiahe formation (Triassic, Sichuan), Niutitang formation (Cambrian, Sichuan), and Doushantuo formation (Ediacaran, Sichuan). Properties of samples are described in Table 1. All samples were ground to pass a sieve size of 60 mesh (250 μm). For outgassing, the pulverized samples were dried and vacuumized at 80°C for 12 hours.

The apparatus used for nitrogen adsorption experiment is Quadrasorb SI surface area and pore size analyzer (manufactured by Quantachrome in USA) which is provided by State Key Laboratory of Oil & Gas Reservoir Geology and Exploitation (China). There are four stations of the experimental instrument. The lower limitation of specific surface area is 0.01 m^2/g for nitrogen. In the aspect of pore size distribution analysis, the minimum pore volume is 0.0001 cc/g (STP), and the pore size range is 0.35~400 nm. In our experiment, nitrogen is used as adsorbate. Measurement is conducted at temperature 77 K, and the minimum P/P_0 is 0.001.

All the experimental data are prepared to analyze nitrogen adsorption processes and determine the value of parameters in shale adsorption isotherm.

3. Adsorption Characteristics of Nitrogen Adsorbed on Shale

3.1. Adsorption Processes. According to BDDT (Brunauer-Deming-Deming-Teller) adsorption isotherm classification [50], nitrogen adsorption isotherms of shale belong to type IV (Figure 1), which indicates that adsorption on shale can be divided into three stages: monolayer adsorption, multilayer adsorption, and capillary condensation [51–54]. The three stages can be specifically expressed as follows: most adsorption isotherms of shale have an inflection point at low relative pressure, which refers to saturated adsorbed content in monolayer adsorption regime. Before this point, only monolayer adsorption takes place. As relative pressure increases, the thickness of adsorbed layers gradually increases and multilayer adsorption occurs. When relative pressure reaches initial capillary condensation pressure (usually around 0.4 P/P_0), adsorption-desorption curve forms hysteresis loop,

TABLE 1: Curve fitting parameters of shale adsorption isotherm for 80 samples.

Number	A	B	K	M	N	R-Square
D1	0.0537	1.9980	2132.0000	1.4890	0.0993	0.9988
D2	6.9810	0.0150	203.5000	0.9568	203.3000	0.9129
D3	0.0780	2.3980	0.0000	0.9168	3.8730	0.7652
D4	2.0550	1.2430	3.7850	0.5093	5274.0000	0.9991
D5	2.3920	0.2903	783.3000	0.7181	538.0000	0.8952
D6	5.2550	1.5350	6.2590	0.1476	191.1000	0.9983
D7	1.6750	1.1630	3.2210	0.2803	144.5000	0.9999
D8	1.9080	1.5610	4.1690	0.3121	1422.0000	0.9994
D9	2.2440	1.3200	6.7910	0.2632	232.6000	0.9993
D10	1.6210	1.2640	5.0760	0.4035	2745.0000	0.9980
D11	0.9914	1.1540	5.6780	0.4159	6902.0000	0.9993
D12	5.4060	1.7330	191.1000	0.4047	201.7000	0.9003
D13	5.3580	0.3832	208.0000	0.4707	202.8000	0.8984
D14	4.2840	1.4770	1896.0000	0.2662	5.7960	0.9995
D15	8.5540	1.3940	59.1100	0.5487	7595.0000	0.9079
D16	3.0770	1.5530	2.5480	0.2783	817.0000	0.9995
D17	7.4640	0.1425	650.6000	0.6635	333.4000	0.9124
D18	3.4710	1.5220	6.1130	0.2741	427.4000	0.9995
D19	3.3230	0.0003	427.4000	0.5498	423.1000	0.9088
D20	3.9300	0.0836	561.5000	0.4809	334.3000	0.9303
D21	3.5090	1.4470	5.5610	0.2747	723.2000	0.9991
D22	1.3000	1.4950	292.4000	0.4657	4.5470	0.9998
D23	4.2450	1.4390	3.9790	0.2875	408.9000	0.9985
D24	2.8740	1.0500	53.0000	0.4740	53.6000	0.9213
D25	2.9540	0.4405	701.8000	0.4577	275.8000	0.9257
D26	3.5920	0.2153	209.1000	0.3085	208.2000	0.9436
D27	5.3050	1.5090	5.4720	0.2243	268.2000	0.9981
D28	5.1140	1.4920	4.6400	0.2279	346.7000	0.9986
D29	9.0830	1.5220	242.3000	0.1936	2.4390	0.9994
D30	7.0750	1.4830	6.5960	0.2288	2220.0000	0.9957
D31	2.7600	1.3680	6.0210	0.2718	156.4000	0.9990
D32	6.3140	1.4600	5.4280	0.2655	6569.0000	0.9966
D33	6.6460	1.5110	4.5510	0.2274	7304.0000	0.9943
D34	4.9060	1.9610	162.9000	0.5029	206.6000	0.9260
D35	6.0750	1.5820	4.5020	0.2520	642.7000	0.9998
D36	10.4800	1.5260	5.6510	0.2440	216.6000	0.9985
D37	5.4940	1.4900	6.4730	0.2477	194.1000	0.9984
D38	5.4510	1.5800	4.8690	0.2569	1299.0000	0.9991
D39	9.2010	1.5080	5.0390	0.2308	174.4000	0.9987
D40	3.6820	1.3920	3.6970	0.3226	2108.0000	0.9997
D41	0.2399	1.3620	7.7860	0.5925	153.7000	0.9969
D42	4.5800	1.5770	4.8860	0.3121	1525.0000	0.9991
D43	4.6990	1.3130	3.8670	0.3180	1697.0000	0.9993
D44	0.8405	1.7690	404.3000	0.3685	3.9270	0.9997
D45	0.5450	1.3380	15.3900	1.2740	2570.0000	0.9852
D46	1.6880	0.6039	176.0000	1.1490	0.0014	0.9834
D47	1.9000	0.5685	206.7000	1.3950	203.6000	0.9780
D48	3.2030	1.5100	3.9550	0.5210	1347.0000	0.9994
D49	0.1630	1.1610	3.9160	0.5117	124.4000	0.9996
D50	0.3986	1.2930	42.2300	1.7860	63.0900	0.9736
D51	3.0890	1.4760	5.7340	0.3389	897.6000	0.9988
D52	2.4900	1.3240	3.8420	0.4336	872.2000	0.9989

TABLE 1: Continued.

Number	A	B	K	M	N	R-Square
D53	3.1810	0.1210	556.3000	1.0080	276.6000	0.9509
D54	2.0150	1.4120	4973.0000	0.5012	4.8210	0.9991
D55	5.4370	1.4850	5.6280	0.2821	1196.0000	0.9994
D56	4.9370	1.5060	5.7860	0.3235	633.8000	0.9993
D57	1.1190	1.5900	2357.0000	0.8201	6.8430	0.9994
D58	2.5580	1.4510	3.8680	0.3408	690.5000	0.9988
D59	2.5710	1.3870	4.9770	0.4927	357.1000	0.9995
D60	0.7093	1.3030	3.2820	0.6943	282.8000	0.9996
D61	0.5753	1.5290	700.2000	0.4714	3.8960	0.9999
D62	0.2555	0.2743	32.7500	3.2820	0.1412	0.9609
D63	4.1530	1.4710	4.0830	0.3707	229.8000	0.9988
D64	0.3084	1.2180	3.3540	0.5845	172.4000	0.9989
D65	0.1870	0.5533	41.5700	1.5160	0.0292	0.9894
D66	0.1506	1.2430	1.9260	0.6289	102.3000	0.9997
D67	0.1322	1.1130	3.5120	0.4219	101.8000	0.9998
D68	1.8820	1.3280	4.1040	0.3971	193.7000	0.9989
D69	1.1050	1.2280	3.9960	0.4548	146.3000	0.9995
D70	3.5390	1.3730	297.5000	1.1020	389.5000	0.9095
D71	0.2066	1.8990	597.9000	0.5551	4.6210	0.9999
D72	1.7930	1.3680	5.0270	0.5813	291.4000	0.9997
D73	3.0830	1.4350	5.1210	0.5572	435.3000	0.9996
D74	1.5910	1.3670	7.1400	0.5368	11840.0000	0.9981
D75	2.7250	2.5950	159.2000	0.9335	174.9000	0.9170
D76	0.7462	0.1465	84.1800	2.1390	0.0039	0.9910
D77	3.8900	0.1753	500.5000	0.7077	264.5000	0.9242
D78	0.1222	1.9200	1187.0000	0.4956	7.9260	0.9999
D79	0.1362	1.8620	126.3000	0.4758	3.7370	0.9998
D80	0.1082	1.8330	96.1900	0.4922	2.8620	0.9996

which demonstrates that capillary condensation exists in the process of nitrogen adsorbed on shale.

3.2. Adsorption Equations and Adsorption Processes. As mentioned above, shale adsorption includes processes of monolayer adsorption, multilayer adsorption, and capillary condensation. Therefore, the generated new shale adsorption isotherm would be capable of depicting all features of these processes.

In terms of monolayer adsorption, Langmuir equation will be an appropriate choice. Langmuir built an adsorption model with the following assumptions: (1) surface of adsorbent has one type of adsorption sites and one site can accommodate only one adsorbate molecule or atom; (2) the surface is homogeneous and there is no lateral interaction between adsorbate molecules; (3) adsorption reaches dynamic equilibrium [55]. Based on these assumptions, the adsorption isotherm can be given as

$$V = \frac{V_m bP}{1 + bP}. \tag{1}$$

All terms used in the equations are defined in the nomenclature section.

For multilayer adsorption, BET adsorption isotherm is a representation of multilayer adsorption model generated by Brunauer, Emmett, and Teller, which assumes the interaction between adsorbate and adsorbent surface is much larger than that between neighboring molecules. The theory is appropriate for adsorption on surface of solid with homogeneous chemical properties, which is frequently applied to calculate specific surface area for porous media. The BET equation can be expressed as [25]

$$V = \frac{V_m cP}{(P_0 - P)\left[1 + (c - 1)(P/P_0)\right]}. \tag{2}$$

Capillary condensation is a process where gas phase transforms into liquid phase. Thus, an adsorption equation applicable to describe liquid adsorption is suitable for this adsorption stage. Among investigated adsorption equations, Freundlich adsorption isotherm is an empirical equation describing equilibrium concentration of solute in solution with respect to concentration of solute adsorbed on the surface of solvent. The adsorption equation is [56]

$$m = kP^{1/n}. \tag{3}$$

Figure 2 points out that if we only apply Langmuir isotherm to shale adsorption in low relative pressure section

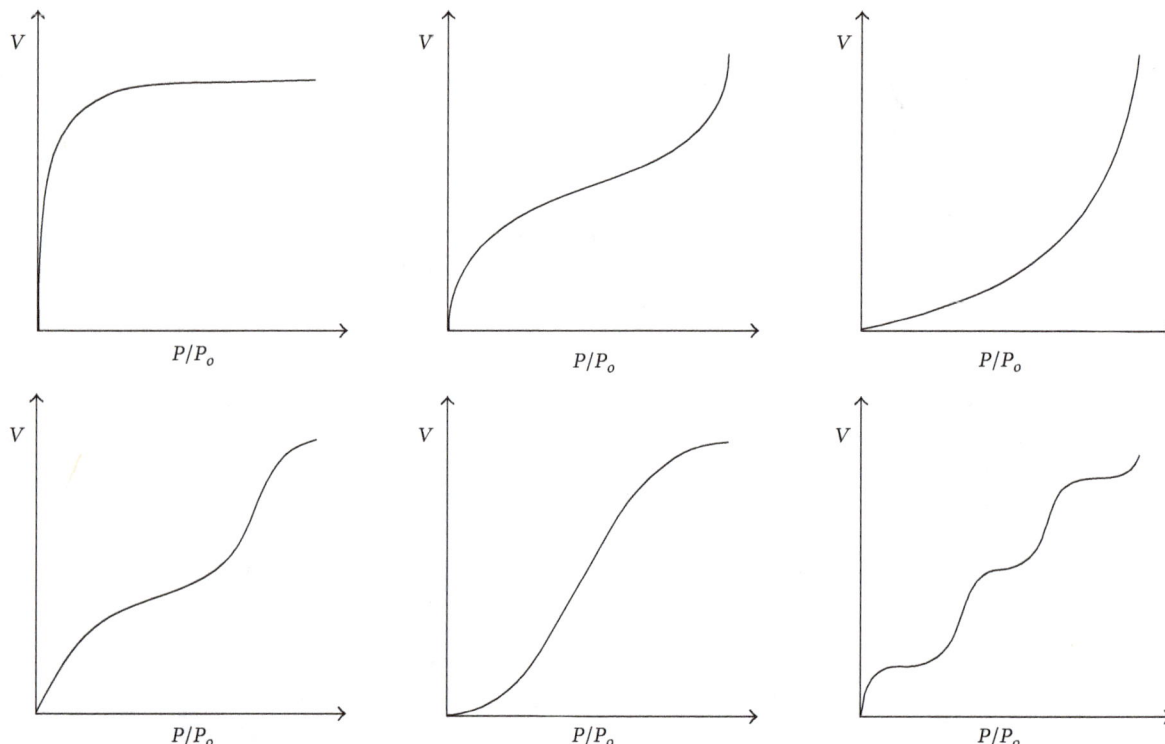

FIGURE 1: Classification of BDDT adsorption isotherms (from Brunauer, Sing et al., 1940).

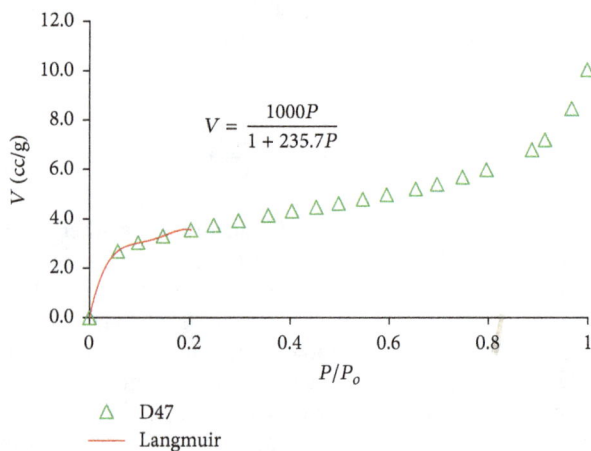

$$V = \frac{1000P}{1 + 235.7P}$$

D47
—— Langmuir

FIGURE 2: Application of Langmuir isotherm to nitrogen adsorption isotherm for shale at low relative pressure section.

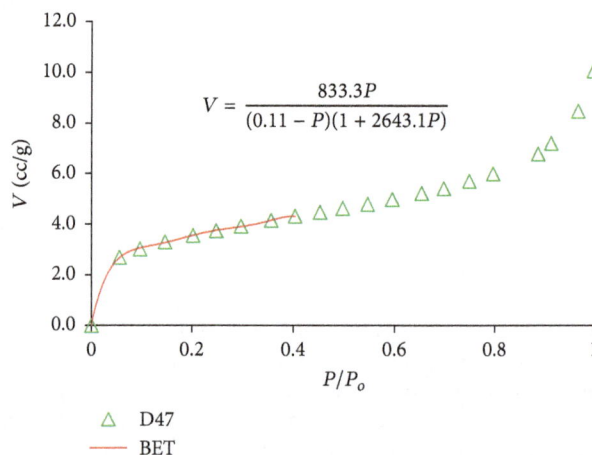

$$V = \frac{833.3P}{(0.11 - P)(1 + 2643.1P)}$$

D47
—— BET

FIGURE 3: Application of BET isotherm to nitrogen adsorption isotherm for shale before capillary condensation.

(before the inflection point where monolayer adsorption switches to multilayer adsorption), Langmuir isotherm can properly match experimental data, which indicates that Langmuir isotherm is suitable for monolayer adsorption in shale and then justifies the analytical result that monolayer adsorption takes place in the process of nitrogen adsorption at low temperature for shale.

As a normal method to acquire surface area of shale, multipoint BET method testifies that BET equation can be applied to describe adsorption on shale at certain conditions (usually relative pressure below 0.4 P/P_0). From curve

fitting result (Figure 3), BET is appropriate for low and medium relative pressure sections, which illustrates that BET adsorption isotherm can depict experimental data before the presence of capillary condensation. This also reveals that multilayer adsorption exists in the process of nitrogen adsorption isotherm for shale.

As shown in Figure 4, Freundlich isotherm fits the medium-high relative pressure section of nitrogen adsorption on shale, especially relative pressure section after occurrence of capillary condensation.

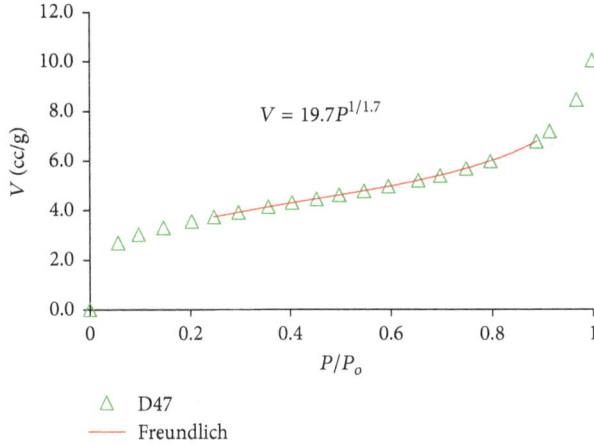

FIGURE 4: Application of Freundlich isotherm to liquefied nitrogen adsorption isotherm for shale for medium-high relative pressure section.

On behalf of potentials of the three equations representing adsorption in different relative pressure sections, the new adsorption equation for shale needs to contain features of Langmuir isotherm, BET isotherm, and Freundlich isotherm.

4. Result

BET and Freundlich adsorption isotherms can be changed to functions which consider relative pressure as an independent variable. Thereafter, Langmuir adsorption isotherm is a case of BET adsorption isotherm, in which the pressure is much lower than saturated vapor pressure.

Rearranging BET adsorption isotherm equation one gets

$$V = \frac{V_m c P_r}{-(c-1) P_r^2 + (c-2) P_r + 1}. \tag{4}$$

Substituting P_r into (3) gives

$$m = k P_r^{1/n} P_0^{1/n}. \tag{5}$$

Freundlich adsorption isotherm describes the relationship between pressure and mass of adsorbate adsorbed on surface of adsorbent per unit of mass. In order to express adsorption capacity in same dimension, (5) is converted to

$$V = \frac{k P_r^{1/n} P_0^{1/n}}{\rho g}. \tag{6}$$

Under experimental conditions, saturated vapor pressure and density of adsorbate are constants. Therefore, setting $k P_0^{1/n}/\rho g = k'$, then (6) becomes

$$V = k' P_r^{1/n}. \tag{7}$$

From (1) and (2), the coefficient and exponent of pressure in Langmuir and BET adsorption isotherm (V_m, b, c) correspond to physical and chemical parameters of monolayer and multilayer adsorption. We apply the function with a

form of BET adsorption isotherm and combine Freundlich adsorption isotherm which can describe characteristic of liquid adsorption to build up shale adsorption isotherm. The coefficient and exponent of relative pressure are variable fitting parameters in the new shale adsorption isotherm expressed as follows:

$$V = \frac{A P_r^M}{(1-B) P_r^N + (B-2) P_r^K + 1}. \tag{8}$$

A, B are undetermined coefficients; M, N, K are undetermined exponent.

5. Discussion

5.1. Physical and Chemical Meaning of Variables in Shale Adsorption Isotherm Equation. From above discussion, coefficient and exponent in the new shale adsorption isotherm equation are related to physical and chemical meanings of coefficients and exponents of Langmuir, BET, and Freundlich adsorption isotherms.

Variable A in (8) can be given as

$$A = V_m c P_0^M. \tag{9}$$

Thus, A is related to maximum amount of monolayer adsorption (V_m), adsorption heat according to formula of c which will be detailed below, saturated vapor pressure at experimental temperature (P_0), and exponent M.

Variable B in (8) is

$$B = c. \tag{10}$$

Since the physical and chemical meaning of variable B in adsorption isotherm equals coefficient c in BET isotherm [57], c is expressed as follows based on BET theoretical derivation [58]:

$$c = \frac{a_1 g e^{((E_1 - E_L)/RT)}}{b_1}. \tag{11}$$

B is related to heat of adsorption (E_1, E_L) and experimental temperature (T).

Thermodynamically, the expression of exponent n in Freundlich adsorption isotherm is [59, 60]

$$n = -\frac{\Delta H_m}{RT}. \tag{12}$$

Compared with exponents of relative pressure in (8), coefficients M, N, and K are relevant to experimental temperature, which indicates that the enthalpy ΔH_m represents the strength of adsorption effect.

After clarifying the physical and chemical meaning of variables in (8), the range of these variables should be determined.

In (9), V_m, c, P_0^M are all positives, and then A should be positive ($A > 0$).

In BET theory, it is assumed that the strength of interaction between adsorbate at first adsorbed layer and adsorbent is much bigger than the strength between adsorbates at subsequent layers. Thus, set heat of adsorption between subsequent adsorbates as E_L. Then, E_1 represents the heat of adsorption between adsorbate at first layer and adsorbent, and $E_1 > E_L$. Thus, c must be larger than 0, and then we can obtain $B > 0$.

In terms of adsorption isotherm system:

$$\Delta H_m = H_2 - H_1 = (U_2 + P_2 V_2) - (U_1 + P_1 V_1). \quad (13)$$

At given temperature, based on ideal gas law, $P_1 V_1 = P_2 V_2$ can be obtained. Thus, in the whole system, change in adsorption enthalpy is equal to change in internal energy; namely, $\Delta H_m = \Delta U$. The adsorption is a process of heat release, $\Delta U < 0$, so $\Delta H_m < 0$ and all variables M, N, K are positive.

For each shale sample, these parameters can be calculated by the new shale adsorption isotherm based on experimental data.

5.2. Specialization of Shale Adsorption Isotherm. When $M = N = 1$; $A = V_m b$; $B = b$, (8) can be simplified to (1) which is Langmuir adsorption isotherm.

When $M = K = 1$; $N = 2$, (8) can be simplified to

$$V = \frac{A P_r}{(1 - B) P_r^2 + (B - 2) P_r + 1}. \quad (14)$$

At the right hand side of equation, multiply both numerator and denominator by saturated vapor pressure P_0 as

$$V = \frac{A P}{(1 - B) P_r^2 P_0 + (B - 2) P_r P_0 + P_0}. \quad (15)$$

The term $(B - 2) P_r P_0$ can be written as $(B - 1) P_r P_0 - P_r P_0$. Equation (15) can be converted to

$$V = \frac{A P}{(P_0 - P)[1 + (B - 1)(P/P_0)]}. \quad (16)$$

Then, $A = V_m * c$; $B = c$, and (16) converts to BET adsorption isotherm as (2).

In (8), if $M \gg 1$; $K \gg 1$, $A = k' P_0^M$, $M = 1/n$, reduces to (7).

By multiplying density of adsorbate (ρ) on both sides of (7) one gets

$$m = k' P^{1/n} \rho. \quad (17)$$

Setting $k' \rho = k$, then (17) can be simplified to Freundlich adsorption isotherm as (3).

5.3. Application of New Adsorption Equation. According to the new shale adsorption isotherm equation and the range of variables, we applied Matlab to perform curve fitting of relative pressure versus amount of adsorption for 80 shale samples. The results of 6 samples are selected randomly and displayed in Figure 5, and the other fitting results are shown in Table 1. It appears that average value of R^2 is 0.9782, maximum value of R^2 is 0.9999, minimum value

of R^2 is 0.7652, and the percentage of shale samples for which the value of R^2 is larger than 0.9 is 96.25%. It indicates that the new generated shale adsorption isotherm can represent a complete process of adsorption including monolayer adsorption, multilayer adsorption, and capillary condensation processes compared with Langmuir, BET, and Freundlich adsorption isotherms individually. In particular, it demonstrates better performance on depicting nitrogen adsorption isotherm at low temperature.

5.4. Coefficient B and the Shape of Adsorption Curve. Taking into account the physical and chemical meaning of coefficient B in the new generated shale adsorption isotherm, it represents coefficient c in BET isotherm (11). According to the research by Kondou et al. [61], the value of c in BET isotherm is related to heat of adsorption. The value of c is bigger, and the heat of adsorption is larger, which indicates that strength of interaction for adsorption is larger and adsorption curve increases more rapidly in low-pressure section shown in Figure 6(a). Focusing on the value of B in shale adsorption isotherm and the shape of adsorption isotherm curve, we figure out that the curve becomes gradually convex as the value of B increases, as shown in Figure 6(b). This illustrates that the generated shale adsorption isotherm can express the difference of heat of adsorption released between different shale samples. Furthermore, the changes in the shape of the curve are relevant to changes in heat of adsorption.

6. Conclusion

(1) The new shale adsorption isotherm is built up based on Langmuir adsorption isotherm, BET isotherm, and Freundlich isotherm, which can offer description for shale adsorption isotherm including monolayer adsorption, multilayer adsorption, and capillary condensation processes. The new shale adsorption isotherm can be converted to Langmuir, BET, and Freundlich isotherms by giving certain values to variables.

(2) The variables in new shale adsorption isotherm are related to coefficients and exponents in Langmuir, BET, and Freundlich adsorption isotherms. The physical and chemical meanings of parameters are figured out and ranges for each parameter are determined, which is used to restrict value of variables in the adsorption isotherm when doing regression analysis to match data from shale samples adsorption experiment.

(3) Based on new shale adsorption isotherm and variable range, curve fitting of relative pressure versus amount of adsorption has been performed. The adsorption isotherms with ability to illustrate the process of monolayer adsorption, multilayer adsorption, and capillary condensation for 80 shale samples from Ordos Basin and Sichuan Basin are obtained. The results of curve fitting are highly accurate.

(4) Variable B in shale adsorption isotherm is related to shape of adsorption curve due to adsorption heat. Variables in shale adsorption isotherm are related to shape of adsorption curve and parameter of heat of adsorption. Adsorption isotherm curve becomes gradually convex as the value of B

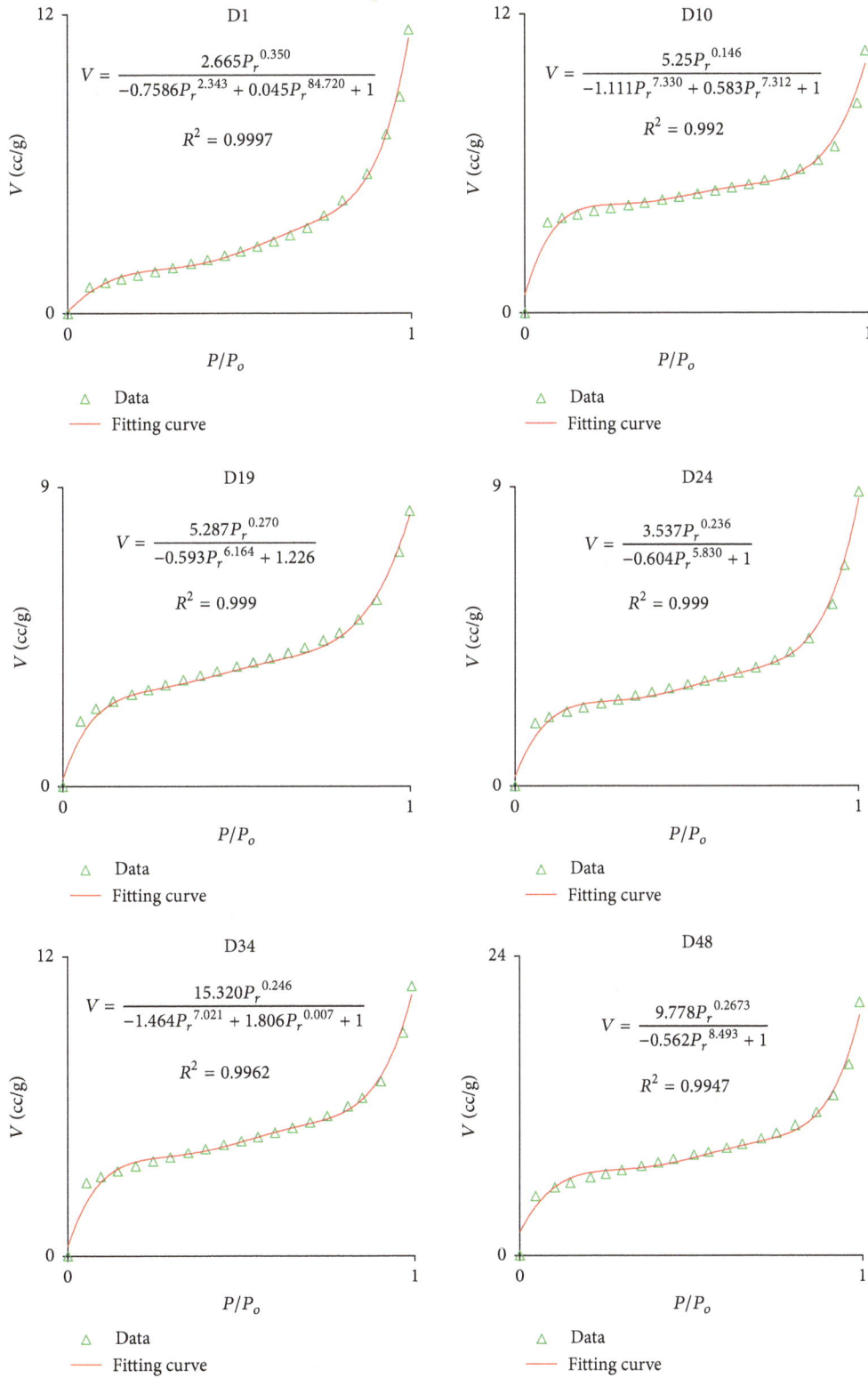

D1

$$V = \frac{2.665 P_r^{0.350}}{-0.7586 P_r^{2.343} + 0.045 P_r^{84.720} + 1}$$

$$R^2 = 0.9997$$

D10

$$V = \frac{5.25 P_r^{0.146}}{-1.111 P_r^{7.330} + 0.583 P_r^{7.312} + 1}$$

$$R^2 = 0.992$$

D19

$$V = \frac{5.287 P_r^{0.270}}{-0.593 P_r^{6.164} + 1.226}$$

$$R^2 = 0.999$$

D24

$$V = \frac{3.537 P_r^{0.236}}{-0.604 P_r^{5.830} + 1}$$

$$R^2 = 0.999$$

D34

$$V = \frac{15.320 P_r^{0.246}}{-1.464 P_r^{7.021} + 1.806 P_r^{0.007} + 1}$$

$$R^2 = 0.9962$$

D48

$$V = \frac{9.778 P_r^{0.2673}}{-0.562 P_r^{8.493} + 1}$$

$$R^2 = 0.9947$$

FIGURE 5: Curve fitting for adsorption isotherm.

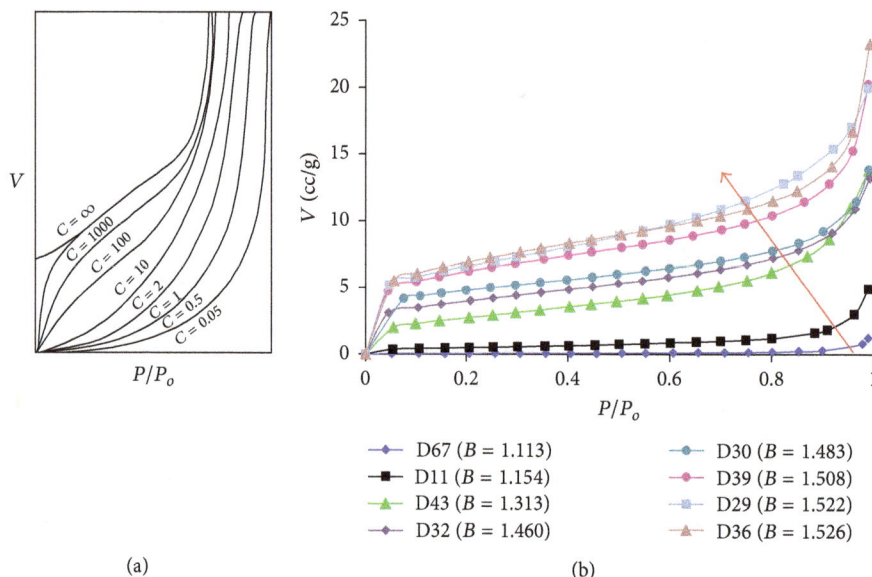

FIGURE 6: Change in shape of adsorption curve versus change in value of B.

increases. Computation according to physical and chemical meaning of coefficient B on heat of adsorption between adsorbate at first layer and adsorbent demonstrates that diversity exists among the heat of adsorption from different shale samples.

Nomenclature

V: Adsorption volume (cc/g)
P: Equilibrium pressure (MPa)
V_m: Saturated adsorption volume (cc/g)
b: Coefficient
P_o: Saturated vapor pressure at certain temperature (MPa)
c: Coefficient
m: Adsorption mass (g/g)
k: Coefficient
n: Coefficient (Mpa)
Pr: The ration of equilibrium pressure P and saturated vapor pressure P_0 (1)
ΔH_m: Adsorption enthalpy (J)
R: Molar gas constant (J/(mol·K))
T: Temperature (K)
ρ_g: Gas density (g/ml)
A: Variable
B: Variable
M: Variable
N: Variable
K: Variable
a_1:   Constant
b_1:   Constant
g: Constant
E_1:   Heat of adsorption between adsorbate at first layer and adsorbent (kJ/mol)
E_L: Heat of adsorption between adsorbate at n layer and adsorbate at $n + 1$ layer ($n \geq 1$) (kJ/mol).

Conflicts of Interest

The authors declare that they have no conflicts of interest.

Acknowledgments

The experimental data support provided by State Key Laboratory of Oil and Gas Reservoir Geology and Exploration is gratefully acknowledged. The authors also thank Dr. Christine Ehlig-Economides from University of Houston for her guidance and assistance.

References

[1] D. J. K. Ross and R. M. Bustin, "Characterizing the shale gas resource potential of Devonian-Mississippian strata in the Western Canada sedimentary basin: application of an integrated formation evaluation," *AAPG Bulletin*, vol. 92, no. 1, pp. 87–125, 2008.

[2] S. Chen, Y. Zhu, H. Wang, H. Liu, W. Wei, and J. Fang, "Shale gas reservoir characterisation: a typical case in the southern Sichuan Basin of China," *Energy*, vol. 36, no. 11, pp. 6609–6616, 2011.

[3] F. Yang, Z. Ning, R. Zhang, H. Zhao, and B. M. Krooss, "Investigations on the methane sorption capacity of marine shales from Sichuan Basin, China," *International Journal of Coal Geology*, vol. 146, pp. 104–117, 2015.

[4] S. R. Etminan, F. Javadpour, B. B. Maini, and Z. Chen, "Measurement of gas storage processes in shale and of the molecular diffusion coefficient in kerogen," *International Journal of Coal Geology*, vol. 123, pp. 10–19, 2014.

[5] M. A. Ahmadi and S. R. Shadizadeh, "Experimental investigation of a natural surfactant adsorption on shale-sandstone reservoir rocks: Static and dynamic conditions," *Fuel*, vol. 159, article no. 9338, pp. 15–26, 2015.

[6] F. Xiong, X. Wang, M. A. Amooie, M. R. Soltanian, Z. Jiang, and J. Moortgat, "The shale gas sorption capacity of transitional

shales in the Ordos Basin, NW China," *Fuel*, vol. 208, pp. 236–246, 2017.

[7] R. Sakurovs, S. Day, S. Weir, and G. Duffy, "Application of a modified Dubinin - Radushkevich equation to adsorption of gases by coals under supercritical conditions," *Energy & Fuels*, vol. 21, no. 2, pp. 992–997, 2007.

[8] T. F. T. Rexer, M. J. Benham, A. C. Aplin, and K. M. Thomas, "Methane adsorption on shale under simulated geological temperature and pressure conditions," *Energy & Fuels*, vol. 27, no. 6, pp. 3099–3109, 2013.

[9] R. Heller and M. Zoback, "Adsorption of methane and carbon dioxide on gas shale and pure mineral samples," *Journal of Unconventional Oil and Gas Resources*, vol. 8, pp. 14–24, 2014.

[10] Y. Dang, L. Zhao, X. Lu et al., "Molecular simulation of CO_2/CH_4 adsorption in brown coal: Effect of oxygen-, nitrogen-, and sulfur-containing functional groups," *Applied Surface Science*, vol. 423, pp. 33–42, 2017.

[11] K. Jessen, G.-Q. Tang, and A. R. Kovscek, "Laboratory and simulation investigation of enhanced coalbed methane recovery by gas injection," *Transport in Porous Media*, vol. 73, no. 2, pp. 141–159, 2008.

[12] I. Langmuir, "The constitution and fundamental properties of solids and liquids, Part I: solids," *Journal of the American Chemical Society*, vol. 38, no. 11, pp. 2221–2295, 1916.

[13] I. Langmuir, "The constitution and fundamental properties of solids and liquids. II. Liquids.1," *Journal of the American Chemical Society*, vol. 39, no. 9, pp. 1848–1906, 1917.

[14] E. C. Markham and A. F. Benton, "The adsorption of gas mixtures by silica," *Journal of the American Chemical Society*, vol. 53, no. 2, pp. 497–507, 1931.

[15] J. A. V. Butler and C. Ockrent, "Studies in electrocapillarity. III," *The Journal of Physical Chemistry*, vol. 34, no. 12, pp. 2841–2859, 1930.

[16] Á. Piñeiro, P. Brocos, A. Amigo, J. Gracia-Fadrique, and M. G. Lemus, "Extended Langmuir isotherm for binary liquid mixtures," *Langmuir*, vol. 17, no. 14, pp. 4261–4266, 2001.

[17] E. Besedovß, D. Bobok, and E. Besedová, "Single-component and binary adsorption equilibrium of 1,2 dichloroethane and 1,2 dichloropropane on activated carbon," *Petroleum & Coal*, vol. 47, no. 2, pp. 47–54, 2005.

[18] J. S. Jain and V. L. Snoeyink, "Adsorption from bisolute systems on active carbon," *Journal of the Water Pollution Control Federation*, vol. 45, no. 12, pp. 2463–2479, 1973.

[19] Y. S. Ho and G. McKay, "Correlative biosorption equilibria model for a binary batch system," *Chemical Engineering Science*, vol. 55, no. 4, pp. 817–825, 2000.

[20] J. Toth, "State equation of the solid-gas interface layers," *Acta Chimica Academiae Scientiarum Hungaricae*, vol. 69, pp. 311–328, 1971.

[21] A. P. Terzyk, J. Chatłas, P. A. Gauden, G. Rychlicki, and P. Kowalczyk, "Developing the solution analogue of the Toth adsorption isotherm equation," *Journal of Colloid and Interface Science*, vol. 266, no. 2, pp. 473–476, 2003.

[22] M. Jaroniec and J. Töth, "Adsorption of gas mixtures on heterogeneous solid surfaces: I. Extension of Tóth isotherm on adsorption from gas mixtures," *Colloid and Polymer Science*, vol. 254, no. 7, pp. 643–649, 1976.

[23] D. M. Ruthven, *Principles of adsorption and adsorption processes*, John Wiley & Sons, 1984.

[24] C. Yao, "Extended and improved Langmuir equation for correlating adsorption equilibrium data," *Separation and Purification Technology*, vol. 19, no. 3, pp. 237–242, 2000.

[25] S. Brunauer, P. H. Emmett, and E. Teller, "Adsorption of gases in multimolecular layers," *Journal of the American Chemical Society*, vol. 60, no. 2, pp. 309–319, 1938.

[26] G. Wang, K. Wang, and T. Ren, "Improved analytic methods for coal surface area and pore size distribution determination using 77 K nitrogen adsorption experiment," *International Journal of Mining Science and Technology*, vol. 24, no. 3, pp. 329–334, 2014.

[27] G. L. Aranovich and M. D. Donohue, "A New Approach to Analysis of Multilayer Adsorption," *Journal of Colloid and Interface Science*, vol. 173, no. 2, pp. 515–520, 1995.

[28] T. L. Hill, "Theory of multimolecular adsorption from a mixture of gases," *The Journal of Chemical Physics*, vol. 14, no. 4, pp. 268–275, 1946.

[29] B. W. Bussey, "Multicomponent gas adsorption of ideal mixtures," *Industrial & Engineering Chemistry Fundamentals*, vol. 5, no. 1, pp. 103–106, 1966.

[30] G. Tiren, "The comparison of different BET-type mixed adsorption equations with some experimental results," *Journal of Chemical Industry and Engineering (China)*, vol. 1, pp. 80–84, 1984.

[31] G. Tiren, "The extended BET adsorption theory in mixture solution," *CHemistry*, vol. 9, pp. 1–7, 1984.

[32] M. Polanyi, "The potential theory of adsorption," *Science*, vol. 141, no. 3585, pp. 1010–1013, 1963.

[33] M. M. Dubinin and L. Radushkevich, "Equation of the characteristic curve of activated charcoal," *Proceedings of the Academy of Sciences USSR*, vol. 55, no. 2, pp. 331–333, 1947.

[34] G. L. Aranovich and M. D. Donohue, "Adsorption isotherms for microporous adsorbents," *Carbon*, vol. 33, no. 10, pp. 1369–1375, 1995.

[35] N. D. Hutson and R. T. Yang, "Theoretical basis for the Dubinin-Radushkevitch (D-R) adsorption isotherm equation," *Adsorption*, vol. 3, no. 3, pp. 189–195, 1997.

[36] M. M. Dubinin and V. A. Astakhov, "Description of Adsorption Equilibria of Vapors on Zeolites over Wide Ranges of Temperature and Pressure," in *Molecular Sieve Zeolites-II*, vol. 102 of *Advances in Chemistry*, pp. 69–85, ACS Publications, 1971.

[37] W. F. L. Brown, "The joule-thermal effect and the equation of state of gases with non-polar molecules," *Russian Journal of Physical Chemistry*, vol. 30, pp. 691–704, 1959.

[38] J. Cortés and P. Araya, "The Dubinin-Radushkevich-Kaganer equation," *Journal of the Chemical Society, Faraday Transactions 1: Physical Chemistry in Condensed Phases*, vol. 82, no. 8, pp. 2473–2479, 1986.

[39] D. Avnir and M. Jaroniec, "An isotherm equation for adsorption on fractal surfaces of heterogeneous porous materials," *Langmuir*, vol. 5, no. 6, pp. 1431–1433, 1989.

[40] R. Sips, "On the structure of a catalyst surface," *The Journal of Chemical Physics*, vol. 16, no. 5, pp. 490–495, 1948.

[41] C. R. Clarkson, R. M. Bustin, and J. H. Levy, "Application of the mono/multilayer and adsorption potential theories to coal methane adsorption isotherms at elevated temperature and pressure," *Carbon*, vol. 35, no. 12, pp. 1689–1705, 1997.

[42] O. Redlich and D. L. Peterson, "A useful adsorption isotherm," *The Journal of Physical Chemistry C*, vol. 63, no. 6, p. 1024, 1959.

[43] K. Y. Foo and B. H. Hameed, "Insights into the modeling of adsorption isotherm systems," *Chemical Engineering Journal*, vol. 156, no. 1, pp. 2–10, 2010.

[44] K. A. Bowker, "Barnett Shale gas production, fort worth basin: issues and discussion," *AAPG Bulletin*, vol. 91, no. 4, pp. 523–533, 2007.

[45] L. Hong et al., "Experimental study of mineral composition and brittle characteristics in longmaxi formation of lower silurian, southeast chongqing," *Science Technology and Engineering*, vol. 13, pp. 8567–8571, 2013.

[46] R. M. Pollastro, D. M. Jarvie, R. J. Hill, and C. W. Adams, "Geologic framework of the Mississippian Barnett Shale, Barnett-Paleozoic total petroleum system, Bend arch-Fort Worth Basin, Texas," *AAPG Bulletin*, vol. 91, no. 4, pp. 405–436, 2007.

[47] D. M. Jarvie, R. J. Hill, T. E. Ruble, and R. M. Pollastro, "Unconventional shale-gas systems: the Mississippian Barnett Shale of north-central Texas as one model for thermogenic shale-gas assessment," *AAPG Bulletin*, vol. 91, no. 4, pp. 475–499, 2007.

[48] S. L. Montgomery, D. M. Jarvie, K. A. Bowker, and R. M. Pollastro, "Mississippian Barnett Shale, Fort Worth basin, north-central Texas: gas-shale play with multi-trillion cubic foot potential," *AAPG Bulletin*, vol. 89, no. 2, pp. 155–175, 2005.

[49] T. Guo, "Evaluation of highly thermally mature shale-gas reservoirs in complex structural parts of the Sichuan Basin," *Journal of Earth Science*, vol. 24, no. 6, pp. 863–873, 2013.

[50] S. Brunauer, L. S. Deming, W. E. Deming, and E. Teller, "On a theory of the van der Waals adsorption of gases," *Journal of the American Chemical Society*, vol. 62, no. 7, pp. 1723–1732, 1940.

[51] U. Kuila and M. Prasad, "Specific surface area and pore-size distribution in clays and shales," *Geophysical Prospecting*, vol. 61, no. 2, pp. 341–362, 2013.

[52] J. C. Groen, L. A. A. Peffer, and J. Pérez-Ramírez, "Pore size determination in modified micro- and mesoporous materials. Pitfalls and limitations in gas adsorption data analysis," *Microporous and Mesoporous Materials*, vol. 60, no. 1-3, pp. 1–17, 2003.

[53] T. Li, H. Tian, J. Chen, and L. Cheng, "Application of low pressure gas adsorption to the characterization of pore size distribution of shales: An example from Southeastern Chongqing area, China," *Journal of Natural Gas Geoscience*, vol. 1, no. 3, pp. 221–230, 2016.

[54] S.-B. Chen, Y.-M. Zhu, H.-Y. Wang, H.-L. Liu, W. Wei, and J.-H. Fang, "Structure characteristics and accumulation significance of nanopores in Longmaxi shale gas reservoir in the southern Sichuan Basin," *Journal of the China Coal Society*, vol. 37, no. 3, pp. 438–444, 2012.

[55] S. Kondou, T. Ishikawa, and I. Abe, "Adsorption Science (Chinese version)," pp. 37–39, Chemical Industry Press, Beijing, China, 2006.

[56] H. Freundlich, "Over the adsorption in solution," *The Journal of Physical Chemistry*, vol. 57, Article ID 385471, pp. 1100–1107, 1906.

[57] S. Kondou, T. Ishikawa, and I. Abe, *Adsorption Science (Chinese version)*, p. 36, Chemical Industry Press, Beijing, China, 2006.

[58] S. Duan, M. Gu, X. Du, and X. Xian, "Adsorption Equilibrium of CO2 and CH4 and Their Mixture on Sichuan Basin Shale," *Energy & Fuels*, vol. 30, no. 3, pp. 2248–2256, 2016.

[59] S. Kondou, T. Ishikawa, and I. Abe, *Adsorption Science (Chinese version)*, p. 42, Chemical Industry Press, Beijing, China, 2006.

[60] M. D. LeVan and T. Vermeulen, "Binary Langmuir and Freundlich isotherms for ideal adsorbed solutions," *The Journal of Physical Chemistry*, vol. 85, no. 22, pp. 3247–3250, 1981.

[61] S. Kondou, T. Ishikawa, and I. Abe, *Adsorption Science (Chinese version)*, p. 44, Chemical Industry Press, Beijing, China, 2006.

Application of Spent Li-Ion Batteries Cathode in Methylene Blue Dye Discoloration

Eric M. Garcia, Hosane A. Taroco, Ana Paula C. Madeira, Amauri G. Souza, Rafael R. A. Silva, Júlio O. F. Melo, Cristiane G. Taroco, and Quele C. P. Teixeira

DECEB, Federal University of São João del-Rei, Campus Sete Lagoas, MG-424, Km 45, 35701-970 Sete Lagoas, MG, Brazil

Correspondence should be addressed to Eric M. Garcia; ericmgmg@hotmail.com

Academic Editor: Carlos Alberto Lberto Martínez-Huitle

This paper aims to present the mechanism study of methylene blue (MB) discoloration using spent Li-ion battery cathode tape and hydrogen peroxide. The recycled cathode used in this work is composed of 72% of $LiCoO_2$, 18% of carbon, and 10% of Al. The value found for surface area is 8.9 m^2/g and the ZCP value occurs in pH = 2.95. Different from what is proposed in the literature, the most likely mechanism of methylene blue discoloration is the oxidation/delitiation of $LiCoO_2$ and the reduction of H_2O_2 forming $OH^\bullet$. Thus, in this paper, an important and promising alternative for discoloration of textile industry dyes using spent Li-ion battery cathode is presented.

1. Introduction

1.1. Li-Ion Battery Recycling: A Brief Overview. The fast popularization of portable devices such as cell phones is due mostly to the introduction of Li-ion batteries (LIBs) by Sony in 1991 [1–3]. The introduction of LIBs to the market was due to characteristics such as high energy density, long lifespan, and lightweight [3, 4]. In general, LIBs have no Cd, unlike Ni-Cd batteries, and have much less Ni than Ni-MH batteries. The cathode material widely used in LIBs is lithium and cobalt oxide, $LiCoO_2$ [1–3, 5]. Due to their widespread use, LIBs' disposal is a topic of great relevance. The waste generated by LIBs will soon be a dramatic problem in countries such as China, where the quantity of discarded LIBs in 2020 can reach 25 billion units [6]. Among many applications, it is reported that the spent $LiCoO_2$ catalyzes methylene blue degradation in the presence of H_2O_2 [3]. Thus, a very audacious project concerns the application of recycled materials from LIBs on waters contaminated with potentially toxic organic molecules. This process is very interesting concerning the environment due to the possibility of promoting the decontamination of organic pollutants using a residue coming from Li-ion battery waste. The substances traditionally used for organic decontamination are O_3 and H_2O_2 [7]. However, the cost of these compounds is relatively high (O_3 is \$2.2 per kg and H_2O_2 is \$1.5 per kg) [7]. The use of $LiCoO_2$ increases the velocity constant of methylene blue (MB) degradation 200 times if compared to pure H_2O_2 with minimum additional cost since this material comes from spent LIB [3]. Thus, the use of recycled LIB cathode is an excellent alternative for degradation of potentially toxic organic molecules. The literature reports hydrogen peroxide oxidation, promoted by $LiCoO_2$ reduction, with intermediate step correspondent to $O_2^{-\bullet}$ radical formation [3]. However, we have reasons to believe that the methylene blue discoloration mechanism, using the H_2O_2 and LIB spent cathode, is different from what is proposed in the literature [3]. The main controversial point is that the reduction potential of $LiCoO_2$ in aqueous solution (see (1)) is around -0.95 V [4] and oxidation of H_2O_2 (see (2)) is -0.89 V [3]. Thus, the global process (see (3)) is thermodynamically unfavorable (-1.84 V). Thus, the pathway to understand the reaction mechanism between $LiCoO_2$ and H_2O_2 involves predominantly the study of the chemical behavior of $LiCoO_2$ in aqueous solution.

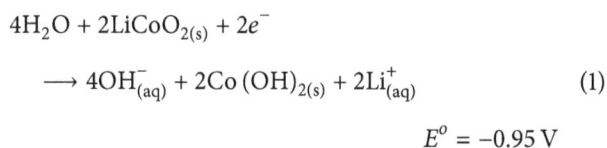

$$4H_2O + 2LiCoO_{2(s)} + 2e^-$$

$$\longrightarrow 4OH_{(aq)}^- + 2Co(OH)_{2(s)} + 2Li_{(aq)}^+ \tag{1}$$

$$E^o = -0.95\,V$$

$$2H_2O_{2(aq)} \longrightarrow 4H^+ + 2e^- + 2O_{2(g)}^{-\bullet} \quad E^o = -0.89\,V \quad (2)$$

$$2LiCoO_{2(s)} + 2H_2O_2$$

$$\longrightarrow 2Co\,(OH)_{2(s)} + 2Li_{(aq)}^+ + 2O_{2(g)}^{-\bullet} \quad (3)$$

$$E^o = -1.84\,V$$

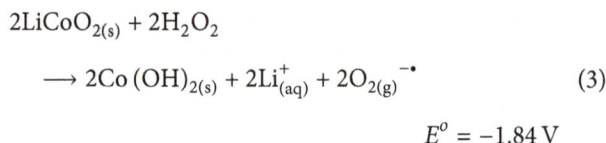

1.2. The Chemical Behavior of LiCoO$_2$ in Aqueous Solution. The LIB operating system basically consists of reversible extraction/insertion of Li$^+$ ions between two electrodes (see (4)) [1, 8, 9]. For LIB with LiCoO$_2$ cathode, in the charge process, the Li$^+$ intercalated in the LiCoO$_2$ structure is forced out by an external applied potential in direction of a carbonaceous matrix (see (4)) [1]. In the discharge (reverse process), the Li$^+$ ions return to the Li$_{(1-x)}$CoO$_2$ matrix due to a concentration gradient [1]. Due to the large LIB operation potential, the electrolyte is a nonaqueous solution composed of lithium salts, such as 1 M LiPF$_6$, in alkyl organic carbonates (ethylene, propylene, and dimethyl carbonates) [9].

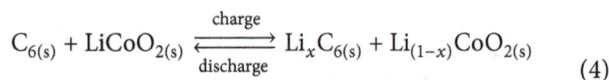

$$C_{6(s)} + LiCoO_{2(s)} \underset{\text{discharge}}{\overset{\text{charge}}{\rightleftharpoons}} Li_xC_{6(s)} + Li_{(1-x)}CoO_{2(s)} \quad (4)$$

$$E \cong 3.7\,V$$

In fact, in aqueous electrolyte with electrode materials LiCoO$_2$ and C$_6$, the reversible intercalation/deintercalation of Li$^+$ happens under potential equal to 1.20 V (99% of Coulomb efficiency) [10]. The LiCoO$_2$ charge in aqueous solution (delitiation) begins around 1.02 V [10] practically independent of the pH value [11].

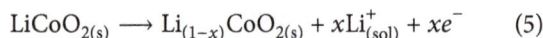

$$LiCoO_{2(s)} \longrightarrow Li_{(1-x)}CoO_{2(s)} + xLi_{(sol)}^+ + xe^- \quad (5)$$

Thus, to force the deintercalation of Li$^+$ ions from LiCoO$_2$ lamellar structure without an external power source, it is necessary to use an oxidant agent with reduction potential higher than 1.02 V. The reduction potential of H$_2$O$_2$/H$_2$O [12] and O$_2$/H$_2$O [12] (considering P_{O2} = 0.21 atm) versus pH is shown in Figure 1. The dotted line represents the limit potential (1.02 V) for LiCoO$_2$ oxidation in aqueous solution (see (5)). The redox reaction represented by LiCoO$_2$ oxidation and the reduction of H$_2$O$_2$ (see (6)) are spontaneous along the pH range 1 to 12. The potential for (6) considering pH =7 is 0.3 V (details in Figure 1). The interference of parallel O$_2$ reduction is observed only in pH <1 (Figure 1).

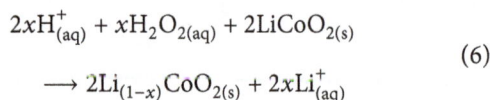

$$2xH_{(aq)}^+ + xH_2O_{2(aq)} + 2LiCoO_{2(s)}$$

$$\longrightarrow 2Li_{(1-x)}CoO_{2(s)} + 2xLi_{(aq)}^+ \quad (6)$$

On the other hand, for acid pH values, the cointercalation of H$^+$ ions in the charged Li$_{(1-x)}$CoO$_2$ (see (7)) [13] normally promotes the obstruction of Li$^+$ diffusion pathways [11].

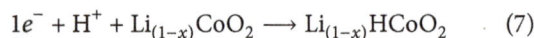

$$1e^- + H^+ + Li_{(1-x)}CoO_2 \longrightarrow Li_{(1-x)}HCoO_2 \quad (7)$$

Thus, this paper aims to present the mechanism study of methylene blue (MB) discoloration using LIB spent cathode tape and hydrogen peroxide. The spent cathode tape

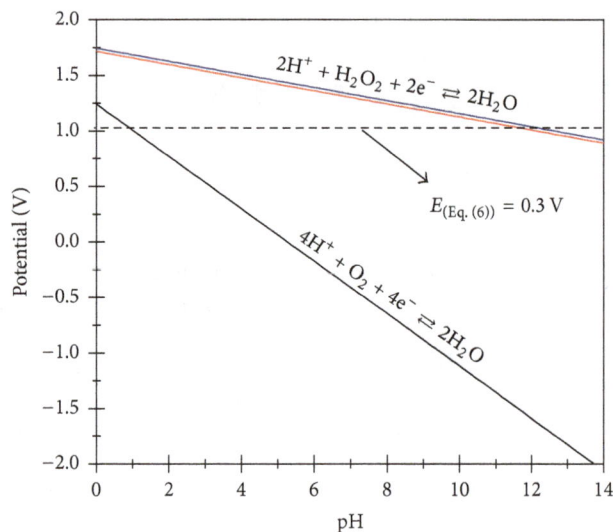

FIGURE 1: Potential versus pH for O$_2$ and H$_2$O$_2$ reduction in water [Haynes]. The dotted line (1.02 V) represents the limit potential for LiCoO$_2$ oxidation (delitiation) in water. It was considered that P_{O2} = 0.21 atm, [H$_2$O$_2$] = 0.10 mol L^{-1} (blue line), and [H$_2$O$_2$] = 0.01 mol L^{-1} (red line).

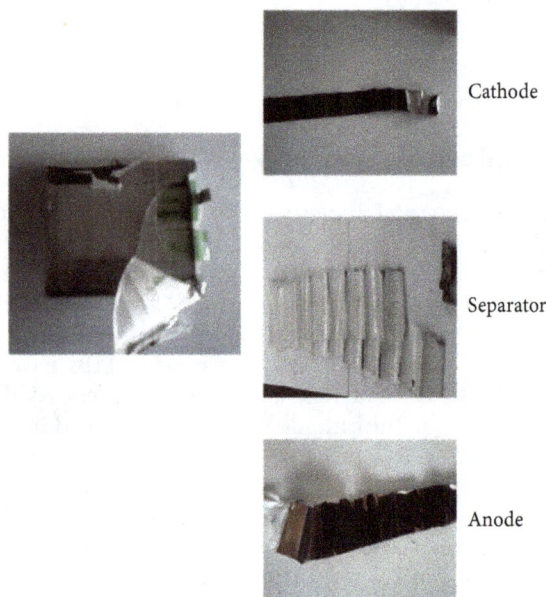

FIGURE 2: Cathode, anode, and separator from a cell phone spent LIB used in this work.

was characterized by X-ray diffraction, atomic absorption spectrophotometry (AAS), zeta potential, SEM, and the MB adsorption technique to determine the surface area.

2. Materials and Methods

2.1. The Characterizations of LIB Spent Cathode. The LIB used in this paper was a spent cell phone battery. The LIB was dismantled using the manual procedure and the components were separated (Figure 2). The spent cathode in tape form was

TABLE 1: The experimental conditions used in the reaction order study of MB discoloration. The variables were spent cathode (SC) area and H_2O_2 and MB concentration.

	Condition 1 SC = 16 cm^2 $[H_2O_2]$ = 0.1 M MB = 12 ppm	Condition 2 SC = 16 cm^2 $[H_2O_2]$ = 0.01 M MB = 12 ppm	Condition 3 SC = 4 cm^2 $[H_2O_2]$ = 0.1 M MB = 12 ppm	Condition 4 SC = 16 cm^2 $[H_2O_2]$ = 0.1 M; MB = 6 ppm
t_{50}	60 s	67 s	242 s	32 s

separated and, to remove the organic solvents, it was heated at 200°C for 5 h according to the literature [1].

2.1.1. The Crystalline Structure.
The spent cathode tape was characterized by X-ray diffraction on a 200 B Rotaflex-Rigaku with Cu Kα irradiation, a Cu filter, and scanning speed of $2°\,min^{-1}$. The spent cathode after degradation of MB was also characterized to verify the possible presence of Co_3O_4 phase.

2.1.2. Atomic Absorption Spectrophotometry (AAS).
The spent cathode powder was scraped from the current collector. 0.20 g of the cathode powder was dissolved in 250 mL of 2 M H_2SO_4 (Sigma-Aldrich, purity ≥ 98%) and 6 mL of 35% w/w H_2O_2 (Sigma-Aldrich). The resulting solution was filtered and analyzed by AAS on a Hitachi Z8200.

2.1.3. Zeta Potential.
The zeta potential was determined by Dynamic Light Scattering (DLS) measurements on a Zeta-Sizer Malvern at 25°C in an aqueous medium after dispersion for 30 min in an ultrasound bath.

2.1.4. Scanning Electron Microscopy (SEM).
The morphology of the spent Li-BCT was observed by Field-Emission Scanning Electron Microscopy on a JEOL JXA model 8900 RL.

2.1.5. Determination of Surface Area.
A successful method to determine the active area was described in the *literature* in recent years. It is based on the adsorption of MB [14]. The measure of the surface area with MB adsorption has the main advantage of measuring the surface accessible to electrolyte. To determine the surface area, the LIB spent cathode was cut into rectangular pieces of 4 cm^2 (~60 mg) and put into a MB (Sigma-Aldrich, dye content ≥95%) solution with different concentrations (1, 2, 3, 5, 7, and 10 ppm) in different pH values (between 3 and 11). After 24 h, the MB concentration was monitored using a UV-spectrophotometer FEMTO Cirrus 80 PR at 659 nm. The MB dye adsorption capacity (q_e) was determined by (8) [15], in which C and C_e are the initial and equilibrium MB concentration, respectively. For the determination of the surface area, the Langmuir isotherm (see (9)) was considered, in which q_{max} is the monolayer adsorption capacity (mg/g) and K_L is a constant related to free energy of adsorption [15]. To convert q_{max} into surface

area (m^2/g), each MB molecule with $1,70 \times 10^{-18}$ m^2 [14] was considered.

$$q_e = \frac{V(C - C_e)}{m}, \qquad (8)$$

$$\frac{C_e}{q_e} = C_e \left(\frac{1}{q_{max}}\right) + \frac{1}{q_{max}K_L}. \qquad (9)$$

2.2. Methylene Blue Discoloration Mechanism.
To study the kinetics of MB discoloration, a 100 mL glass cell containing 50 mL of MB solution, H_2O_2, and LIB spent cathode tape (cut into rectangular pieces) was stirred magnetically (10 rpm) in a bath with controlled temperature at 25°C. Before the H_2O_2 addition, the MB solution was put in contact with LIB spent cathode tape for 24 hours to establish the adsorption equilibrium. The pH of the solution was adjusted at around 7 (exactly 7.02). The discoloration process was monitored by analyzing aliquots at 659 nm. The reaction order was evaluated by variation of spent cathode area and H_2O_2 and MB concentration. The variations of the parameters are shown in Table 1. Considering t_{50} as the interval of time to promote 50% of discoloration, the reaction order (a_R) in relation to generic reactant "R" can be obtained by (10) [3] in which t_{50} and t'_{50} are the interval of time to promote 50% of discoloration in two different concentrations represented by $[R]$ and $[R]'$, respectively.

$$a_R = \frac{\log\left(t'_{50}/t_{50}\right)}{\log\left([R]/[R]'\right)}. \qquad (10)$$

The active radical was identified by the addition of 2 M isopropanol (i-PrOH) as a scavenger for OH$^\bullet$ [16]. The cyclic experiment was performed using the same spent cathode in 5 cycles of MB discoloration (600 min). All measurements were made in the absence of visible light to avoid the possible electron excitation of cobalt lithium oxide conduction band to the valence band generating the electron (BC) and hole (BV) pair. After MB discoloration, the resultant solution was analyzed with AAS aiming to determine the possible lixiviation of Co and Li.

3. Results and Discussion

3.1. The Characterizations of LIB Spent Cathode Powder.
The characterization of LIB spent cathode is very important since its composition can present Fe, Mn, and Ni. Thus, X-ray

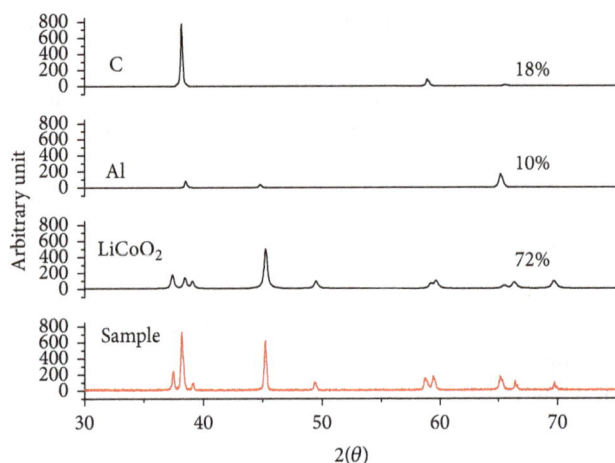

FIGURE 3: The X-ray diffractogram of LIB spent cathode tape before MB discoloration reaction. The proportion of phase was analyzed with Rietveld method using the free software FullProf©.

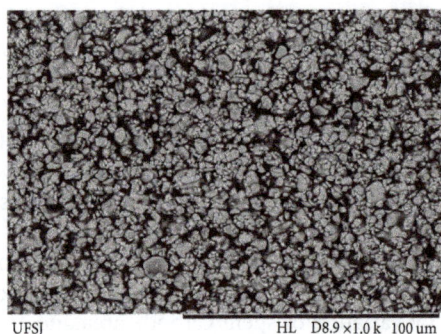

FIGURE 4: Scanning Electron Microscopy (SEM) of LIB spent cathode tape.

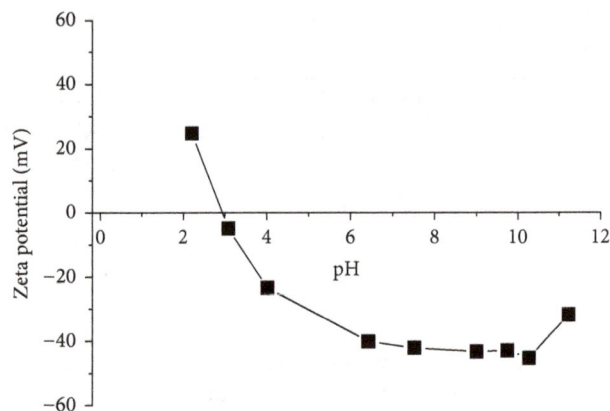

FIGURE 5: Zeta potential of LIB spent cathode powder.

diffraction and AAS were used to elucidate the chemical composition of LIB spent cathode. The spent cathode powder was dissolved (Materials and Methods) and the resulting solution was analyzed by AAS. The presence of Fe, Mn, and Ni was not detected (detection limit: 0.04 ppm). The Li and Co concentrations were 55 ppm and 418 ppm, respectively. Considering the atomic mass of Co and Li (58.933 and 6.941 g/mol [12]), these concentrations are compatible with a stoichiometric proportion of approximately 1 : 1. The X-ray of LIB spent cathode tape was analyzed with Rietveld method using the free software FullProf© [17] (Figure 3). The X-ray diffraction showed the proportion of phases present in the cathode. The recycled cathode is composed of 72% of $LiCoO_2$, 18% of carbon, and 10% of Al. Carbon is added to increase the electrical conductivity of the cathodic material [1]. The Al presence is due to the current collector composition and its detection in X-ray diffraction is possible due to the porosity of the LIB cathode [1]. It is worth pointing out that the presence of other Li_xCoO_2 phases (with x varying from 1 to 0.5) is also possible, but their detection through X-ray diffraction is complex when $0.5 < x < 1$ [11].

Figure 4 shows the Scanning Electron Microscopy (SEM) of the spent cathode from Li-ion battery. The values found

for the porosity and the average size of grain were 42% and 31 μm, respectively [18].

To clarify the adsorption process, the zeta potential was assessed (Figure 5). The zeta potential remains practically constant ($\sim$−40 mV) in the pH ranging from 11 to 6. Considering the values of pH ranging from 6 to 3, the zeta potential varies considerably until it reaches the zero charge point (ZCP). The spent cathode has relatively acidic ZCP (close to pH = 2.95).

To determine the surface area, Figure 6(a) shows the C_e/q_e versus C_e plot for MB adsorption onto the spent LIB cathode. The adjustment of experimental points in Langmuir isotherm (see (9)) showed good concordance ($R^2 = 0.99$). Similar to the zeta potential measurement, the surface area measured by MB adsorption undergoes a variation with pH (Figure 6(b)). For values of pH ranging from 11 to 6, the surface area changes minimally. The value found for surface area based on MB adsorption was 8.9 m^2/g in pH = 7. With the pH ranging from 6 to 3, the MB adsorption decreases quite probably due to the H^+ adsorption competition.

3.2. Methylene Blue Discoloration Mechanism. To obtain the reaction order and the reaction determinant step (RDS), MB discoloration was performed in different conditions as shown in Figure 7. Table 1 shows t_{50} in different situations. The reaction orders obtained in relation to H_2O_2, SC (spent cathode), and MB are, respectively, $a_{H_2O_2} = 0.04 \approx 0$, $a_{SC} = 1,008 \approx 1$, and $a_{MB} = 1.102 \approx 1$. Thus, the reaction determinant step can be written as $v_{rds} = k[SC][MB]$.

To prove that the reaction mechanism can be represented by (6), the pH and Co and Li concentration were measured before and after the discoloration reaction. The experimental design is given by Table 1 in condition 3. In the previous solution, the concentrations of Co and Li were below the detection limit (0.04 ppm). After MB discoloration, the Co concentration remained below 0.04 ppm and the Li concentration reached 1.2 ppm. Moreover, after the discoloration reaction, the pH value ranged from 7.01 to 7.25. It is important to note that, for charged $Li_{(1-x)}CoO_2$ phase in aqueous solution, the Li^+ intercalation (which corresponds to the LIB discharge process) only must be considered favorable in

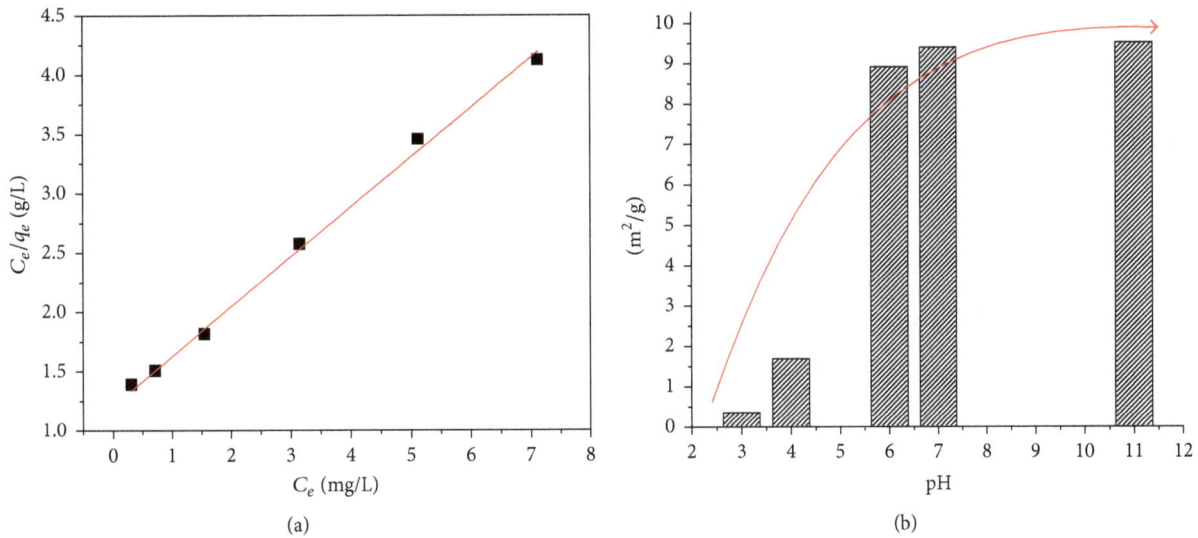

FIGURE 6: (a) The C_e/q_e versus C_e plot for MB adsorption onto spent LIB cathode. (b) The surface area measured by MB adsorption with pH variation.

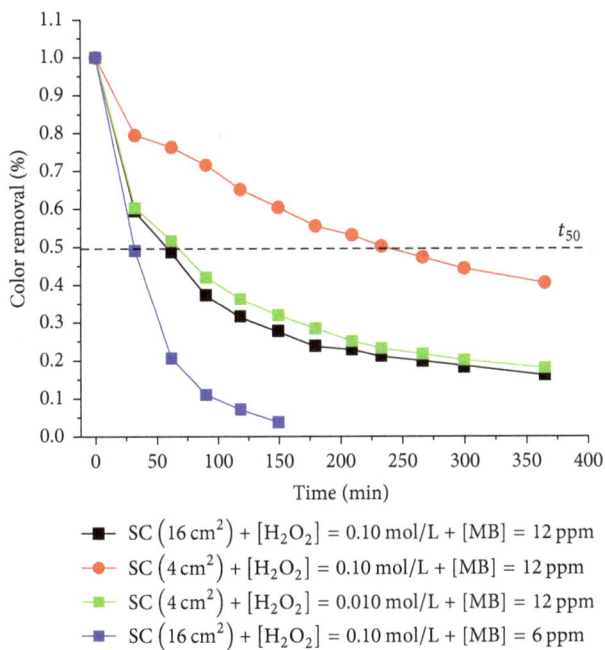

FIGURE 7: Process of MB discoloration (color removal) versus time for pH = 7.02 with different conditions of H_2O_2, spent cathode (SC), and MB solution.

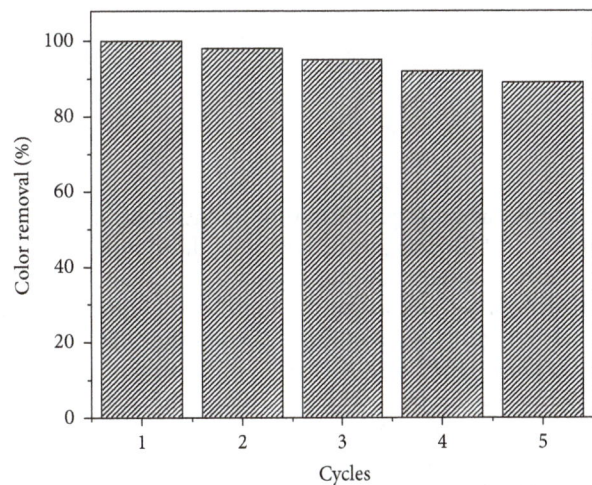

FIGURE 8: The result of five cyclic tests of MB discoloration. The experimental conditions used were MB 12 ppm, spent cathode 4 cm², and 0.1 M H_2O_2.

special conditions as reported in the literature (e.g., in 5 M $LiNO_3$) [10]. Thus, the only spontaneous process for $LiCoO_2$ under the experimental conditions used in this work is the oxidation process shown by (6). With the decrease of Li content in the cathode ($LiCoO_2$), there is also a decrease in the kinetics of MB discoloration. This is easily seen in the cyclic test performed with the same LIB spent cathode tape (Figure 8). A slight drop of discoloration capacity is noticed, which can be evidence that the spent cathode is a reagent

and not a catalyst as reported previously in the literature [3]. Thus, the charged phase $Li_{(1-x)}CoO_2$ reacts in the oxidation direction by

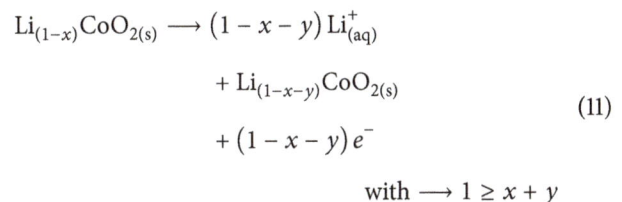

$$Li_{(1-x)}CoO_{2(s)} \longrightarrow (1 - x - y)\, Li_{(aq)}^{+}$$
$$+ Li_{(1-x-y)}CoO_{2(s)}$$
$$+ (1 - x - y)\, e^{-}$$

$$\text{with} \longrightarrow 1 \geq x + y$$

(11)

The X-ray diffraction was performed on the spent cathode after degradation of methylene blue. The main objective was to verify the presence of Co_3O_4. The charged Li_xCoO_2 (with $1 > x$) is metastable and, after the charge/discharge cycles,

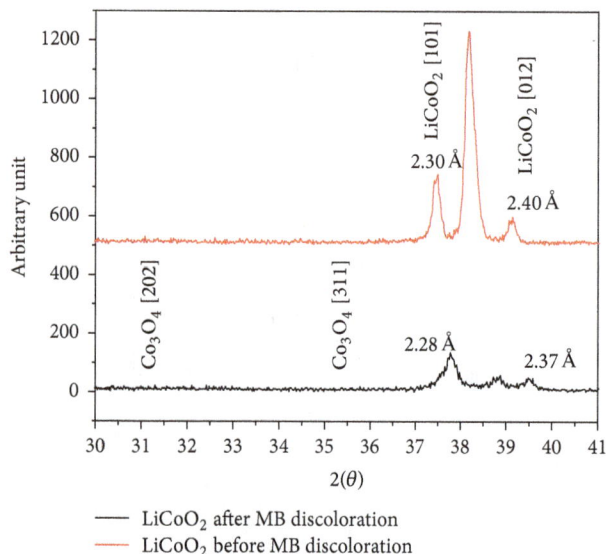

FIGURE 9: The X-ray diffractogram of LIB spent cathode tape before and after MB discoloration reaction (condition 3 in Table 1).

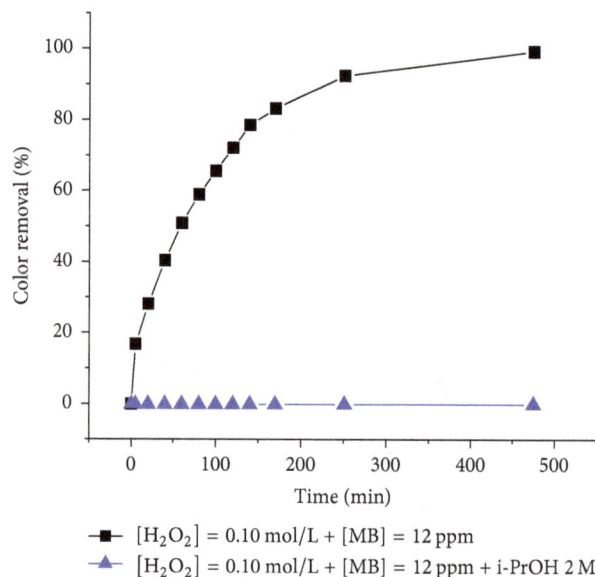

FIGURE 10: The discoloration of 12 ppm MB with 0.10 M H_2O_2 and spent cathode with and without 2 M i-PrOH.

the Co_3O_4 phase can be formed through the decomposition solid-state reaction (see (12)) [19].

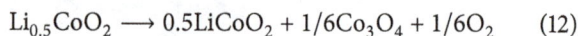

$$Li_{0.5}CoO_2 \longrightarrow 0.5LiCoO_2 + 1/6Co_3O_4 + 1/6O_2 \quad (12)$$

This reaction takes place at around 190°C [19]. In this work, the most probable process involving $LiCoO_2$ is delitiation (LIB charge); in addition, the temperature of the reaction system did not exceed 25°C. Thus, after the methylene blue discoloration reaction, Co_3O_4 presence is not expected. Figure 9 shows the positions of the main peaks of Co_3O_4 phase (JCPDS 78-1970 [13]). In Figure 9, we observe only the presence of $LiCoO_2$ crystallographic planes. Thus, the presence of Co_3O_4 phase in the spent cathode after degradation of methylene blue is not confirmed by X-ray diffraction. It is noted that, in the spent cathode after MB discoloration reaction, the interplanar distance undergoes a subtle decrease, showing that the H^+ cointercalation also is not evidenced [11].

To confirm the mechanism, the active radicals were identified by the addition of isopropanol (i-PrOH) as a scavenger for $OH^\bullet$ [16]. Figure 10 shows the discoloration of 12 ppm MB with 0.1 M H_2O_2 and spent cathode with and without 2 M i-PrOH. The measure shows that $OH^\bullet$ is the predominant radical formed by H_2O_2 reduction (see (13)) promoted by $LiCoO_2$ oxidation.

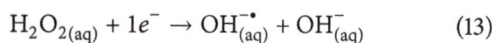

$$H_2O_{2(aq)} + 1e^- \longrightarrow OH^\bullet_{(aq)} + OH^-_{(aq)} \quad (13)$$

4. Conclusion

In this paper, we presented an important and promissory alternative for discoloration of textile industry using LIB recycled cathode. The model molecule used was methylene blue. The recycled cathode is composed of 72% of $LiCoO_2$, 18% of carbon, and 10% of Al. The value found for surface area of the spent cathode was $8.9\,m^2/g$ and the ZCP value occurs in pH = 2.95. Different from what is proposed in the literature, the more probable mechanism of methylene blue discoloration is the oxidation/delitiation of $LiCoO_2$ and the reduction of H_2O_2 forming the $OH^\bullet$.

Conflicts of Interest

The authors declare that they have no conflicts of interest regarding the publication of this paper.

Acknowledgments

This work was supported by CNPq, DECEB, Federal University of São João del-Rei (UFSJ), Sete Lagoas. This work is a collaboration research project of members of the Rede Mineira de Química (RQ-MG) supported by FAPEMIG (Project CEX-RED-00010-14).

References

[1] M. B. J. G. Freitas and E. M. Garcia, "Electrochemical recycling of cobalt from cathodes of spent lithium-ion batteries," *Journal of Power Sources*, vol. 171, no. 2, pp. 953–959, 2007.

[2] E. M. Garcia, V. D. F. C. Lins, H. A. Tarôco, T. Matencio, R. Z. Domingues, and J. A. F. Dos Santos, "The anode environmentally friendly for water electrolysis based in LiCoO2 recycled from spent lithium-ion batteries," *International Journal of Hydrogen Energy*, vol. 37, no. 22, pp. 16795–16799, 2012.

[3] M. C. Abreu Gonçalves, E. M. Garcia, H. A. Taroco et al., "Chemical recycling of cell phone Li-ion batteries: Application in environmental remediation," *Waste Management*, vol. 40, pp. 144–150, 2015.

[4] Y. Xu, D. Song, J. Li et al., "Electrochemical properties of LiCoO2 + x% S mixture as anode material for alkaline secondary battery," *Electrochimica Acta*, vol. 85, pp. 352–357, 2012.

[5] X. Zeng, J. Li, and B. Shen, "Novel approach to recover cobalt and lithium from spent lithium-ion battery using oxalic acid," *Journal of Hazardous Materials*, vol. 295, pp. 112–118, 2015.

[6] X. Chen, C. Luo, J. Zhang, J. Kong, and T. Zhou, "Sustainable Recovery of Metals from Spent Lithium-Ion Batteries: A Green Process," *ACS Sustainable Chemistry and Engineering*, vol. 3, no. 12, pp. 3104–3113, 2015.

[7] Y. Yao, Y. Cai, G. Wu et al., "Sulfate radicals induced from peroxymonosulfate by cobalt manganese oxides ($Co_xMn_{3-x}O_4$) for Fenton-Like reaction in water," *Journal of Hazardous Materials*, vol. 296, pp. 128–137, 2015.

[8] B. Scrosati and J. Garche, "Lithium batteries: status, prospects and future," *Journal of Power Sources*, vol. 195, no. 9, pp. 2419–2430, 2010.

[9] S. Goriparti, E. Miele, F. De Angelis, E. Di Fabrizio, R. P. Zaccaria, and C. Capiglia, "Review on recent progress of nanostructured anode materials for Li-ion batteries," *Journal of Power Sources*, vol. 257, pp. 421–443, 2014.

[10] R. Ruffo, F. La Mantia, C. Wessells, R. A. Huggins, and Y. Cui, "Electrochemical characterization of LiCoO2 as rechargeable electrode in aqueous LiNO3 electrolyte," *Solid State Ionics*, vol. 192, no. 1, pp. 289–292, 2011.

[11] E. E. Levin, S. Y. Vassiliev, and V. A. Nikitina, "Solvent effect on the kinetics of lithium ion intercalation into $LiCoO_2$," *Electrochimica Acta*, vol. 228, pp. 114–124, 2017.

[12] W. M. Haynes, *CRC Handbook of Chemistry and Physics*, 92nd edition, 2011.

[13] Y. Liu, C. Chang, D. Zhang, and Y. Wu, "Improved electrochemical properties by lithium insertion into Co3O4 in aqueous LiOH solution," *Progress in Natural Science: Materials International*, 2013.

[14] J. Cho, "Correlation of capacity fading of LiMn2O4 cathode material on 55 ∘C cycling with their surface area measured by a methylene blue adsorption," *Solid State Ionics*, vol. 138, no. 3-4, pp. 267–271, 2001.

[15] T. Akar, I. Tosun, Z. Kaynak, E. Kavas, G. Incirkus, and S. T. Akar, "Assessment of the biosorption characteristics of a macrofungus for the decolorization of Acid Red 44 (AR44) dye," *Journal of Hazardous Materials*, vol. 171, no. 1-3, pp. 865–871, 2009.

[16] F. Chai, K. Li, C. Song, and X. Guo, "Synthesis of magnetic porous Fe3O4/C/Cu2O composite as an excellent photo-Fenton catalyst under neutral condition," *Journal of Colloid and Interface Science*, vol. 475, pp. 119–125, 2016.

[17] FullProf Suite, https://www.ill.eu/sites/fullprof/.

[18] http://imagej.nih.gov/ij/download.html.

[19] Y. Baba, S. Okada, and J.-I. Yamaki, "Thermal stability of LixCoO2 cathode for lithium ion battery," *Solid State Ionics*, vol. 148, no. 3-4, pp. 311–316, 2002.

In Vitro Study of Adsorption Kinetics of Dextromethorphan Syrup onto Activated Charcoal in Simulated Gastric and Intestinal Fluids

Shobha Regmi,[1] **Balmukunda Regmi,**[1] **Sajan Lal Shyaula,**[2] **Shiva Pathak,**[1] **Bishnu Prasad Bhattarai,**[1] **and Saroj Kumar Sah**[1]

[1]*Department of Pharmacy, Institute of Medicine, Tribhuvan University, Maharajgunj Medical Campus, Kathmandu, Nepal*
[2]*Nepal Academy of Science and Technology (NAST), Khumaltar, Lalitpur, Nepal*

Correspondence should be addressed to Saroj Kumar Sah; sarozsah@iom.edu.np

Academic Editor: Pranav S. Shrivastav

Adsorption kinetics of dextromethorphan (DXM) syrup in simulated gastric and intestinal fluids onto activated charcoal (AC) were investigated in an in vitro model. The adsorption studies were performed as a function of time, initial concentration, and temperature. The quantification of DXM adsorbed onto AC was obtained from the Langmuir adsorption isotherms using HPLC. The maximum adsorption capacities (at 95% confidence limits) of AC for DXM were 111.615 [106.38; 126.85] mg in simulated intestinal environment (pH 6.8) and 78.314 [86.206; 70.422] mg in simulated gastric environment (pH 1.2). The adsorption capacity of AC for DXM in simulated gastric fluid (pH 1.2) was not significantly different from the adoption capacity of AC for DXM in simulated intestinal fluid (pH 6.8). Moreover, the adsorption kinetics behavior of dextromethorphan onto AC followed pseudo-second-order kinetics. Our results show that AC in therapeutically acceptable doses can be beneficial in the majority of oral overdose of DXM.

1. Introduction

In acute oral drug overdose, the drug should be removed as soon as possible, before it is significantly absorbed from the gastrointestinal tract. A nonspecific measure for gastrointestinal decontamination of a poisoned patient is the oral administration of activated charcoal (AC). Oral ingestion of AC has long been known to be effective in reducing the systemic absorption of many drugs due to its adsorptive properties. Thus, it is taken as a useful agent in the management of acute oral drug overdoses.

Dextromethorphan (DXM) is an antitussive drug. It is one of the active ingredients in many over-the-counter (OTC) cold and cough medicines [1]. Cases of recreational abuse of DXM [2] have been reported in USA, Sweden, Australia, Germany, and Korea, primarily among adolescents and young adults [3]. In a study in the USA during a 2-year period from 2004 and 2005, the emergency departments treated over 1,500 children for adverse effects related to OTC

cough and cold medication use. A 6-year retrospective study from 1999 to 2004 of the California Poison Control System showed a 10-fold increase in the rate of DXM abuse cases in all ages and a 15-fold increase in the rate DXM abuse cases in adolescents [3]. The risk of overdose, incorrect dosing, and adverse events are increasing in young children due to the greater number of colds they acquire each year [4]. Due to its fashionable color, sweet taste, and careless storage of DXM syrup nearby the access of child population, they seem to be more susceptible toward its overdose. In a study conducted in the USA to monitor trends in DXM abuse by using the National Poison Data System: 2000–2010, the mean annual prevalence of DXM cases reported to poison control centers was 13.4 cases per million population for all ages and 113.0 cases per million for 15–19-year-old people. The prevalence of DXM cases for all ages increased steadily until 2006 to a peak of 17.6 calls/million and subsequently plateaued at 15.7 cases per million in 2010. A preponderance of male adolescents was noted throughout the study period [5].

It has been an important impulse for a resurgence of interest in the use of AC in toxicology [6]. The adsorption patterns differ with different drugs. So, there is a need to understand this variability to recognize the proper adsorption patterns of AC for different drugs in order to calculate the appropriate amount of AC to administer during acute poisoning [7, 8]. In addition, there is a need to study the maximum adsorption capacity of activated charcoal for many drugs and establish a drug-activated charcoal dose relationship to reduce the systemic toxicity of ingested drugs and achieve the maximal efficacy of activated charcoal. Studies on in vitro drug adsorption onto AC have been carried out at different pH values to simulate in vivo conditions in the gastrointestinal tract [7–10]. An acidic pH (1.2) has been used to simulate the gastric environment, and basic pH (6.8) has been used to simulate the environment in the small intestine.

DXM, an OTC drug, is easily available and have a potential of abuse [11]. Although the frequent use of AC in an asymptomatic patient with DXM poisoning has been practiced, data regarding safety and efficacy of AC in the management of DXM poisoning remain largely unknown [12]. Thus, the need of this study was anticipated to investigate in vitro efficacy of activated charcoal in adsorption of DXM. Based on available data from in vivo and in vitro studies, the recommended dose regimen of AC varies between 25 and 100 g in adults [13], 1 g/kg body weight in children [14], or 0.5–2 g/kg body weight [15]. For many drugs (or poison as well as pesticides), the adsorption onto AC varies with a change in pH [16–20]. The main purpose of this study was to elucidate the difference in maximum adsorption capacity of DXM onto activated charcoal in simulated in vitro environments at two extremities of pH inside gastrointestinal tract. It is well known that oral AC should not only adsorb toxic compounds effectively, but also adsorb the particular intoxicants as fast as possible. This leads to the statement that the investigations on the kinetics of adsorption of different drugs are still under exploration [21]. In our study, we have investigated the in vitro adsorption of DXM at two different pH values (pH 1.2 and pH 6.8) of simulated gastric and intestinal fluids onto AC and elucidated the kinetics of the adsorption.

2. Material and Methods

2.1. Chemicals and Reagents. Standard dextromethorphan was obtained from Lomus pharmaceuticals private limited, Gothatar, Bhaktapur, Nepal. Clinical grade activated charcoal powder (D-Tox powder; Asian Pharmaceuticals Pvt. Ltd, Nepal) was purchased from a community pharmacy. All the reagents such as potassium dihydrogen phosphate (KH_2PO_4), sodium hydroxide (NaOH), and orthophosphoric acid were of analytical grade. Methanol and water of HPLC grade were used in the preparation of the mobile phase for HPLC. The simulated gastric and intestinal fluid environments (pH 1.2 and pH 6.8) were prepared as per United States Pharmacopoeia, 2002 [22]. Simulated gastric fluid pH 1.2 (SGF) was prepared by taking 2 g NaCl and 7 mL concentrated HCl in 1000 mL distilled water. Simulated intestinal fluid, pH 6.8, was prepared by taking 68.05 g potassium dihydrogen phosphate (KH_2PO_4) and 8.96 g sodium hydroxide

(NaOH) in 10 L water. Similarly, 0.1 M phosphate buffer of pH 3 for mobile phase was prepared according to European pharmacopoeia; 13.6 g of potassium dihydrogen phosphate was dissolved in 900 mL of water, pH being adjusted to 3.0 with phosphoric acid, and diluted to 1000 mL.

2.2. Calibration Curves. DXM stock solution of 1000 μg/mL was prepared in simulated gastric and intestinal fluids. Then, different concentrations of standard solutions of DXM were obtained by dilution. The desired concentrations were 500 μg/mL, 250 μg/mL, 125 μg/mL, 62.5 μg/mL, 31.25 μg/mL, and 15.625 μg/mL. Each concentration was filtered with the 0.45 μm filter and injected in HPLC. The separation was achieved by C18 column using an isocratic mobile phase consisting of methanol–dihydrogen phosphate buffer at pH 3 (50:50, % v/v). The analysis was performed at a flow rate of 1 mL/min and at a detection wavelength of 220 nm. The analysis was performed at oven temperature 32°C. Prior to any analysis, the mobile phase was degassed and filtered using 0.45 μm filters. The system was equilibrated with the mobile phase before injection. The area obtained was then plotted against the known concentration and calibration curve was obtained [23].

2.3. Adsorption Study. To obtain Langmuir's adsorption isotherm, AC and dextromethorphan (DXM syrup 10 mg/mL) were mixed in different proportions of 1:1, 3:1, 6:1, and 15:1 in 50 mL volumetric flasks. The amount of dextromethorphan in all the experiments was 30 mg. The sample mixtures in simulated gastric and intestinal fluids were bathsonicated for 15 min at 37°C. The solution was filtered through a filter paper. Experiments were performed in triplicate. The filtrate was then analyzed by HPLC. Before injecting the sample into HPLC, it was refiltered through 0.45 μm filter paper. The separation was achieved on LC18 column using a mobile phase constituting 50:50 (% v/v) mixture of methanol and water. The analysis was performed at a flow rate of 1 mL/min [23].

2.4. Kinetics Study. In order to perform kinetics study of the drug adsorption over the AC, the suspension of DXM and the AC at ratios varying from 1:1 up to 1:12 was kept in 6 different flasks and mixed. Each flask was taken out of the sonicator in 5, 10, 15, 20, 25, and 30 min. After filtration, solutions were analyzed by the method as described previously. The data obtained were examined with pseudo-first-order and pseudo-second-order reaction equations [24]. The kinetics study of drug adsorption was performed in both simulated gastric and intestinal fluids.

2.5. DXM Analysis. The concentrations of DXM, before and after the attainment of equilibrium during adsorption phenomenon, were estimated using HPLC. The data obtained using HPLC were plotted on the calibration curve to obtain the exact amount of DXM adsorbed in the simulated gastrointestinal environments.

2.6. Estimation of the Langmuir Parameter. A derived equation of the Langmuir adsorption isotherm was used to estimate q_m (monolayer capacity) that reflects the maximal quantity of the drugs (in mg) adsorbed per gram AC. The q_m was calculated from the following equation.

$$\frac{C_e}{q_e} = \frac{1}{q_m K} + \frac{C_e}{q_m},$$ (1)

where C_e (mg/L) is the drug concentration in the liquid phase at equilibrium and "K" is Langmuir constant. q_e is equilibrium adsorbed quantity (mg of adsorbate per gram of adsorbent) calculated as $q_e = V(C_0 - C_e)/W$, where "V" is total volume in litre and W is the quantity of AC in grams; C_0 (mg/L) is the initial quantity of dextromethorphan and C_e is the concentration after the adsorption.

The Langmuir isotherm adsorption parameter "q_m" was obtained by linear least square fitting of the experimental data. A plot of C_e/q_e versus C_e yields a straight line with the slope $1/q_m$ and intercept $1/q_m K$.

2.7. Calculation of Adsorption Energy. Langmuir constant "K" calculated from the Langmuir equation can be related to the adsorption free energy, by the following equation: $G = RT \ln K$, where "R" is the universal gas constant (1.987 cal mol^{-1} K^{-1}) and "T" is the absolute temperature (Kelvin) taken as 303 K. Free energy is a measure of the capacity of the system to do work. If the value is negative, the system will have a tendency to do work spontaneously (exothermic reaction). Free energy is measured in kilojoules per mole, also called Gibb's free energy. The negative values of free energy change show that the solute is more concentrated on the adsorbent than in the bulk solution; that is, the system has reached equilibrium [25].

2.8. Calculation of Suitability of Adsorbent to Adsorbate/ Constant Separation Factor. To determine the characteristic behavior of the adsorption, constant separation factor (R) was used. The equation is given as; $R = 1/(1 + KX_0)$, where X_0 is the initial amount of the adsorbate. "R" is a direct function of "K" and is related to the slope $1/q_m$ of the linearized Langmuir expression. The slope of the linearized Langmuir isotherm can be used to interpret the type of adsorption by using the value of "R" as follows: $R > 1.0$, not suitable; $R = 1.0$, linear; $0 < R < 1.0$ suitable (good adsorbent); and $R = 0$, irreversible.

2.9. Statistical Analysis. Data analysis was conducted by the Microsoft excel 2010 and IBM SPSS Statistics for Windows, Version 20.0 (Armonk, NY: IBM Corp.). Statistical values were calculated using unpaired t-tests. Differences with p values of less than 0.05 were considered statistically significant.

3. Results

3.1. Adsorption Saturation of AC with Dextromethorphan. The final concentration of DXM (C_e; mg/L) was plotted against the amount of DXM adsorbed per gram of charcoal q_e (mg/g of AC) for each sample in both SGF and SIF. Plots are as shown in Figure 1.

TABLE 1: Maximal adsorption capacity, q_m (mg DXM adsorbed/g of AC) in SGF and SIF.

Simulated fluid	q_m (mg of DXM adsorbed/g of AC)
Simulated gastric fluid (pH 1.2)	78.314 [86.206; 70.422]
Simulated intestinal fluid (pH 6.8)	116.615 [106.38; 126.85]

TABLE 2: Effect of pH of simulated gastrointestinal fluids on MAC. This shows that there is no significant effect of pH in the MAC.

		p value
pH 6.8	pH 1.2	0.097

3.2. Maximal Adsorption of DXM from Simulated Gastrointestinal Fluids. Langmuir adsorption isotherm was used to calculate the maximal adsorption capacity (q_m) (mg DXM adsorbed/g of AC). The maximum adsorption capacities (MAC) of AC for DXM in SGF and SIF were different. Each gram of AC adsorbed 111.615 mg [106.38; 126.85] DXM at intestinal pH and 78.314 [86.206; 70.422] mg DXM at gastric environment. In other words, AC at its standard dose of 50 g can eliminate 5.8307 g [5319–6342.5] DXM at the intestinal environment and 3.9157 g [4310.3–3521.1 g] DXM at gastric environment (Table 1).

3.3. Effect of pH of Simulated Gastrointestinal Fluids on Maximum Adsorption Capacity of AC. There was no significant effect of pH of simulated gastrointestinal fluids in the MAC (Table 2). The significant difference was measured at 95% confidence interval (CI) using independent sample t-test and was considered significant if $p \leq 0.05^*$.

3.4. Equilibrium Isotherm Model for Adsorption of DXM onto AC. Equilibrium isotherm equations were used to describe the experimental adsorption data. The two most common isotherms for describing solid–liquid adsorption systems are Langmuir and Freundlich adsorption isotherms. Langmuir's plots showed the excellent coefficient of determination (R^2) at both pHs (1.2 and 6.8) for simulated fluid, indicating the excellent fitting model to the experiment data (Figures 2 and 3).

3.5. Comparison of Linear Coefficient of Determination, R^2, for Different Adsorption Isotherm Model. The data were fitted on both Langmuir and Freundlich adsorption isotherms (Table 3). The Langmuir isotherm plot showed the excellent coefficient of determination compared with that of Freundlich isotherm for each trial of study. Thus, results were analyzed by plotting the data on Langmuir isotherm equation.

3.6. Calculation of Adsorption Energy ($G = RT \ln K$) and Suitability of Adsorbent to Adsorbate/Constant Separation Factor ($R = 1/(1 + KX_0)$). The negative values of change in Gibb's free energy indicated that the solute was more concentrated on the adsorbent rather than in the bulk solution, and equilibrium condition was attained (Table 4).

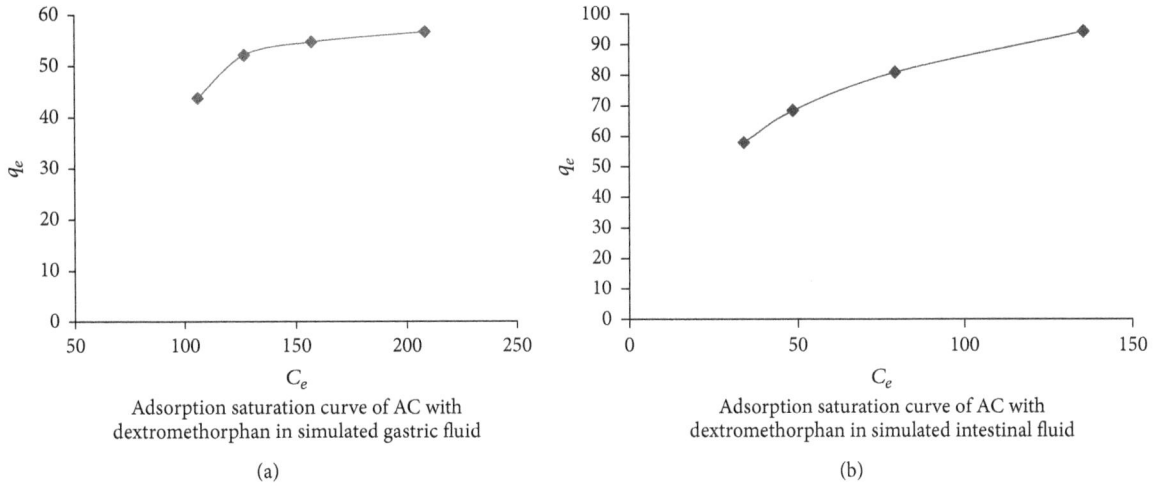

Adsorption saturation curve of AC with
dextromethorphan in simulated gastric fluid

(a)

Adsorption saturation curve of AC with
dextromethorphan in simulated intestinal fluid

(b)

FIGURE 1: Adsorption saturation curve of AC with DXM in (a) simulated gastric fluid and (b) simulated intestinal fluid.

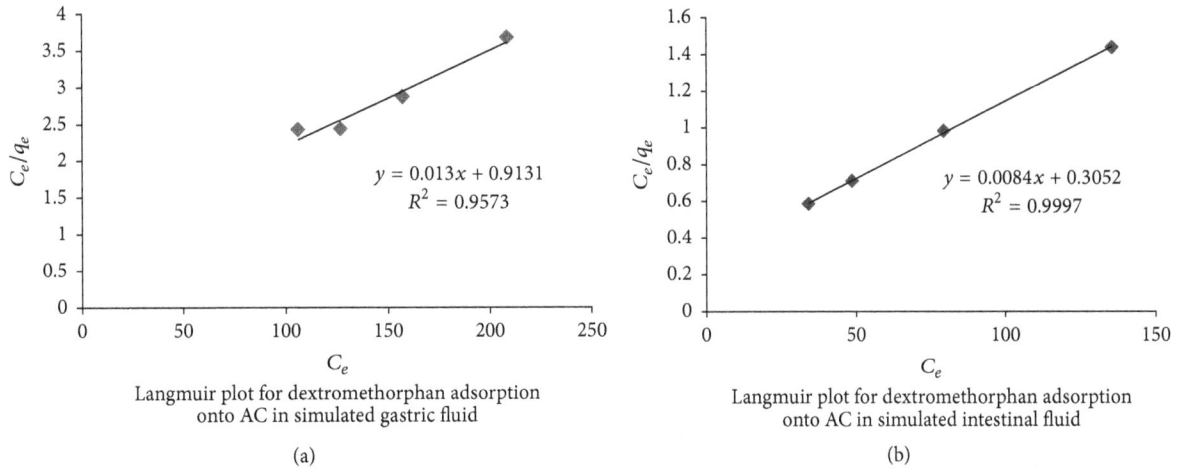

Langmuir plot for dextromethorphan adsorption
onto AC in simulated gastric fluid

(a)

Langmuir plot for dextromethorphan adsorption
onto AC in simulated intestinal fluid

(b)

FIGURE 2: Langmuir plot of AC with DXM in (a) simulated gastric fluid and (b) simulated intestinal fluid.

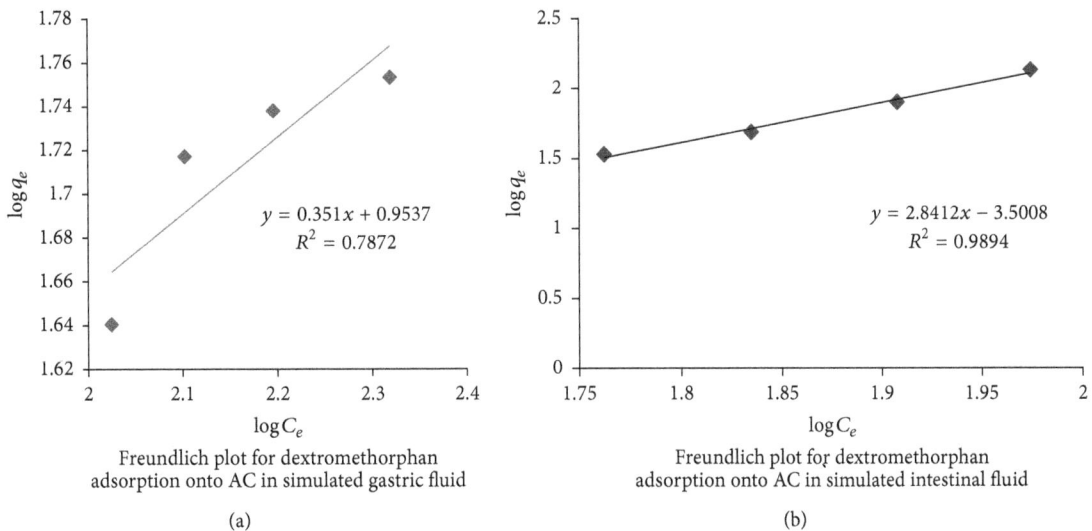

Freundlich plot for dextromethorphan
adsorption onto AC in simulated gastric fluid

(a)

Freundlich plot for dextromethorphan
adsorption onto AC in simulated intestinal fluid

(b)

FIGURE 3: Freundlich plot of AC with DXM in (a) simulated gastric fluid and (b) simulated intestinal fluid.

TABLE 3: Comparison of coefficient of determination, R^2.

Adsorption isotherm	R^2 SGF	R^2 SIF
Langmuir isotherm	0.9573	0.9997
Freundlich isotherm	0.7872	0.9894

TABLE 4: Calculation of adsorption energy ($G = RT \ln K$).

K	G (pH 1.2) Kcal/mole	K	G (pH 6.8) Kcal/mole	Inference
0.01498	−2.5286	0.03185	−2.075	Positive

TABLE 5: Calculation of suitability of adsorbent to adsorbate/constant separation factor/suitability of adsorption ($R = 1/(1 + KX_0)$).

R (pH 1.2)	Inference	R (pH 6.8)	Inference
0.672	Suitable	0.327	Suitable

TABLE 6: Comparison of coefficient of determination, R^2.

Adsorption kinetics	R^2 SGF	R^2 SIF	K_2 SGF	K_2 SIF
Pseudo-first-order	0.437	0.6599		
Pseudo-second-order	0.9968	1.000	0.02433	0.0809

The value of constant separation factor "R" is less than 1 in all of the above cases. This showed that AC is a good adsorbent for DXM (Table 5).

3.7. Adsorption Kinetics Study. Kinetic studies for the adsorption of DXM onto AC were studied and data were plotted to two most used kinetic model equations; pseudo-first-order and pseudo-second-order model. Results showed that the adsorption kinetics behavior of DXM onto AC followed pseudo-second-order kinetics model rather than pseudo-first-order kinetics. For pseudo-first-order kinetics, the plot of log ($q_e - q_t$) versus "t" should be a straight line with negative slope. When we plotted this value, we obtained a line with negative slope but R^2 value was less than that of the pseudo-second-order plot. When we plotted a graph for t/q_t versus "t" according to pseudo-second-order model, we obtained a straight line with positive slope and good R^2 value (>0.9968). So, the adsorption studies of DXM on AC followed the pseudo-second-order kinetic model. This is represented in Figures 4 and 5.

3.8. Comparison of R^2 for Different Adsorption Kinetic Models. The data were plotted on both pseudo-first- and pseudo-second-order equation. The pseudo-second-order model yielded an excellent coefficient of determination for each trial of study compared to that of pseudo-first-order (Table 6). Thus, all the results were analyzed by fitting the data on pseudo-second-order equation.

4. Discussion

The adsorption studies were performed in simulated environments of pH 1.2 and 6.8 to mimic the human gastrointestinal tract. To achieve the saturation of AC with study drug, the proportion of the mass of AC : DXM was varied from 1 : 1 to 15 : 1. DXM in liquid form from the cough syrup was taken in order to simulate in vivo intoxication conditions when patients are intoxicated with the DXM syrup available in the market.

Equilibrium isotherms were used to describe the experimental adsorption data. The data were plotted onto the two most common adsorption isotherms; Langmuir isotherm and Freundlich isotherm, and their coefficient of determination (R^2) was compared to get the best adsorption isotherm equation. Based on the value of R^2, Langmuir isotherm was selected. To fit the data into the Langmuir adsorption isotherm, the value R^2 should be 0.87–1, indicating that the experimental data fits strongly into the individual regression. In the current study, all the R^2 values satisfied this condition and greater values of R^2 were found in Langmuir adsorption isotherm (0.9573 at simulated gastric fluid and 0.9997 in simulated intestinal fluid, resp.). Although Freundlich isotherm also fitted in SIF, a better value of R^2 was obtained with Langmuir isotherm. The study showed the suitability of Langmuir adsorption isotherm. The R^2 value in the current study was higher in simulated intestinal fluid compared with that in simulated gastric fluid. This can be explained on the basis of the excellent logarithmic relationship observed between adsorption affinity and solubility of DXM in simulated intestinal fluid.

Adsorption kinetics models were used to describe the experimental data of adsorption of DXM. The data were plotted onto two common adsorption kinetics models; pseudo-first-order and pseudo-second-order model, and their coefficients of determination (R^2) were compared. The pseudo-second-order model showed the best results in both SGF and SIF. To fit the data, the R^2 value should range from 0.87 to 1.00. The pseudo-second-order kinetics model fitted for this study with good R^2 (>0.9961) value, both in gastric and intestinal pH. The adsorption rate was higher in SIF than SGF.

The difference in maximum adsorption capacity of AC for DXM in SGF and SIF was insignificant ($p = 0.097$). AC (50 g) adsorbed 5.319 g to 6.3425 g of DXM (mean value: 5.8307 g; i.e., 19 bottles of 100 mL capacity in which each 5 mL contains 15 mg of DXM) at intestinal environment and 4.3103 g to 3.5211 g (mean value: 3.9157 g; i.e., 13 bottles of 100 mL capacity in which each 5 mL contains 15 mg of DXM) at gastric environment.

The calculation of Gibb's free energy and calculation of suitability of adsorption factor showed that the reaction was spontaneous, and equilibrium was reached. Thus, these data indicated that charcoal is suitable for adsorption of this drug [25]. A study performed by American Association of Poison Control Centers has concluded that activated charcoal could be administered as part of the management of a DXM poisoned patient despite the lack of specific data to support its use [12]. In the current study, the DXM showed

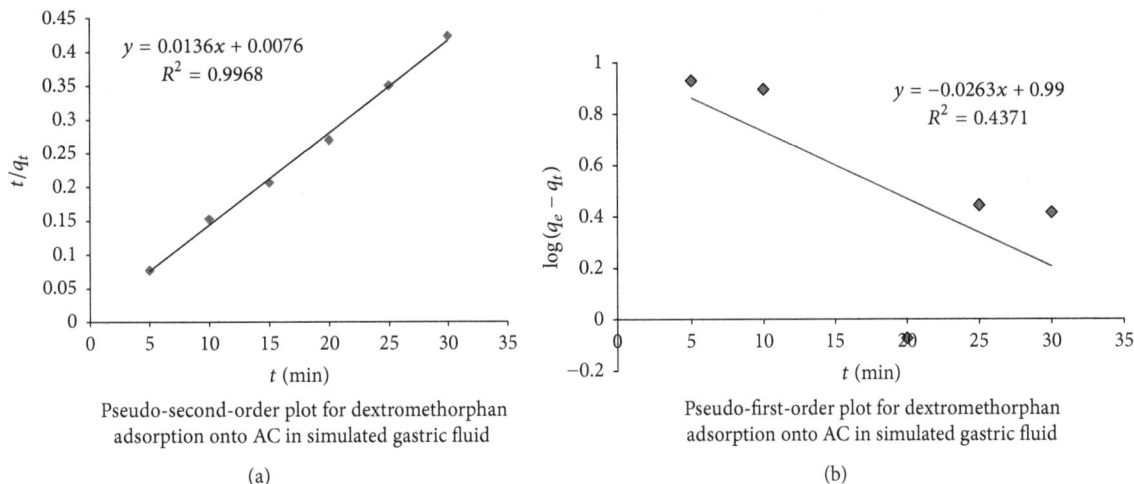

FIGURE 4: (a) Pseudo-second-order plot for adsorption of DXM on AC in SGF and (b) pseudo-first-order plot for adsorption of DXM on AC in SGF.

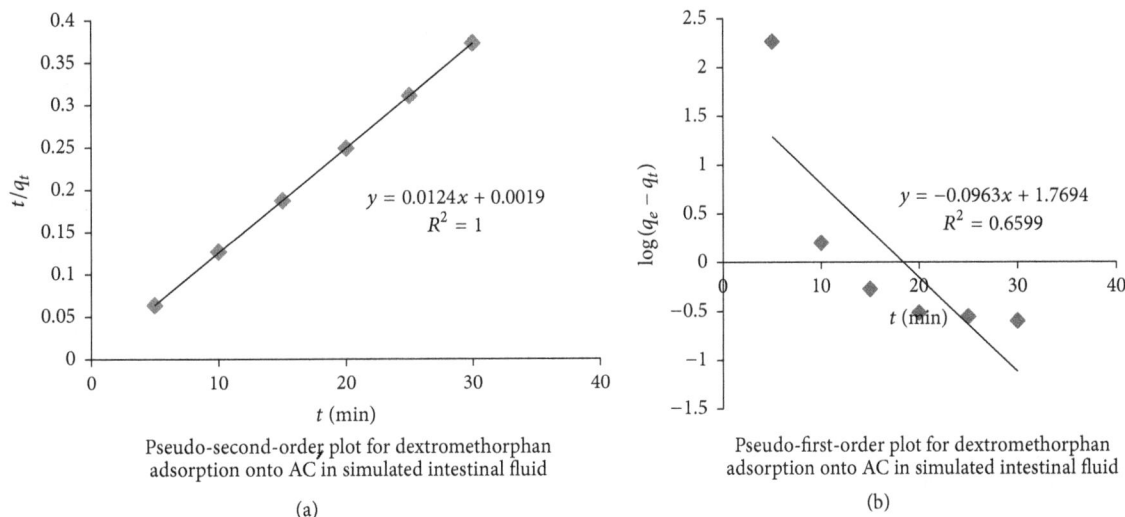

FIGURE 5: (a) Pseudo-second-order plot for adsorption of DXM on AC in SIF and (b) pseudo-first-order plot for adsorption of DXM on AC in SIF.

good adsorption; 78.314 mg/g of charcoal was adsorbed in simulated gastric fluid and 116.615 mg/g of charcoal was adsorbed in simulated intestinal fluid (p = 0.097). Our results further support the use of activated charcoal for DXM poisoning in clinical settings.

5. Conclusion

The main objective of our research was to investigate the adsorption capacity of activated charcoal for DXM syrup. Further studies need to be conducted to correlate our in vitro data with the in vivo conditions. Nonetheless, the present in vitro study shows that AC can adsorb sufficient amount of DXM. This in vitro study showed that sufficient amount of DXM can be adsorbed by the activated charcoal in its

standard treatment dose. The use of charcoal might be the most effective therapeutic approach for DXM poisoning.

Competing Interests

The authors declare that they have no competing interests.

References

[1] P. Amaratunga, M. Clothier, B. L. Lemberg, and D. Lemberg, "Determination of dextromethorphan in oral fluid by LC-MS-MS," *Journal of Analytical Toxicology*, vol. 40, no. 5, pp. 360–366, 2016.

[2] P. J. Perry, K. Fredriksen, S. Chew et al., "The effects of dex-tromethorphan on driving performance and the standardized

field sobriety test," *Journal of Forensic Sciences*, vol. 60, no. 5, pp. 1258–1262, 2015.

[3] T. Tartu and E. M. L. De Lima, "WHO Expert Committee on Drug Dependence".

[4] T. Ryan, M. Brewer, and L. Small, "Over-the-counter cough and cold medication use in young children," *Pediatric Nursing*, vol. 34, no. 2, pp. 174–184, 2008.

[5] M. D. Wilson, R. W. Ferguson, M. E. Mazer, and T. L. Litovitz, "Monitoring trends in dextromethorphan abuse using the National Poison Data System: 2000–2010," *Clinical Toxicology*, vol. 49, no. 5, pp. 409–415, 2011.

[6] C. M. A. Rademaker, A. van Dijk, M. H. de Vries, F. Kadir, and J. H. Glerum, "A ready-to-use activated charcoal mixture: adsorption studies in vitro and in dogs: its influence on the intestinal secretion of theophylline in a rat model," *Pharmaceutisch Weekblad Scientific Edition*, vol. 11, no. 2, pp. 56–60, 1989.

[7] S. Panthee and S. P. Lohani, "In vitro adsorption studies of paracetamol to activated charcoal capsule, powder and suspension," *The Open Toxicology Journal*, vol. 2, no. 1, pp. 22–25, 2008.

[8] L. C. G. Hoegberg, H. R. Angelo, A. B. Christophersen, and H. R. Christensen, "Effect of ethanol and pH on the adsorption of acetaminophen (paracetamol) to high surface activated charcoal, in vitro studies," *Journal of Toxicology: Clinical Toxicology*, vol. 40, no. 1, pp. 59–67, 2002.

[9] D. N. Bailey and J. R. Briggs, "The effect of ethanol and pH on the adsorption of drugs from simulated gastric fluid onto activated charcoal," *Therapeutic Drug Monitoring*, vol. 25, no. 3, pp. 310–313, 2003.

[10] J. M. Valente Nabais, B. Ledesma, and C. Laginhas, "Removal of amitriptyline from simulated gastric and intestinal fluids using activated carbons," *Journal of Pharmaceutical Sciences*, vol. 100, no. 12, pp. 5096–5099, 2011.

[11] M.-Y. Lo, M. W. Ong, J.-G. Lin, and W.-Z. Sun, "Codeine consumption from over-the-counter anti-cough syrup in Taiwan: a useful indicator for opioid abuse," *Acta Anaesthesiologica Taiwanica*, vol. 53, no. 4, pp. 135–138, 2015.

[12] P. A. Chyka, A. R. Erdman, A. S. Manoguerra et al., "Dextromethorphan poisoning: an evidence-based consensus guideline for out-of-hospital management," *Clinical Toxicology*, vol. 45, no. 6, pp. 662–677, 2007.

[13] D. Cooney, *Activated Charcoal in Medical Applications*, Marcel Dekker, New York, NY, USA, 2nd edition, 1995.

[14] P. A. Chyka and D. Seger, "Position statement: single-dose activated charcoal," *Journal of Toxicology - Clinical Toxicology*, vol. 35, no. 7, pp. 721–741, 1997.

[15] M. Smilkstein, "Techniques used to prevent gastrointestinal absorption of toxic compounds," in *Goldfrank's Toxicologic Emergencies*, pp. 35–51, Appleton & Lange, Stamford, Conn, USA, 6th edition, 1998.

[16] C. A. Bainbridge, E. L. Kelly, and W. D. Walkling, "In vitro adsorption of acetaminophen onto activated charcoal," *Journal of Pharmaceutical Sciences*, vol. 66, no. 4, pp. 480–483, 1977.

[17] P. J. Neuvonen, K. T. Olkkola, and T. Alanen, "Effect of ethanol and pH on the adsorption of drugs to activated charcoal: Studies *In Vitro* And In Man," *Acta Pharmacologica et Toxicologica*, vol. 54, no. 1, pp. 1–7, 1984.

[18] E. M. Sellers, V. Khouw, and L. Dolman, "Comparative drug adsorption by activated charcoal," *Journal of Pharmaceutical Sciences*, vol. 66, no. 11, pp. 1640–1641, 1977.

[19] A. T. Sitoura, J. Atta-Politou, and M. A. Koupparis, "In vitro adsorption study of fluoxetine onto activated charcoal at gastric and intestinal pH using high performance liquid chromatography with fluorescence detector," *Journal of Toxicology - Clinical Toxicology*, vol. 35, no. 3, pp. 269–276, 1997.

[20] H. Guven, Y. Tuncok, S. Gidener et al., "in vitro adsorption of dichlorvos and parathion by activated charcoal," *Journal of Toxicology: Clinical Toxicology*, vol. 32, no. 2, pp. 157–163, 1994.

[21] A. P. Terzyk and G. Rychlicki, "The influence of activated carbon surface chemical composition on the adsorption of acetaminophen (paracetamol) in vitro: the temperature dependence of adsorption at the neutral pH," *Colloids and Surfaces A: Physicochemical and Engineering Aspects*, vol. 163, no. 2-3, pp. 135–150, 2000.

[22] J. Pharmacopoeia, "United States' Pharmacopeia," 2002.

[23] M. R. Louhaichi, S. Jebali, M. H. Loueslati, N. Adhoum, and L. Monser, "Simultaneous determination of pseudoephdrine, pheniramine, guaifenisin, pyrilamine, chlorpheniramine and dextromethorphan in cough and cold medicines by high performance liquid chromatography," *Talanta*, vol. 78, no. 3, pp. 991–997, 2009.

[24] E. K. Putra, R. Pranowo, J. Sunarso, N. Indraswati, and S. Ismadji, "Performance of activated carbon and bentonite for adsorption of amoxicillin from wastewater: mechanisms, isotherms and kinetics," *Water Research*, vol. 43, no. 9, pp. 2419–2430, 2009.

[25] A. M. Aljeboree, A. N. Alshirifi, and A. F. Alkaim, "Kinetics and equilibrium study for the adsorption of textile dyes on coconut shell activated carbon," *Arabian Journal of Chemistry*, 2014.

18

Effect of Bentonite on the Pelleting Properties of Iron Concentrate

Hao Liu,[1,2] Bing Xie,[1] and Yue-lin Qin[2]

[1]College of Materials Science and Engineering, Chongqing University, Chongqing 400044, China
[2]School of Metallurgy and Materials Engineering, Chongqing University of Science and Technology, Chongqing 401331, China

Correspondence should be addressed to Yue-lin Qin; qinyuelin710@163.com

Academic Editor: María D. Alba

The physical and chemical properties such as particle size, montmorillonite content, swelling degree, water absorption, and blue absorption of A, B, and C bentonites were studied under laboratory conditions. The effects of adding different quality and different proportion of bentonite on falling strength, compression strength, and shock temperature of green pellet were investigated. The experimental results show that the montmorillonite content, water absorption, and methylene blue absorption of bentonite-B are the highest. And the quality of bentonite-B is the best, followed by bentonite-C and bentonite-A poor quality. When the amount of bentonite-B reduced from 1.5% to 1.0%, the strength of green pellets and the shock temperature both decrease. As the same proportion of A, B, and C bentonites, the green-ball strength and shock temperature are as follows: bentonite-A > bentonite-B > bentonite-C.

1. Introduction

Pellets should have the properties of high grade, good intensity, and uniform granularity as beneficial furnace burden in the ironmaking process. To increase yield and reduce coke consumption and to improve economic benefits, acidic pellet can be combined with highly basic pellet to form appropriate burden structure [1]. Pellet production has developed significantly as the appropriate burden design of "high basicity sinter + acidic pellet" has become widely adopted throughout the country.

Bentonite is a traditional metallurgical pellet binder, which ensures that the dry and roasted pellets have certain strength properties that meet transportation requirements [2]. Bentonite is a hydrated clay mineral with montmorillonite as its main constituent. The chemical structure of montmorillonite is $(Al_2, Mg_3) (Si_4O_{10})(OH)_2 \cdot nH_2O$. Montmorillonite also contains small amounts of illite, kaolinite, halloysite, chlorite, zeolite, quartz, feldspar, and calcite. Bentonite is either white or yellowish in color and has a wax-like, soil-like, or grease-like luster that depends on iron content. Bentonite can be classified as Na-bentonite, natural bleaching earth, and Ca-bentonite on the basis of exchangeable cation type and content, as well as the size of the montmorillonite layer charge.

Bentonite has strong hygroscopicity and expansibility. It can absorb water by as much as 8–15 times its dry mass. When wet, it can expand by even up to 30-fold of its dry mass. Bentonite can be dispersed in aqueous medium as a viscous and suspended material. Bentonite solutions have viscous, variable, and lubricant properties. Different types of bentonites influence pellet properties, as well as the interaction among heterogeneous or homogeneous bentonites and iron concentrate [3–6]. In this study, we utilized three different types of bentonites for a pelletizing test. We investigated the effects of different granularity, components, and content of bentonite on the falling strength, compression strength, and shock temperature on the produced green pellet. The results of this study will provide an experimental basis for the industrial production of bentonite pellets.

2. Experimental

2.1. Materials. Iron concentrates and bentonites produced by an iron and steel company in southwest China were used. The main chemical constituents of the iron concentrate and the

TABLE 1: Physical and chemical analysis of the iron concentrate (%).

Constituents	TFe	SiO_2	Al_2O_3	CaO	TiO_2	MgO	FeO	S	−0.074 mm	Moisture
Content	57.57	5.93	1.20	1.10	1.80	4.80	19.78	0.413	73.28	4.5

TABLE 2: Chemical analysis of the bentonites (%).

Constituents	MgO	CaO	Al_2O_3	P_2O_5	SiO_2	Fe_2O_3	K_2O	Na_2O	MnO_2
Bentonite-A	4.10	2.83	13.80	0.034	61.68	2.41	0.830	0.156	0.014
Bentonite-B	3.58	3.34	15.93	0.040	71.12	1.97	0.527	1.070	0.020
Bentonite-C	3.49	1.96	14.66	0.010	63.64	1.38	0.751	0.465	0.050

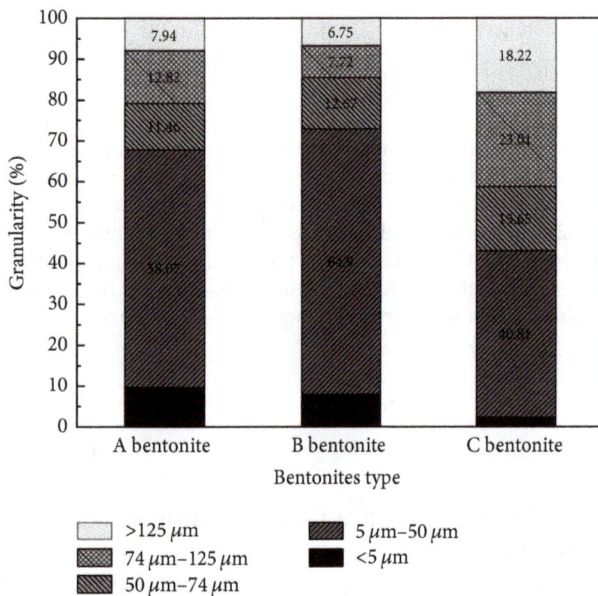

FIGURE 1: Size distribution of bentonite.

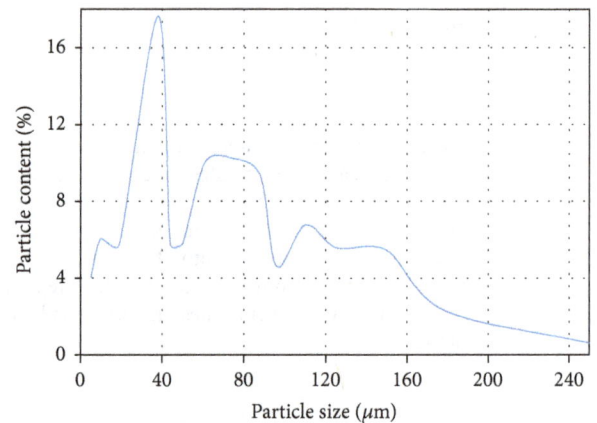

FIGURE 2: The full size distribution of the iron ore concentrate.

bentonites were analyzed by X-ray Fluorescence (XRF). Both the granularity and moisture content of the iron concentrate were also determined. The results are presented in Tables 1 and 2. The granularity of three different bentonites types A, B, and C were analyzed and determined by Winner 3002 particle size analyzer, and it is a laser diffraction instrument. The bentonite was analyzed as a dry powder, and it was ground, as shown in Figure 1.

As indicated in Figure 1, the particle size distribution of bentonite types A, B, and C was mainly concentrated in the 5–50 μm range at 58.07%, 64.90%, and 40.81%, respectively. Type B bentonite had the smallest particle size, followed by type A. Type C bentonite had the largest particle size and the most even particle distribution. Figure 2 shows the full size distribution of the iron ore concentrate.

2.2. Experimental Method. The parameters that represent bentonite properties include colloid value, degree of swelling, water absorption, methylene blue absorption, and montmorillonite content. (1) Colloid value, or colloid content or colloid degree, indicates the montmorillonite content of

bentonite. Colloid value is expressed by the total volume percentage of colloidal mixture in a sample to a certain proportion of water after standing for 24 hours. (2) Expansion factor, or degree of expansion, is related to the dispersion ability of bentonite in water. (3) Methylene blue absorption refers to the grams of methylene blue absorbed by 100 g bentonite in water. (4) Montmorillonite content is measured on the basis of the absorption of methylene blue by montmorillonite. Metallurgical bentonite quality standard of China (GB/T20973-2007) is adopted in the process of experiment to test and determine the main property of the bentonite.

This study utilized a disc pelletizer with 45° incline and 1 m diameter to conduct the pelletizing experiment. The pelletizing process consists of three stages: (1) mixing of the raw material to produce the raw pellet, (2) pellet enlargement, and (3) pellet hardening. During the pelletizing, the atomized water was added in the area of disc pelletizer. The water was deionized to avoid the effects of water chemistry on pelletization performance. The total weight of raw material was 4 kg. During the first stage, which occurred for 3 min, raw pellets were produced from 1 kg of raw material. After 9 min, more raw material was added to enlarge the raw pellets. During the last stage, the enlarged pellet was hardened for 3 min. The total time of pelletizing process was 15 min. Afterwards, the green pellets were removed and continually sprayed with water to maintain moisture content at 7.0±0.5%.

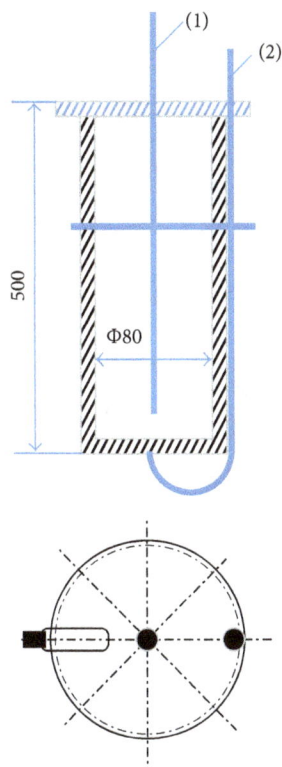

(1) Thermocouple
(2) Compressed air into the pipe

FIGURE 3: Burst temperature measuring device.

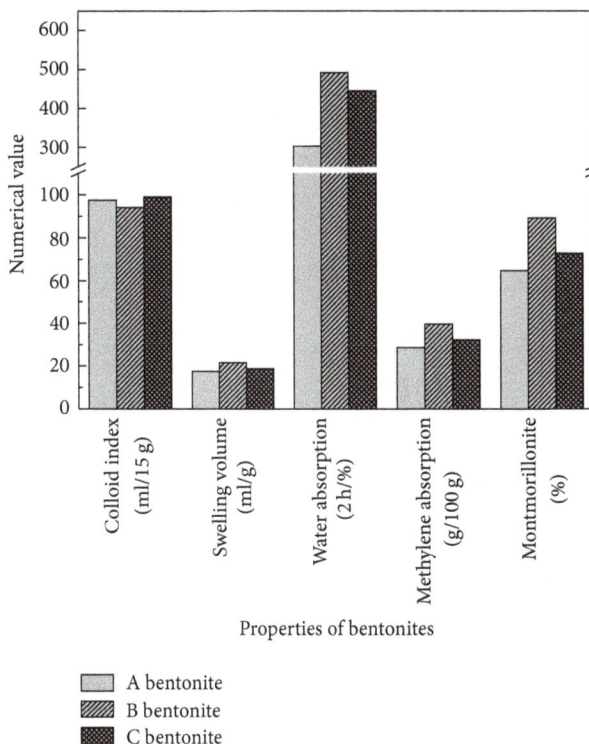

FIGURE 4: Results of bentonite pellet test.

The dry configuration proportions were 1.5% (bentonite-A), 1.5% (bentonite-B), 1.3% (bentonite-C), 1.0% (bentonite-B), and 1.0% (bentonite-C).

Green pellet properties include compression strength, moisture content, falling strength, and shock temperature. (1) To test compression strength, the green pellet was placed on an electronic scale and pressed at constant speed until the green pellet broke. The pressure value was recorded when the pellet broke. The same process was repeated 10 times, and the recorded pressure values were averaged over the 10 repeats. (2) To test the moisture content, the green pellet was pulverized. Then, a 20–25 g sample was placed in a rapid moisture measuring instrument to measure and record the value of moisture content. The same process was repeated thrice. The moisture content value was averaged over the three attempts. (3) To test falling strength, the green pellet was dropped from a height of 0.5 m until the ball broke for certain number of times. The test was repeated 10 times. Falling strength values were averaged over the 10 attempts. (4) Shock temperature was assessed with the Dynamic Method developed by AC Company, USA (Figure 3). The shock temperature test was conducted in the special vertical tube furnace. Ambient-temperature air was sent from an air compressor into the tube furnace at a speed controlled by a rotary flowmeter. The tube furnace was equipped with a silicon carbide bar. Its temperature was controlled by an automatic silicon temperature controller. A $\varphi 80 \times 500$ mm

stainless steel hot air blast pipe, which contained a $\varphi 10$ mm alumina ball 200 mm in height, was installed in the device. The electric furnace heated the alumina ball, which instantly heated up the air from the compressor. The thermocouple for real-time detection of the hot air temperature was inserted in a hole from the top of the furnace. As illustrated in Figure 3, the drying container used to collect the green pellet had a heat-resistant nickel chromium wire frame, an inner diameter of 50 mm, and a height of 120 mm. Thus, air flow can easily pass through the green pellet layer. Then, 25 green pellets were loaded into the nickel chromium wire frame and hoisted into the reaction drum from the top of the drum. The air speed inside the drum was 1.5 m/s under cold conditions. After 3 minutes, the green pellets were removed. The highest temperature required to break 4% of the green pellets was recorded as the shock temperature.

3. Results and Discussion

3.1. Physical and Chemical Properties of Bentonite. Figure 4 illustrates the properties of the three bentonite types utilized by this experiment. The montmorillonite content of type B bentonite is 89.25%, which is higher than those of type A (64.7%) and type C (72.74%). Type B has the highest values for swelling volume, water absorption, and methylene blue absorption, followed by type C and type A. Montmorillonite absorbs methylene blue when bentonite is dispersed in water. Hence, bentonite with a high montmorillonite content absorbs more methylene blue. Methylene blue absorption is also an indicator for absorptivity and water absorption

TABLE 3: Effect of bentonite type on the pellet index.

Exp. number	Moisture content/%	Falling strength/times·pellet^{-1}	Compression strength/N·pellet^{-1}	Shock temperature/°C	Bentonite content/%	Weight of qualified pellets (>10 mm)/kg
1	7.28	2.53 ± 0.01	9.02 ± 0.02	550 ± 3	1.5 (A)	0.94
2	6.79	3.00 ± 0.02	9.76 ± 0.03	550 ± 2	1.5 (B)	1.43
3	6.64	2.60 ± 0.01	9.42 ± 0.02	550 ± 2	1.3 (B)	0.50
4	7.46	2.53 ± 0.02	8.23 ± 0.02	500 ± 3	1.0 (B)	0.67
5	7.13	2.07 ± 0.02	7.97 ± 0.02	500 ± 2	1.0 (C)	1.41

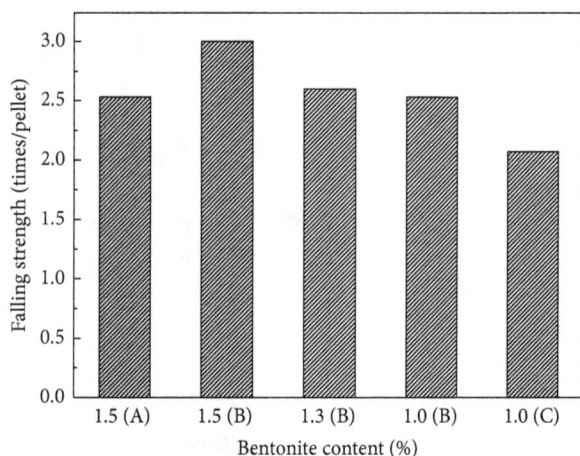

FIGURE 5: Effect of different types and proportions of bentonite on the falling strength of the ball.

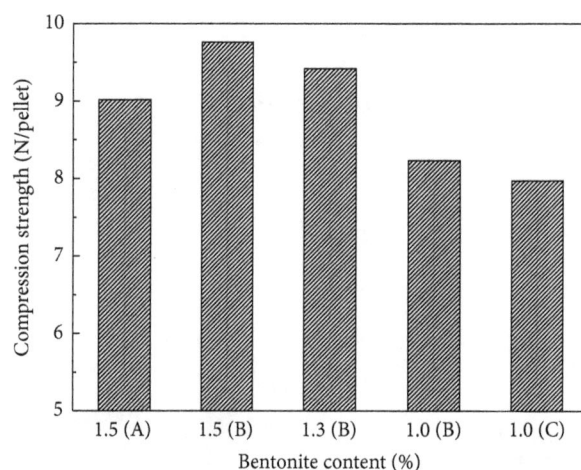

FIGURE 6: Effect of different types and proportions of bentonite on the compression strength of the ball.

[3]. A higher methylene blue absorption value indicates that the bentonite has higher absorptivity, water absorption capacity, and larger swelling volume. Methylene blue absorption, swelling volume, montmorillonite content, and water absorption are important indices for evaluating bentonite quality. Therefore, type B bentonite has the highest quality, followed by type C and finally type A.

3.2. Effect of Different Types and Content of Bentonite on the Properties of Green Pellets.

Table 3 present the properties of green pellets that contain different types and proportions of bentonites. Type A and type B bentonite content is set at 1.5% in experiments 1 and 2. Bentonite-B content is set at 1.3% in experiments 3. Type B and bentonite-B and bentonite-A content is set at 1.0% in experiments 4 and 5.

Figure 5 shows the effect of different types and proportions of bentonite on the falling strength of the green pellet. When bentonite content is 1.5% and 1.3%, the falling strength of the green pellet improved compared with those of green pellets that contain 1.0% bentonite. Increasing homogeneous bentonite content from 1.5% to 1.0% gradually decreases the falling strength from 3.00 times/pellet to 2.53 times/pellet. When the bentonite content is 1.5%, the falling strength of the green pellets of type B is better than that of the green pellets of type A. When the bentonite content is 1.0%, the falling strength of green pellets of type B is better than that of green pellets of type C. The falling strength of the three different

types of bentonites decreases in the following order: type B > type C > type A.

Figure 6 presents the effect of different types and proportions of bentonite on the compression strength of the green pellet. With increasing homogeneous bentonite content from 1.5% to 1.0%, the compression strength decreases from 9.76 N/pellet to 8.23 N/pellet. When bentonite contents are 1.5% and 1.3%, the compression strength of the green pellet improved compared with those of green pellets that contain 1.0% bentonite. When the bentonite content is 1.5%, the compression strength of the green pellets of type B is better than that of the green pellets of type A. When the bentonite content is 1.0%, the compression strength of green pellets of type B is better than that of green pellets of type C. The compression strength of the three different types of bentonites decreases in the following order: type B > type C > type A.

Figure 7 shows that when the bentonite contents are 1.5% and 1.3%, the shock temperature of the pellet is 823 K, compared with the 773 K shock temperature of the pellet with 1.0% added bentonite. The pellet explodes during the drying process because moisture transfers towards loosely structured areas of the pellet, which increases internal pressure. The pellet explodes when the internal pressure exceeds the pellet's tensile strength. Given that bentonite has a greater capacity for water absorption, the crystal layers of bentonite can firmly absorb moisture onto the surface

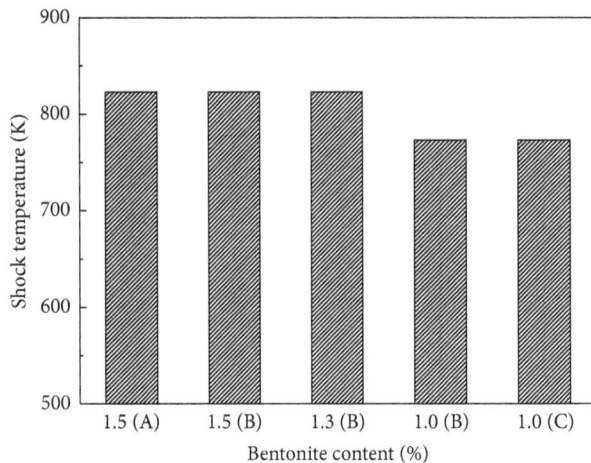

FIGURE 7: Effect of different types and proportions of bentonite on the shock temperature of the ball.

of montmorillonite because of the effect of negative field force. Therefore, free water does not easily evaporate and transfer. During the drying and evaporation process, the water absorption property of bentonite slows down water evaporation from the pellet. Thus, moisture is released slowly from the pellet, which reduces the internal steam pressure and increases the shock temperature of the pellet [7–10]. Therefore, increasing bentonite content increases the shock temperature of the pellet.

The results indicate that the quality of the three different types of bentonites decreases in the following order: type B > type C > type A. Therefore, at a constant content of heterogeneous bentonites, falling strength and compression strength are dependent on bentonite quality because bentonite particles are fine with a specific surface area of $100\,\mathrm{m^2/g}$. Hence, the dispersion property of bentonite is not only acceptable, but its water absorption and expansion properties also enable fine particles to fill and absorb between the ore particles. The nature of the ore surface changes accordingly to form microcapillaries when a solid bridge and liquid bridge are added. When capillary force increases, the compression strength of the green pellet also improves [11]. Montmorillonite content dictates the physicochemical properties of bentonite and montmorillonite arranges into structured layers of aqueous aluminum silicate with expansion property [7, 12]. Bentonite begins to swell upon absorbing water. Under mechanical shear stress, the swollen bentonite can improve the relative sliding between the ore particles via plastic deformation [13, 14]. Therefore, bentonite is not easily broken and the falling strength of the green pellet is enhanced.

4. Conclusion

This study investigated the influence of type and proportion of bentonite on falling strength, compression strength, and shock temperature of green pellets by adding bentonite of different types and proportions to iron concentrates. The following conclusions were made:

(1) Montmorillonite content is directly correlated with the degree of swelling, water absorption, and methylene blue absorption of bentonite. The degree of swelling, water absorption, and ethylene blue absorption increase as montmorillonite content increases. These characteristics indicate the good quality of bentonite.

(2) Adding bentonite improves the strength of the green pellet. The falling strength and compression strength of green pellet increase as bentonite content increases. For homogeneous bentonites, the strength of the green pellet increases as bentonite content increases. For heterogeneous bentonites, the strength of green pellet depends on the quality of bentonite at a constant bentonite content.

(3) Adding a higher proportion of high-quality bentonite to the iron concentrates increases the bursting temperature resistance of the green pellet.

Conflicts of Interest

The authors declare that there are no conflicts of interest regarding the publication of this paper.

Acknowledgments

The authors would like to acknowledge the National Natural Science Foundation of China (no. 51604054), Scientific and Technological Research Program of Chongqing Municipal Education Commission (nos. KJ1501307 and KJ1601331), and Chongqing Research Program of Basic Research and Frontier Technology (no. cstc2016jcyjA0647).

References

[1] X. B. Zhang, "Effect of bentonite additive on pelletizing operation of our country," *Sintered pellets*, vol. 28, no. 6, pp. 3–6, 2003.

[2] X. B. Zhang, "Discussion on performance index of bentonite used for metallurgical pelletizing," *Sintered pellets*, vol. 35, no. 4, pp. 12–16, 2010.

[3] Y. Z. Zhang, "Influence of bentonite properties on pellet performances," *Sintered pellets*, vol. 31, no. 2, pp. 4–21, 2006.

[4] Y. R. Li, M. S. Zhou, L. W. Zhai, L. G. Zhang, and W. Ren, "Effect of types of bentonites on green pellet propert," *Anshan Iron and Steel Technology*, vol. 3, no. 1, pp. 15–19, 2009.

[5] Z. C. Huang, Y. Jiang, Z. G. Han, and G. X. Huang, "Study on the mechanism of using bentonite as binder in spheric agglomeration of iron ores," *Chemical Minerals and Processing*, vol. 8, no. 1, pp. 13–15, 2005.

[6] Z. Y. Lv, M. S. Shen, and W. G. Kang, "Experimental research on restraining reduced expansion of pellets by adding bentonite," *Baotou Steel Technology*, vol. 37, no. 4, pp. 33–35, 2011.

[7] B. Alince, F. Bednar, and T. G. M. Van de Ven, "Deposition of calcium carbonate particles on fiber surfaces induced by cationic polyelectrolyte and bentonite," *Colloids and Surfaces A: Physicochemical and Engineering Aspects*, vol. 190, no. 1-2, pp. 71–80, 2001.

[8] T. C. Eisele and S. K. kawatra, "A review of of binders in iron ore palletization," *Minerals Processing and Extractive Metallurgy Review*, vol. 24, no. 1, pp. 1–47, 2003.

[9] Y. M. Zhang, *Pellet theory and technology*, Metallurgical Industry Press, Beijing, China, 2008.

[10] Z. L. Xue, "Study on the shock temperature of iron ore green pellets," *Journal of Wuhan Iron and Steel Institute*, vol. 15, no. 3, pp. 249–255, 1992.

[11] H. Q. Zhang, "Application of bentonite in iron ore oxidation pellets," *China Mining*, vol. 18, no. 8, pp. 99–102, 2009.

[12] S. K. Kawatra and S. J. Ripke, "Effects of bentonite fiber formation in iron ore pelletization," *International Journal of Mineral Processing*, vol. 65, no. 3-4, pp. 141–149, 2002.

[13] C. X. Li, R. C. Ren, and M. Y. Liao, "Study on binder of bentonite from western Liaoning for iron pellet," *Non-metallic mineral*, vol. 32, no. 3, pp. 42–46, 2009.

[14] H. M. Feng and Y. H. Wang, "Explore mechanism of benonite in iron ore pelletting," *China Non-metallic Mineral Industry Guide*, vol. 6, no. 1, pp. 15–18, 2009.

19

Thermal Stability and Luminescent Properties of Tri-cellulose Acetate Composites Containing Dy(SSA)₃Phen Complex

Xiangwei Sun, Feiyue Wu, Yan Luo, Mengjun Huang (iD), Yuntao Li, Xiang Liu, and Kunpeng Liu

College of Materials and Chemical Engineering, Chongqing University of Arts and Sciences, Honghe Road 319, Chongqing 402160, China

Correspondence should be addressed to Mengjun Huang; huangmj806@gmail.com

Academic Editor: Marco Anni

Dysprosium (Dy) ternary complex was prepared using 5-sulfosalicylic acid (SSA) as the first ligand and 1,10-phenanthroline (Phen) as the second ligand, denoted as Dy(SSA)₃Phen. The complex was blended with tri-cellulose acetate (TCA) via a cosolvent method to obtain polymer luminescent materials. The composition and structure of the rare-earth complex were characterized by means of elemental analysis, infrared (IR) spectra, and thermogravimetric analysis (TGA). The fluorescence spectra displayed this pure Dy(SSA)₃Phen complex, and the TCA/Dy(SSA)₃Phen composites all emit blue light. The (90/10) composite possesses fine luminescent properties with quantum yield of 33.5% and thermal stability for potential usage as blue fluorescent materials.

1. Introduction

Polymer/rare earth (RE) composites possess excellent optical, electrical, and spectroscopic properties stemming from the additive rare-earth compounds and meantime have fine processability and flexibility arising from polymer materials, which can be potentially applied in light-emitting diodes, detector systems, X-ray phosphors, solid-state lasers and bioinorganic sensors, etc. [1–5]. Considerable attention has been paid to design and synthesize polymer/RE luminescent materials with good luminescent properties and high internal quantum efficiencies [6–8]. RE-based emitters which are usually doped or bonded with macromolecules can generate characteristically sharp and narrow emissions, because of the distinctive excitation mechanism of antenna effect and f-f radiative transition [9–14]. A considerable number of research reports have focused on four kinds of RE metals, i.e., europium (Eu), terbium (Tb), samarium (Sm), and dysprosium (Dy), especially for the Eu or Tb ions based complexes, because their resonance energy levels are close to the triplet levels of organic ligands, and so the RE ions could emit strong luminescent properties via the efficient energy transfer between ligands and RE ions [15–19].

Notably, doping luminescent RE organic complexes into polymer matrices is one of the most effective approaches to obtain the polymer luminescent materials via a simple, convenient, and high efficient blending method. However, some doped-polymer/RE composites exhibited poor luminescent properties and poor mechanical properties, mainly owing to the nonuniform dispersion of RE complexes in polymer matrices and the poor chemical affinity between the doped complex and polymer matrix [9, 20]. In order to solve these problems, organic ligands of RE complexes play a significant role in boosting the chemical affinity between the added complexes and macromolecules, and especially the organic compounds possessing multifunctional groups which could be potential hydrogen-bonding acceptors or donors are primarily selected as ligands for RE complexes [21–23]. Conjugated acids were usually used as the first ligand as a result of their strong coordination ability with RE ions and the potential hydrogen-bonding acceptors from carbonyl groups, e.g., salicylic acid [13], phthalic acid [24], benzoic acid and its derivatives [25], and acrylic acid [26]. It is worthwhile pointing out that the molecular structure of polymer matrices is another significant factor for promoting the chemical affinity between the added RE complexes and polymer chains.

In this paper, tri-cellulose acetate (TCA) was selected as the polymer matrix, because of the potential hydrogen-bonding donors from the substantial hydroxyl groups. Furthermore, as a result of good filming performance, excellent transparency, fine glossiness, well solubility, and good processability of TCA, it had been widely used in the protective film of the polarizing film of the liquid-crystal display (LCD) which required a high transparency [27]. TCA could be easily soluble in a variety of solvents, favorable to prepare polymer/RE composites by means of solution blending. Additionally, sulfosalicylic acid (SSA) was chosen as the first ligand, not only because it could sensitize the luminescent property of Dy ions but also since its oxygen of carbonyl groups provides hydrogen-bonding acceptors [13, 19]. Therefore, the hydrogen bonding is most probably formed between the ligands of the added RE complex and the TCA matrix, favorable to improve the uniform dispersion of the complex in the matrix. As a series of studies on polymer luminescent materials emitting three-color light (red, green, and blue) [10, 28], Dy-based complexes blended with TCA have been highly concerned. There are few reports of Dy^{3+} complex-doped polymer materials with fine luminescence, and thus, we synthesized Dy^{3+} ternary complexes using SSA and 1,10-phenanthroline (Phen) as the first and second ligands, respectively. The $Dy(SSA)_3Phen$ complex can retain fine luminescent properties dispersing into the TCA matrix, and finally, the $TCA/Dy(SSA)_3Phen$ composites emitting blue light with good thermal stability were obtained.

2. Materials and Methods

2.1. Materials. Tri-cellulose acetate (TCA), with the weight-average molecular weight of 200,000, was supplied by our collaborating laboratory (Xiuli Wang's Research Group, Sichuan University) and purified by acetone. Dysprosium oxide (Dy_2O_3) at 99.99% purity was provided by Shanghai Yuelong Rare Earth New Materials Co., Ltd., China. 5-sulfosalicylic acid (SSA, AR), 1,10-phenanthroline (Phen, AR), dimethyl sulfoxide (DMSO, AR), acetone (EA, AR), and absolute ethyl alcohol (AR) were purchased from Chengdu Kelong Chemical Reagent Factory, China. Sodium hydroxide (NaOH, AR) was purchased from Chongqing Chuandong Chemical Co., Ltd., China. Concentrated hydrochloric acid (HCl, AR) was supplied by Chongqing Inorganic Chemical Reagent Factory, China.

2.2. Preparation of $Dy(SSA)_3Phen$ Complex. Weighed 3 mmol Dy_2O_3 and 80 ml concentrated hydrochloric acid in a 100 ml bottom beaker immersed in an oil bath with magnetic stirrer, heated to 100°C to dissolve Dy_2O_3 powders absolutely, kept at the temperature for 10 hours to evaporate the solution to dryness and then cooled to room temperature. Then dissolved the remains using absolute EA in a 250 ml volumetric flask, added absolute EA to the mark, and finally got 0.1 mol/L $DyCl_3$ ethanol solution. About

15 mmol SSA, 5 mmol Phen, and 30 ml absolute EA were successively added in 100 ml beaker immersed in an oil bath with a magnetic stirrer to dissolve all components absolutely, and then, 5 mmol Dy^{3+} ethanol solution was added in dropwise, keeping the reaction temperature at 80°C. The pH value of the mixture was adjusted to about 6.5 with NaOH aqueous solution (6 mol/L), and the mixed solution was kept at 80°C, stirred for 4 hours, and cooled to room temperature. The precipitates were repeatedly washed with EA and deionized water to remove chloride ions, dried in a vacuum oven at 60°C for 24 hours, and then stored in a damp-proof cabinet. The ternary complex, $Dy(SSA)_3Phen$, was prepared.

2.3. Preparation of $TCA/Dy(SSA)_3Phen$ Composites. Dissolved 0.50 g solid mixtures of the purified TCA and Dy$(SSA)_3Phen$ complex powders using 30.0 ml DMSO in a 100 ml flat bottom beaker immersed in an oil bath with stirring, heated the mixed solution to 60°C with stirring, then cooled to room temperature, and underwent a natural evaporation process until the solution became semisolid samples. The prepared semisolid samples were transferred to an evaporating dish with flat bottom and dried in a vacuum oven at 60°C for 10 hours. Finally thin film-like TCA/Dy$(SSA)_3Phen$ composites were obtained and stored in a damp-proof cabinet. The weight ratios of TCA to Dy$(SSA)_3Phen$ were 100/0, 96/4, 93/7, 90/10, and 87/13.

2.4. Characterization. Elemental analysis of C, H, and N in the pure $Dy(SSA)_3Phen$ complex was conducted on an Elemental analyzer (Vario EL III, Elementar Co., Germany).

Infrared transmittance spectrum was recorded on a Nicolet 6670 FTIR spectrophotometer (Thermo Fisher Corp., USA). Infrared attenuated total reflection (IRATR) was implemented and measured by averaging 32 scans at a resolution of $4 \, cm^{-1}$ over the range $4000–450 \, cm^{-1}$.

Thermogravimetric analysis (TGA) was implemented by a thermogravimeter (Q500, TA Instruments Co., USA) in the nitrogen atmosphere using a heating rate of 10°C/min from room temperature to 1000°C.

The luminescent properties of the pure $Dy(SSA)_3Phen$ complex and all $TCA/Dy(SSA)_3Phen$ composites were characterized by a fluorescence spectrophotometer (Hitachi F-7000, Hitachi High-Technologies Corp., Japan). The adopted mode was luminescence. The supplied PMT voltage and scan speed were set to 400 V and 1200 nm/min, respectively. The excitation (EX) and emission (EM) slit widths were fixed at 2.5 and 5.0 nm.

3. Results and Discussion

3.1. Elemental Analysis. The results of elemental analysis for the $Dy(SSA)_3Phen$ complex are, respectively, listed as follows: measured value (theoretical value) C 41.6% (40.7%), H 2.3% (2.5%), N 4.8% (5.0%), and S 6.5% (5.7%). These results suggest that the molar ratio of Dy^{3+} to SSA to Phen is $1:3:1$ in the prepared $Dy(SSA)_3Phen$ complex.

3.2. FTIR Analysis. Figure 1 shows infrared spectra of the two ligands and the Dy(SSA)$_3$Phen complex, in the wavelength range of 3550 to 450 cm^{-1}. As for the spectrum of Phen, the peak at 1561 cm^{-1} is assigned to the stretching vibration of -N=C group, and the peaks at 852 cm^{-1} and 739 cm^{-1} are ascribed to the out-of-plane bending vibration of -C-H group, and for the spectrum of SSA, the peak at 1656 cm^{-1} belongs to the stretching vibration of the carbonyl group. Obviously, compared with the spectrum of Phen, the peak of the spectrum of the Dy(SSA)$_3$Phen complex corresponding to the $v_{N=C}$ vibration shifts from 1561 cm^{-1} to 1536 cm^{-1}, and the peaks belonging to the r_{C-H} vibration shift from 852 cm^{-1} and 739 cm^{-1} to 843 cm^{-1} and 719 cm^{-1}, respectively, which indicates that there are coordination bonds between two nitrogen atoms from Phen ligand and the central ion (Dy^{3+} ion) in the complex [29]. Moreover, by comparison with the spectrum of SSA ligand, the peak of the spectrum of the complex assigned to the $v_{C=O}$ vibration shifts from 1656 cm^{-1} to 1621 cm^{-1}, indicative of the formation of the coordination bonds between carbonyl group from SSA ligand and Dy^{3+} ion [13, 30]. Furthermore, it is worth noting that as for the spectrum of the complex two new absorption peaks appearing at 549 cm^{-1} and 461 cm^{-1} are attributed to the stretching vibration of O $\longrightarrow$ RE and N $\longrightarrow$ RE, respectively [31].

Infrared spectra of TCA and the TCA/Dy(SSA)$_3$Phen (95/5) and (90/10) composites are displayed in Figure 2, ranging from 4000 cm^{-1} to 450 cm^{-1}. It can be seen that the two peaks of the spectrum of TCA occurring at 3480 cm^{-1} and 1740 cm^{-1} are assigned to the v_{O-H} and $v_{C=O}$ vibrations, respectively. Clearly, the spectrum of the (95/5) composite is similar to that of the pure TCA except that the peak corresponding to the v_{O-H} vibration of the composite shifts from 3480 cm^{-1} to 3415 cm^{-1}. The possible reason is that the amount of the added Dy(SSA)$_3$Phen complex is comparatively low so that the characteristic peaks of the complex are almost covered by the strong peaks of TCA matrix. Nevertheless, as for the spectrum of the (90/10) composite, the peaks at 3393 cm^{-1} and 1740 cm^{-1} are, respectively, attributed to the v_{O-H} and $v_{C=O}$ vibrations of the TCA matrix, and the peak at 1625 cm^{-1} is assigned to the $v_{C=O}$ vibration of SSA from the added complex. Furthermore, the main characteristic peaks of Dy(SSA)$_3$Phen complex at 846 cm^{-1}, 717 cm^{-1}, and 470 cm^{-1} also appear in the spectrum of the (90/10) composite. Compared with the spectra of the pure TCA and Dy(SSA)$_3$Phen complex as shown in Figures 1 and 2, it can be found that the v_{O-H} peak for the (90/10) composite is red-shifted about 90 cm^{-1} and the $v_{C=O}$ peak for the composite is red-shifted 4 cm^{-1} from the 1621 cm^{-1}, indicative of the existence of hydrogen bonds between the oxygen atom from carbonyl groups from SSA and the hydroxyl groups from TCA.

3.3. Thermal Analysis. Thermal properties of pure Dy (SSA)$_3$Phen complex and TCA/Dy(SSA)$_3$Phen composites were characterized via TG analysis, and the TGA curves are shown in Figure 3. As for the plot of the Dy(SSA)$_3$Phen complex, there is a two-stage decomposition process,

corresponding to the removal processes of two ligands. The total mass loss of the complex is 79.2%, which is in good agreement with the calculated value (80.9%) when the ligands of the Dy(SSA)$_3$Phen complex are completely decomposed and the final residues are rare-earth oxide [32]. The weight loss of the first stage is 58.6% which is considered as the removal process of SSA ligand, and that of the second stage is 20.4% belonging to the decomposition process of Phen ligand. Furthermore, at the first decomposition stage of the complex, the extrapolated onset (T_e) and final (T_c) temperatures are about 273°C and 325°C, and at the second stage, the T_e and T_c are about 557°C and 669°C, suggesting that the Dy(SSA)$_3$Phen complex possesses fitting thermal stabilities for the usage of luminescent materials. Based on the above-mentioned elemental analysis, FTIR analysis, and TGA results, the molecular structure of the Dy(SSA)$_3$Phen complex can be ascertained, as shown in Figure 4.

As for TCA and the TCA composites, the TGA thermograms are displayed in Figure 3, and the T_e and T_c values as well as the mass-loss values are listed in Table 1. Obviously, there is only one stage of decomposition occurring on the TGA curves of TCA and the composites, which suggests that the decomposition process of the composites is similar to that of the pure TCA, and during the processes, each chain scission reaction has an analogous energy barrier [33]. It can be seen that with the increase of the content of the added complex, the T_e, T_i, and T_c values show a descending trend. This is probably caused by the fact that the degradation process of TCA macromolecules is related to the rupture of the anhydroglucose ring, i.e., the backbone rupture as well as the carbonization process and the volatile substances released from organic ligands of the added complex are probably favorable to boost the rupture process of TCA macromolecular chains [34]. The more the additive content of the complex, the more the released substances and the faster the degradation processes of the TCA composites. However, it is particularly worth pointing out that the T_e and T_i values of all composites are nearly over 300°C, much higher than the onset temperature. As a consequence, the TCA/Dy(SSA)$_3$Phen composites still possess practicable thermal properties for materials.

3.4. Photoluminescent Properties. Figures 5 and 6 show the excitation and emission spectra of pure Dy(SSA)$_3$Phen complex and all TCA/Dy(SSA)$_3$Phen composites, respectively. It is obvious that the profile of the spectra of all composites is similar to that of the pure complex, i.e., there are broad and strong bands ranging from 260 nm to 360 nm in all excitation spectra, which suggests that the pure complex and the TCA composites can be well excited in the given wavelength range, as shown in Figure 5. The maximal values of the excitation spectra are the optimum excitation wavelengths for emission spectra, i.e., 326 nm is the optimum excitation wavelength for the pure Dy(SSA)$_3$Phen complex; 309 nm, 309 nm, 312 nm, and 312 nm are corresponding to the TCA/Dy(SSA)$_3$Phen (96/4, 93/7, 90/10, and 87/13) composites, respectively. Meanwhile, it can be seen that the excitation spectra of the composites represent

(a)   (b)

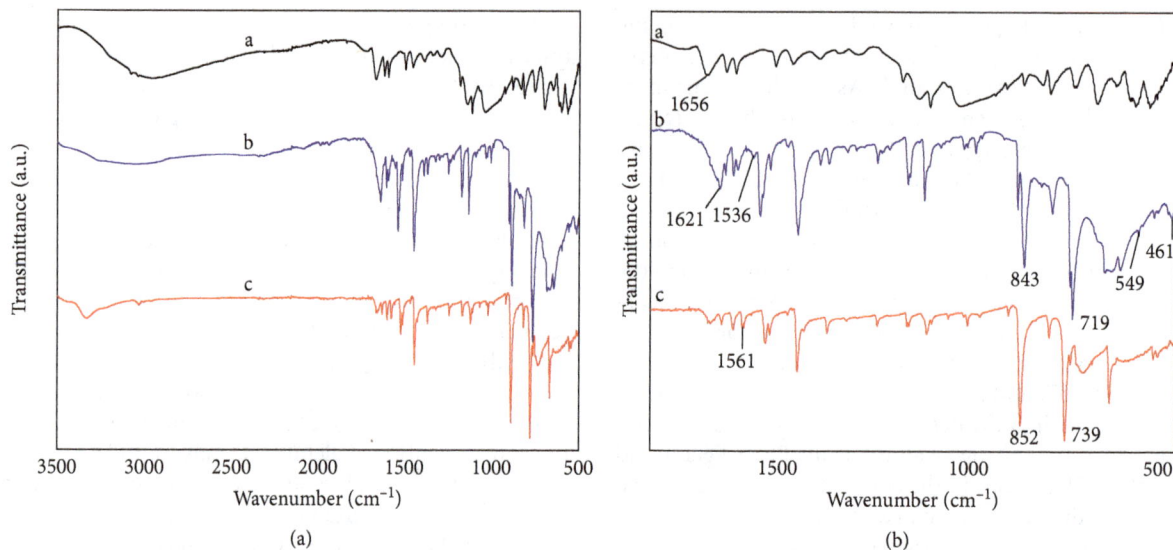

FIGURE 1: FTIR spectra ((a) 3550–450 cm^{-1}; (b) 1800–450 cm^{-1}) of ligands and the complex: a, SSA; b, Dy(SSA)$_3$Phen; c, Phen.

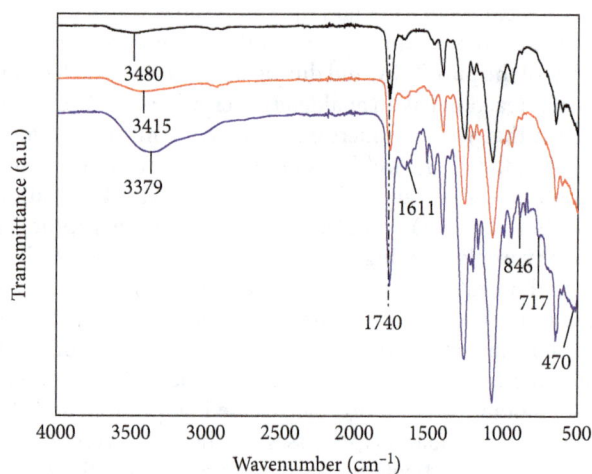

FIGURE 2: FTIR spectra of TCA and the TCA/Dy(SSA)$_3$Phen composites: a, TCA; b, (96/4); c, (90/10).

FIGURE 4: Molecular structure of the Dy(SSA)$_3$Phen complexes.

TABLE 1: The T_e, T_i, T_c, and mass-loss values of pure TCA and the TCA/Dy(SSA)$_3$Phen composites.

TCA/Dy(SSA)$_3$Phen	T_e (°C)	T_i (°C)	T_c (°C)	Mass loss (%)
(100/0)	347.0	366.9	380.1	92.1
(96/4)	325.9	342.1	364.3	89.9
(93/7)	313.4	331.3	355.2	88.6
(90/10)	297.7	316.9	342.6	86.6

T_e, T_i, and T_c denote extrapolated onset, onset, and final temperatures, respectively.

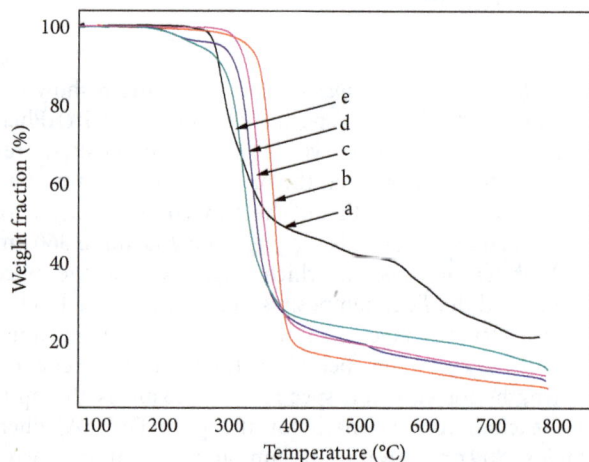

FIGURE 3: TGA thermograms of pure Dy(SSA)$_3$Phen complex and TCA/Dy(SSA)$_3$Phen composites: a, Dy(SSA)$_3$Phen; b, TCA; c, (96/4); d, (93/7); e, (90/10).

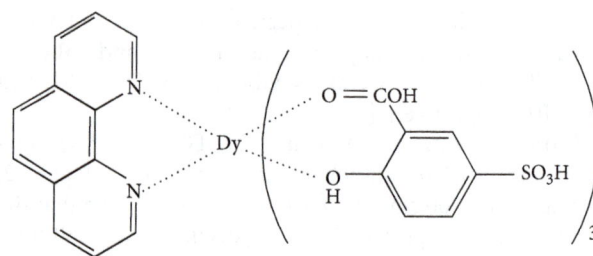

obvious blueshift compared with the 326 nm peak value for the pure complex, and especially for the (96/4) and (93/7) composites, the peaks are blueshifted by 17 nm, indicating that TCA macromolecules may influence the excitation process of the added complex. The main reason is that the interaction between TCA and the added Dy(SSA)$_3$Phen complex as discussed in FTIR analysis probably affects the energy transfers of the SSA ligands during the excitation process of the doped complex in TCA composites.

Emission spectra of TCA/Dy(SSA)$_3$Phen (0/100, 96/4, 93/7, 90/10, and 87/13) composites were obtained by exciting at 326 nm, 312 nm, 312 nm, 309 nm, and 309 nm, respectively, as shown in Figure 6. For the pure Dy (SSA)$_3$Phen complex, the strong emission band with a maximum at 436 nm is assigned to $\pi \longrightarrow \pi^*$ electron transition of the SSA ligands and the weak band at around

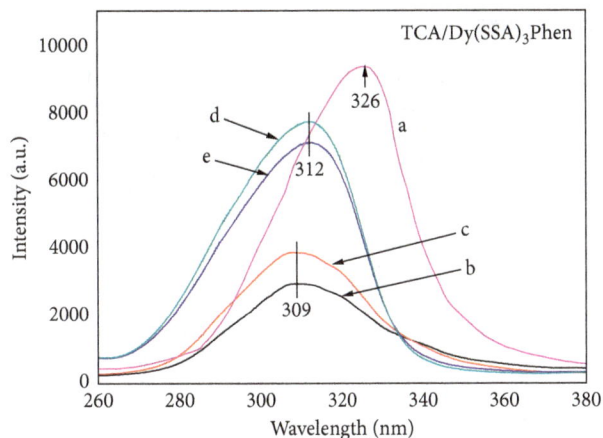

FIGURE 5: Excitation spectra of Dy(SSA)$_3$Phen complex and TCA/Dy(SSA)$_3$Phen composites: a, 0/100; b, 96/4; c, 93/7; d, 90/10; e, 87/13.

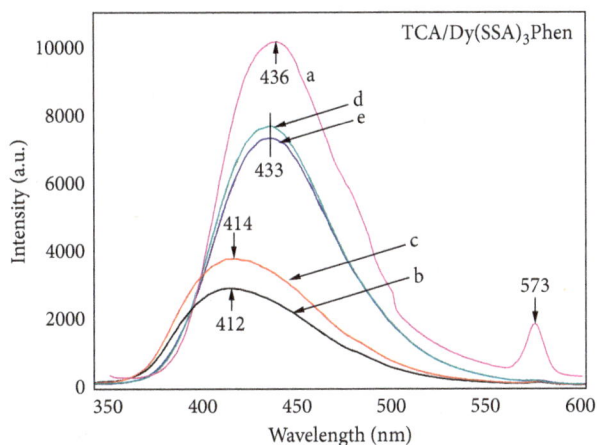

FIGURE 6: Emission spectra of Dy(SSA)$_3$Phen complex and TCA/Dy(SSA)$_3$Phen composites: a, 0/100; b, 96/4; c, 93/7; d, 90/10; e, 87/13, under excitation wavelengths at 326 nm, 312 nm, 312 nm, 309 nm, and 309 nm, respectively.

573 nm is ascribed to the $^4F_{9/2} \longrightarrow {}^6H_{13/2}$ transition of Dy^{3+} ions [2, 35, 36]. This is probably because the introduction of the central Dy^{3+} ions is propitious to the energy transfers of the SSA ligands, due to the larger planar structure of the Dy (SSA)$_3$Phen complex (Figure 4) [36]. The Commission Internationale de L'Eclairage (CIE) coordinate of the pure complex is calculated from its emission spectrum excited at 326 nm. The coordinates (0.15, 0.09) indicate that the complex emits blue light, as shown in Figure 7. Obviously, as for the TCA/Dy(SSA)$_3$Phen composites, there arc intense broad emission bands in the 360 nm–500 nm wavelength range in the emission spectra, with the maxima at 412 nm, 414 nm, 434 nm, and 434 nm corresponding to the (96/4), (93/7), (90/10), and (87/13) composites, respectively, upon the optimum excitation wavelengths. This suggests that the doped Dy(SSA)$_3$Phen complexes in the composites can still exhibit the characteristic emission of the SSA ligands similar to that of the pure complex. Dissimilarly, the weak bands at around 573 nm are absent in the emission spectra of all

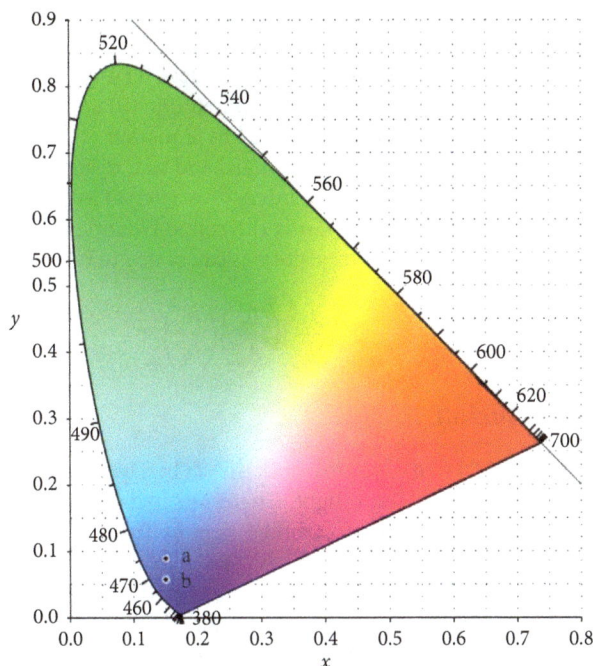

FIGURE 7: CIE coordinate photograph of the Dy(SSA)$_3$Phen complex and the TCA/Dy(SSA)$_3$Phen (95/5) composite: a, (0.15, 0.09) for the complex; b, (0.15, 0.06) for the composite.

composites. Furthermore, the emission spectra of the composites are also blueshifted, similar to the blueshift of the excitation spectra, which suggests that the TCA matrix has an impact on the excitation and emission processes of the additive Dy(SSA)$_3$Phen complex in the TCA composites.

It should be noted that a comparison of the spectral intensity of all samples indicates that the excitation and emission spectra of the pure Dy(SSA)$_3$Phen complex shows the strongest spectral intensity, and the spectral intensity of the TCA/Dy(SSA)$_3$Phen composites represents a nonlinear relationship with increasing the concentration of the added complex. When the content of Dy(SSA)$_3$Phen complex in the composites is less than 10 wt.%, the spectral intensity of the composites gradually increases with the increase of the additive content of the complex, and the intensity of the (87/13) composite is slightly lower than that of the (90/10) composite. The possible reason is that in the (96/4) and (93/7) composites, the concentration of the doped complexes is relatively low so that the complexes might be completely diluted, leading to weak spectral intensity. Nevertheless, when the additive content exceeds 10%, superfluous Dy(SSA)$_3$Phen complexes maybe aggregate in the TCA matrix, probably causing the concentration quenching existing in the (87/13) composite. Another possible reason is the inhomogeneity of the film. These caused the excitation, and emission intensity of the composite decreases. Clearly, when the concentration of the Dy(SSA)$_3$Phen complex is appropriate, namely, 10%, the spectral intensity of the composite is strong and close to that of the pure complex. The coordinates (0.15, 0.06) indicate that the (90/10) composite can also emit blue light, as shown in Figure 7.

A comparison method was adopted to calculate quantum yield (QY), and quinine sulfate (in 0.1 M H_2SO_4) was the standard with literature quantum yield 0.54 at 350 nm [37]. The QY of the TCA/Dy(SSA)$_3$Phen (90/10) composite in dimethyl sulfoxide was calculated by Equation (1), where subscripts "s" and "x" denote standard and test, respectively; I is the integrated fluorescence intensity excited at 312 nm; A is the absorption at 312 nm; n is the refractive index of the solvent. The QY of the composite was calculated to be 33.5%.

$$QY_x = QY_s \left(\frac{I_x}{I_s}\right)\left(\frac{A_s}{A_x}\right)\left(\frac{n_x}{n_s}\right). \qquad (1)$$

4. Conclusion

The Dy(SSA)$_3$Phen complex and the TCA/Dy(SSA)$_3$Phen composites all emit blue light. There are hydrogen bonds between oxygen atom of carbonyl groups from SSA and hydroxyl groups from TCA, which is favorable for the added complex to retain strong luminescence in the TCA composites. When the content of the added Dy(SSA)$_3$Phen complex is more than 10%, the concentration quenching or the inhomogeneity of the film probably exists in the (87/13) composite. The TCA/Dy(SSA)$_3$Phen (90/10) composite possesses fine luminescent properties with the quantum yield of 33.5% and thermal stability for potential usage as blue fluorescent materials.

Conflicts of Interest

The authors declare that they have no conflicts of interest to disclose in relation to the current research.

Authors' Contributions

Xiangwei Sun carried out the data analysis, wrote the draft, and revised the manuscript. Feiyue Wu participated in the design of the study and revised the manuscript. Yan Luo participated in the data analysis. Yuntao Li and Xiang Liu participated in the statistical analysis and revised the manuscript. Kunpeng Liu participated in the design of the study. Mengjun Huang conceived the study and revised the manuscript. All authors read and approved the final manuscript.

Acknowledgments

This work was financially supported by the Scientific and Technological Research Program of Chongqing Municipal Education Commission (No: KJZD-K201800301, KJ1711281, KJ1501128, KJ1601108, KJ1601121, and KJ1711266), Natural Science Foundation of Chongqing Municipal Science and Technology Commission (No: cstc2017jcyjAX0140, cstc2017jcyjAX0244), the Research Program of Yongchuan District Water Authority of Chongqing Municipal (No: 2016-04), and the Research Program of Chongqing University of Arts and Sciences (No: R2015CH10, R2015CH12, and XSKY2016064).

References

[1] K. S. V. Krishna Rao, H. G. Liu, and Y. I. Lee, "Fluorescence spectroscopy of polymer systems doped with rare-earth metal ions and their complexes," *Applied Spectroscopy Reviews*, vol. 45, no. 6, pp. 409–446, 2010.

[2] N. Xu, C. Wang, W. Shi, S. P. Yan, P. Cheng, and D. Z. Liao, "Magnetic and luminescent properties of Sm, Eu, Tb, and Dy coordination polymers with 2-hydroxynicotinic acid," *European Journal of Inorganic Chemistry*, vol. 2011, no. 15, pp. 2387–2393, 2011.

[3] M. Ge, X. Guo, and Y. Yan, "Preparation and study on the structure and properties of rare-earth luminescent fiber," *Textile Research Journal*, vol. 82, no. 7, pp. 677–684, 2012.

[4] A. Kozłowska, M. Nakielska, J. Sarnecki et al., "Spectroscopic investigations of rare-earth materials for luminescent solar concentrators," *Optica Applicata*, vol. 41, no. 2, pp. 359–365, 2011.

[5] M. Flores, U. Caldiño, G. Córdoba, and R. Arroyo, "Luminescence enhancement of Eu^{3+}-doped poly(acrylic acid) using 1,10-phenanthroline as antenna ligand," *Optical Materials*, vol. 27, no. 3, pp. 635–639, 2004.

[6] Y. Zhang, Q. Wang, C. Wang, X. Pei, and T. Wang, "Reversible emission color change of Eu-polymer complex induced by acetone," *Journal of Macromolecular Science, Part B*, vol. 51, no. 2, pp. 235–241, 2012.

[7] C. Yang, J. Xu, Y. Zhang et al., "Efficient monochromatic red-light-emitting pleds based on a series of nonconjugated Eu-polymers containing a neutral terpyridyl ligand," *Journal of Materials Chemistry C*, vol. 1, no. 32, p. 4885, 2013.

[8] M. Bortoluzzi, G. Paolucci, M. Gatto et al., "Preparation of photoluminescent PMMA doped with tris(pyrazol-1-yl) borate lanthanide complexes," *Journal of Luminescence*, vol. 132, no. 9, pp. 2378–2384, 2012.

[9] Y. Mei and B. Yan, "White luminescent hybrid soft materials of lanthanide (Eu^{3+}, Sm^{3+}) beta-diketonates and Ag/Ag$_2$S nanoparticles based with thiol-functionalized ionic liquid bridge," *Inorganic Chemistry Communications*, vol. 40, pp. 39–42, 2014.

[10] H. Y. Li, J. Wu, W. Huang et al., "Synthesis and photoluminescent properties of five homodinuclear lanthanide (Ln^{3+}=Eu^{3+}, Sm^{3+}, Er^{3+}, Yb^{3+}, Pr^{3+}) complexes," *Journal of Photochemistry and Photobiology A: Chemistry*, vol. 208, no. 2-3, pp. 110–116, 2009.

[11] Y. Ma and Y. Wang, "Recent advances in the sensitized luminescence of organic europium complexes," *Coordination Chemistry Reviews*, vol. 254, no. 9-10, pp. 972–990, 2010.

[12] X. J. Sun, W. X. Li, W. J. Chai, T. Ren, and X. Y. Shi, "The studies of enhanced fluorescence in the two novel ternary rare-earth complex systems," *Journal of Fluorescence*, vol. 20, no. 2, pp. 453–461, 2010.

[13] B. Gao, W. Zhang, Z. Zhang, and Q. Lei, "Preparation of polymer-rare earth complex using salicylic acid-containing polystyrene and its fluorescence emission property," *Journal of Luminescence*, vol. 132, no. 8, pp. 2005–2011, 2012.

[14] Z. Hong, W. L. Li, D. Zhao et al., "White light emission from OEL devices based on organic dysprosium-complex," *Synthetic Metals*, vol. 111-112, no. 4, pp. 43–45, 2000.

[15] W. G. Quirino, C. Legnani, R. M. B. dos Santos et al., "Electroluminescent devices based on rare-earth tetrakis β-diketonate complexes," *Thin Solid Films*, vol. 517, no. 3, pp. 1096–1100, 2008.

[16] S. Biju, L.-J. Xu, M. A. H. Alvesc, R. O. Freirec, and Z.-N. Chen, "Bright orange and red light-emitting diodes of

new visible light excitable tetrakis-ln-β-diketonate (Ln = Sm^3$^+$, Eu^{3+}) complexes," *New Journal of Chemistry*, vol. 41, no. 4, pp. 1473–1484, 2013.

[17] T. Cantat, F. Jaroschik, F. Nief, L. Ricard, N. Mezailles, and P. Le Floch, "New mono- and bis-carbene samarium complexes: synthesis, X-ray crystal structures and reactivity," *Chemical Communications*, vol. 41, no. 41, pp. 5178–5180, 2005.

[18] J. Priya, N. K. Gondia, A. K. Kunti, and S. K. Sharma, "Pure white light emitting tetrakis β-diketonate dysprosium complexes for OLED applications," *ECS Journal of Solid State Science and Technology*, vol. 5, no. 10, pp. R166–R171, 2016.

[19] P. P. Lima, S. A. Junior, O. L. Malta et al., "Synthesis, characterization and luminescence properties of Eu^{3+} 3-phenyl-4-(4-toluoyl)-5-isoxazolonate based organic-inorganic hybrids," *European Journal of Inorganic Chemistry*, vol. 2006, no. 19, pp. 3923–3929, 2006.

[20] J. G. Kang, H. G. Cho, J. G. Kim, and K. S. Choi, "Thermal and luminescent properties of polymeric acidic organophosphorus complexes of Eu(III) and Tb(III) undoped and doped with 1,10-phenanthroline," *Materials Chemistry and Physics*, vol. 91, no. 1, pp. 172–179, 2005.

[21] D. B. Raj, B. Francis, M. L. Reddy, R. R. Butorac, V. M. Lynch, and A. H. Cowley, "Highly luminescent poly(methyl methacrylate)-incorporated europium complex supported by a carbazole-based fluorinated beta-diketonate ligand and a 4,5-bis(diphenylphosphino)-9,9-dimethylxanthene oxide co-ligand," *Inorganic Chemistry*, vol. 49, no. 19, pp. 9055–9063, 2010.

[22] Q. F. Li, D. Yue, W. Lu, X. Zhang, C. Li, and Z. Wang, "Hybrid luminescence materials assembled by [Ln(DPA)$_3$]$^{3-}$ and mesoporous host through ion-pairing interactions with high quantum efficiencies and long lifetimes," *Scientific Reports*, vol. 5, p. 8385, 2015.

[23] J. Zhang, L. Zhang, Y. Chen, X. Huang, L. Wang, and Q. Zhang, "Influence of different carboxylic acid ligands on luminescent properties of Eu(Lc)$_3$phen (Lc=MAA, AA, BA, SA) complexes," *Journal of Nanomaterials*, vol. 2013, Article ID 768535, 8 pages, 2013.

[24] T. Zhang, R. F. Li, X. Feng, S. W. Ng, and R. F. Bai, "Synthesis, crystal structure and characterization of a lanthanide complex based on 4-(4-oxypyridinium-1-yl) phthalic acid and 1,10-phenanthroline," *Inorganic and Nano-Metal Chemistry*, vol. 47, no. 3, pp. 375–379, 2016.

[25] L. Liu, Z. Xu, Z. Lou, F. Zhang, B. Sun, and J. Pei, "Photoluminescence and electroluminescence mechanism on a novel complex Eu(o-BBA)$_3$(phen)," *Journal of Luminescence*, vol. 122-123, pp. 961–963, 2007.

[26] X. Liu, J. Zhu, and Z. Xu, "Luminescent properties of tri-cellulose acetate composites doped with rare earth terbium (iii) complex," *Nanomaterials and Energy*, vol. 3, no. 1, pp. 25–29, 2014.

[27] A. K. Mann, D. Varandani, B. R. Mehta, L. K. Malhotra, G. Mangamma, and A. K. Tyagi, "Synthesis and luminescence properties of alumina encapsulated InN nanorods," *Journal of Nanoscience and Nanotechnology*, vol. 8, no. 12, pp. 6290–6296, 2008.

[28] X. Liu, J. Zhu, H. Ni, B. Ma, and L. Liu, "Luminescent properties of a polymer photoluminescent composite containing the binuclear (Eu, Tb) complex as an emitter," *Journal of Macromolecular Science, Part B*, vol. 55, no. 1, pp. 20–32, 2015.

[29] L. Yuguang, Z. Jingchang, C. Weiliang, Z. Fujun, and X. Zheng, "Syntheses and characterization of binuclear complexes Tb$_{1-x}$Eu$_x$(TTA)$_3$phen," *Journal of Rare Earths*, vol. 25, no. 3, pp. 296–301, 2007.

[30] T. V. Koksharova, S. V. Kurando, and I. V. Stoyanova, "Coordination compounds of 3d-metals salicylates with thiosemicarbazide," *Russian Journal of General Chemistry*, vol. 82, no. 9, pp. 1481–1484, 2012.

[31] C. Yang, J. Luo, J. Ma, M. Lu, L. Liang, and B. Tong, "Synthesis and photoluminescent properties of four novel trinuclear europium complexes based on two tris-β-diketones ligands," *Dyes and Pigments*, vol. 92, no. 1, pp. 696–704, 2012.

[32] X. W. Yang, W. J. Sun, C. Y. Ke, H. G. Zhang, X. Y. Wang, and S. L. Gao, "Thermochemical properties of rare earth complexes with salicylic acid," *Thermochimica Acta*, vol. 463, no. 1-2, pp. 60–64, 2007.

[33] N. Španić, "Chemical and thermal properties of cellulose acetate prepared from white willow (Salix alba) and black alder (Alnus glutinosa) as a potential polymeric base of biocomposite materials," *Chemical and Biochemical Engineering Quarterly*, vol. 29, no. 3, pp. 357–365, 2015.

[34] M. C. C. Lucena, A. E. V. Alencar, S. E. Mazzeto, and S. A. Soares, "The effect of additives on the thermal degradation of cellulose acetate," *Polymer Degradation and Stability*, vol. 80, no. 1, pp. 149–155, 2003.

[35] J. Y. Rodrigues Silva, L. Lourenco da Luz, F. G. Martinez Mauricio et al., "Lanthanide-organic gels (LOGs) as a multifunctional supramolecular smart platform," *ACS Applied Materials & Interfaces*, vol. 9, no. 19, pp. 16458–16465, 2017.

[36] R. X. Wang, C. B. Liu, and G. Y. Zhu, "Study and application of a new L∗-L type fluorescence enhancement system of lanthanum-phenyl salicylate-trioctyl phosphine oxide-triton X-100," *Journal of Shandong University*, vol. 40, no. 2, pp. 92–95, 2005.

[37] J. E. Gill, "The fluorescence excitation spectrum of quinine bisulfate," *Photochemistry and Photobiology*, vol. 9, no. 4, pp. 313–322, 1969.

Metal-Organic Framework-101 (MIL-101): Synthesis, Kinetics, Thermodynamics, and Equilibrium Isotherms of Remazol Deep Black RGB Adsorption

Vo Thi Thanh Chau,[1] Huynh Thi MinhThanh,[2] Pham Dinh Du,[3] Tran Thanh Tam Toan,[4] Tran Ngoc Tuyen,[4] Tran Xuan Mau ⓘ,[4] and Dinh Quang Khieu ⓘ[4]

[1]*Tran Quoc Tuan High School, Quảng Ngãi 570000, Vietnam*
[2]*Department of Chemistry, Qui Nhon University, Bình Định 590000, Vietnam*
[3]*Faculty of Natural Sciences, Thu Dau Mot University, Thủ Dầu Một 820000, Vietnam*
[4]*University of Sciences, Hue University, Hue 530000, Vietnam*

Correspondence should be addressed to Dinh Quang Khieu; dqkhieu@hueuni.edu.vn

Academic Editor: Hossein Kazemian

In the present paper, the synthesis of metal-organic framework-101 (MIL-101) and Remazol Deep Black RGB (RDB) adsorption on MIL-101 were demonstrated. The kinetics of RDB adsorption on MIL-101 was studied using Weber's intraparticle diffusion model and the pseudo-first- and pseudo-second-order kinetic models. Particularly, the statistical method of piecewise linear regression and multi-nonlinear regression was employed to analyse the adsorption data according to the previously mentioned kinetic models. The results indicated that the adsorption process followed the three-step pseudo-first-order kinetic equation, which was consistent with the results of the intraparticle diffusion model with three linear segments. This model best described the experimental data. In addition, the adsorption isotherm data were studied using five adsorption models, namely, Langmuir, Freundlich, Redlich–Peterson, Toth, and Sips in nonlinear forms, and the Langmuir model is the most appropriate for the experimental data. The values of energies of activation of adsorption were calculated, and they revealed that the adsorption process was of endothermic chemical nature. A statistical comparison using Akaike information criterion to estimate the goodness of fit of the kinetic and isotherm models was presented.

1. Introduction

It is well known that wastewater from textile industries, pulp mills, and dyestuff manufacturing has been a potential threat to environment [1, 2]. Various treatment processes such as physical separation, chemical oxidation, and biological degradation have been widely investigated to remove dyes from wastewaters [3]. Among these processes, adsorption technology is considered as one of the most competitive methods because it does not require high operating temperature and has a simple operation as well as low cost, and several coloring materials can be removed simultaneously [4].

MIL-101 has demonstrated good performance in storage and adsorption of gas such as hydrogen storage [5], adsorption of CO_2 and CH_4 [6], and long-chain alkanes [7]. However, there are very few reports studying the adsorption

of dyes from aqueous solutions [8–10]. MOFs in general and MIL-101 in particular exhibit high efficiency for adsorption of dyes from aqueous solutions due to their unique structures such as large surface areas, ordered porosity, and high density of adsorption sites (anion and cation). MIL-101 demonstrates excellent adsorption properties for dyes, such as methyl orange (MO), xylenol orange (XO), and uranine. For expanding applications of MIL-101, studying the adsorption of this material for Remazol Deep Black RGB (denoted as RDB) used widely in dye industry will be a concern of this work.

Azo dyes with an azo group are widely used in many textile industries due to their low cost, high solubility, and stability. These dyes and their intermediate products are toxic, carcinogenic, and mutagenic to aquatic life. Remazol Deep Black RGB is a common diazo reactive dye, widely used in textile industries [11]. It is stable and hard to degrade biologically due

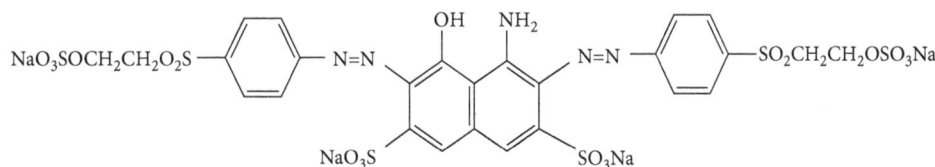

$$NaO_3SOCH_2CH_2O_2S \underset{}{} \text{—} \underset{NaO_3S}{\overset{OH}{\underset{}{}}} \overset{NH_2}{\underset{SO_3Na}{}} \text{—} SO_2CH_2CH_2OSO_3Na$$

SCHEME 1: Molecular structure of RDB.

to the presence of aromatic rings. Thus, RDB removal from textile wastewater has drawn much attention among researchers. Several techniques including adsorption, electrochemistry, and biosorption for RDB treatment have been reported. Soloman et al. [11] applied the electrochemical treatment to hydrolyze Remazol Black. They demonstrated that the performance of the batch recirculation system was better than other reactor configurations studied in terms of capacity utilization and energy consumption. Brazilian pine-fruit shells (*Araucaria angustifolia*) in their natural form were used as an adsorbent for the removal of RDB dye from aqueous effluents [12]. A biosorption process to discard azo dyes by fungi (*Aspergillus flavus*) was investigated in batch reactors [13]. Ninety percent of the dye in a 100 mg/L solution was removed. Recently, Thanh et al. [14] reported that iron doping to ZIF-8 significantly enhances RDB adsorption capacity. Fe–ZIF-8 also exhibits photocatalytic degradation of RDB under visible light [15].

Batch adsorption studies focus on two main trends: (i) designing and optimizing experiments with the evaluation of the influence of the experimental variables—this approach enables to estimate the magnitude of the influence of the factors affecting the process and their interactions [16]—and (ii) kinetics, thermodynamics, and equilibrium isotherm adsorption studies [14, 17]. For the latter, several models are used to study adsorption kinetics and isotherms. The parameters in these models are calculated with linear or nonlinear regression approaches. However, the number of parameters in each mode is different. For example, the Langmuir isotherm model contains two parameters, while the Redlich–Peterson isotherm model has three parameters. It is obvious that the greater the number of model parameters, the lower the relative errors (REs) or the sum of squared errors (SSEs). Therefore, the model compatibility needs to be evaluated including SSEs or REs and the number of model parameters as well as the experimental points. However, in the majority of current publications, the goodness of fit for models is estimated based on only the REs or SSEs. To the best of our knowledge, the research on this issue is limited.

In the present study, MIL-101 was employed as an adsorbent for removing RDB dye. The effects of initial concentration, adsorbent particle size, agitation speed, temperature, and pH on the adsorption behavior of RDB onto MIL-101 were investigated. The adsorption kinetic and isothermal studies and the goodness of fit for models were addressed.

2. Experimental

2.1. Materials.
Chromium (III) nitrate nonahydrates $(Cr(NO_3)_3 \cdot 9H_2O$, Merck, Germany), terephthalic acid, $(C_6H_4(COOH)_2$, Merck, Germany) (H_2BDC), and hydrofluoric acid (HF, 40%, Merck, Germany) were utilized in this study. Remazol Deep Black RGB $(C_{26}H_{21}N_5Na_4O_{19}S_6)$, (molecular weight = 991.82) was supplied by Thuy Duong Textile Company, Vietnam. The molecular structure of RDB is shown in Scheme 1.

2.2. Apparatus.
The powder X-ray diffraction (XRD) pattern was recorded by means of a D8 Advance (Bruker, Germany) with CuKα radiation ($\lambda = 1.5406$ Å). The morphology of the obtained samples was determined using transmission electron microscope (TEM) on JEOL JEM-2100F. The specific surface area of the samples was determined by means of nitrogen adsorption/desorption isotherms using a Micromeritics 2020 volumetric adsorption analyzer system. Visible spectrophotometry was measured by using Lambda 25 Spectrophotometer (PerkinElmer, Singapore) at λ_{max} of RDB dye (600 nm).

2.3. Preparation of MIL-101.
MIL-101 was synthesized from chromium (III) nitrate nonahydrates, H_2BDC, and HF using the hydrothermal method [18]. Three samples of MIL-101 with different molar ratios of HF/$H_2BDC = 0.00$, 0.25, and 0.75 were prepared. In a typical procedure, a mixture of 10 mmol of H_2BDC, 12.5 mmol of $Cr(NO_3)_3 \cdot 9H_2O$, x mmol of HF ($x = 0.00$, 0.25, and 0.75), and 350 mmol of H_2O was heated in a Teflon-lined stainless steel autoclave at 200°C for 8 h. The resulting green solid was filtered and then washed with ethanol in a Soxhlet apparatus for around 12 h to completely remove the unreacted amount of H_2BDC. The obtained MIL-101 was denoted as MHF0 for the molar ratio of HF/$H_2BDC = 0$, MHF0.25 for 0.25, and MHF0.75 for 0.75.

2.4. Adsorption Kinetic and Thermodynamic Studies.
Kinetic studies were carried out in a 3 L plastic beaker. This beaker was equipped with a stainless steel flat-blade impeller using an electric motor to stir the dye solution and a tap near the bottom to withdraw the liquid at any time. MIL-101 (0.50 g) was mixed thoroughly with 1000 mL of the dye solution in the beaker at room temperature. 10 mL of the mixture was withdrawn periodically, and MIL-101 was removed by using a centrifuge. The final dye concentration was determined using the spectrophotometric method. The adsorption capacity of the adsorbent was calculated according to the following formula:

$$q_t = \frac{V(C_0 - C_t)}{m}, \quad (1)$$

where q_t is the adsorption capacity (mg·g^{-1}) at t time, C_0 is the initial dye concentration (mg·L^{-1}), C_t is the dye concentration ($mg \cdot L^{-1}$) at time t, V is the volume of dye solution (L), and m is the mass of the adsorbent (g).

For the adsorption isotherm study, an amount of 30, 40, 50, 60, 70, 80, 90, and 110 mg of MIL-101 was added to 8 stopper 250 mL Erlenmeyer flasks containing 100 mL of 100 mg/L RDB solution. The flasks were then placed into a shaker bath at $28 \pm 1°C$ for 24 hours. Thereafter, the supernatant liquid was separated by centrifugation and the final dye concentration was determined with the method mentioned above.

In order to study formal and diffusion kinetics, Weber's intraparticle diffusion model and pseudo-first- and pseudo-second-order kinetic models were used. Weber's intraparticle diffusion model is described as in the following equation [19]:

$$q_t = k_i \cdot t^{1/2} + I, \tag{2}$$

where k_i is intraparticle diffusion rate constant ($mg \cdot g^{-1} \cdot min^{-0.5}$) and I is the intercept which represents the thickness of the boundary layer.

If intraparticle diffusion is the rate-limiting step, then the plot of q versus $t^{0.5}$ will give a straight line with a slope that equals k_i and an intercept equal to zero.

The pseudo-first- and pseudo-second-order kinetic models in the nonlinear form are expressed as follows [20]:

$$q_t = q_e \cdot \left(1 - e^{-k_1 \cdot t}\right),$$

$$q_t = q_e \cdot \frac{q_e \cdot k_2 \cdot t}{1 + q_e \cdot k_2 \cdot t}, \tag{3}$$

where q_t and q_e are the adsorption capacity at time t (min) and at equilibrium time, respectively, k_1 is the rate constants of the pseudo-first-order model (min^{-1}), and k_2 is the rate constant of the pseudo-second-order kinetic model ($g \cdot mg^{-1} \cdot min^{-1}$).

For the thermodynamic study, the experiments were conducted in the same way as in the adsorption kinetics study. However, the adsorption temperature was fixed at 298, 308, 318, and 328 K. The activation energy, E_a, was determined using the Arrhenius equation [21]:

$$k = A \cdot e^{-\left(E_a/R \cdot T\right)}, \tag{4}$$

where k is the rate constant, A is the frequency factor, R is the gas constant ($8.314 J \, mol^{-1} \cdot K^{-1}$), and T is the absolute temperature in Kelvin. Taking the natural logarithm of both sides of (4), one obtains

$$\ln k = -\frac{E_a}{R \cdot T} + \ln A. \tag{5}$$

By linearly plotting $\ln k$ versus $1/T$, one could obtain E_a from the slope ($-E_a/R$).

In order to evaluate whether the adsorption process is spontaneous, the adsorption thermodynamic parameters are needed. The standard Gibbs free energy of adsorption (ΔG^0) is given by the following expression:

$$\Delta G^0 = \Delta H^0 - T \Delta S^0, \tag{6}$$

where ΔG^0, ΔH^0, and ΔS^0 are the change of standard Gibbs energy, enthalpy, and entropy, respectively.

ΔG^0 is given by the van't Hoff equation:

$$\Delta G^0 = -R \cdot T \cdot \ln K_d, \tag{7}$$

where K_d is the distribution coefficient of the solute ions and equal to q_e/C_e [22, 23] and the others are described earlier.

By replacing (6) to (7), one obtains

$$\ln K_d = -\frac{\Delta H^0}{R \cdot T} + \frac{\Delta S^0}{R}. \tag{8}$$

The value of ΔH^0 and ΔS^0 is calculated from the slope and intercept of the linear plot of $\ln K_d$ versus $1/T$.

2.5. Adsorption Isotherm Study. Experimental data were analysed according to five isotherm models by Langmuir, Freundlich, Redlich–Peterson, Sips, and Toth.

Langmuir isotherm: Langmuir model is valid for monolayer sorption onto the surface. It could be expressed as follows [24]:

$$q_e = q_m \cdot \frac{K_L \cdot C_e}{1 + K_L \cdot C_e}, \tag{9}$$

where q_m is the maximum monolayer capacity amount ($mg \cdot g^{-1}$), K_L is the equilibrium constant, q_e is the equilibrium adsorption capacity ($mg \cdot g^{-1}$), and C_e is the equilibrium concentration of adsorbate ($mg \cdot L^{-1}$).

Freundlich isotherm: The Freundlich equation is an empirical relation based on the sorption onto a heterogeneous surface. It is commonly represented as [25]

$$q_e = K_F \cdot C_e^{1/n}, \tag{10}$$

where K_F $((mg \cdot g^{-1}) \cdot (mg \cdot L^{-1})^n)$ and n are the Freundlich parameters related to adsorption capacity and adsorption intensity, respectively.

The maximum capacity, q_m ($mg \cdot g^{-1}$), can be calculated from the following equation [26]:

$$q_m = K_F \cdot C_0^{1/n}, \tag{11}$$

where C_0 is the initial concentration.

Redlich–Peterson isotherm: Redlich–Peterson isotherm [27] contains three parameters and involves the features of both Langmuir and Freundlich isotherms. It can be described as follows:

$$q_e = \frac{K_R \cdot C_e}{1 + a_R \cdot C_e^{b_R}}, \tag{12}$$

where K_R ($L \cdot g^{-1}$) and a_R ($L \cdot mg^{-1}$) are the Redlich–Peterson isotherm constants and b_R is the exponent which lies between 0 and 1.

When b_R approaches 1, (12) becomes the Langmuir equation. Then, the maximum adsorption capacity, q_m ($mg \cdot g^{-1}$), can be determined by the following equation:

$$q_m = \frac{K_R}{a_R}. \tag{13}$$

Sips isotherm: Sips isotherm is a combination of the Langmuir and Freundlich isotherms and expected to describe heterogeneous surfaces much better [24]. The model can be written as [28]

$$q_e = \frac{K_s \cdot C_e^{1/n}}{1 + a_s \cdot C_e^{1/n}}, \tag{14}$$

where K_s and a_s are the Sips constants related to the energy of adsorption. The maximum monolayer adsorption capacity could is given by K_a/a_s.

Toth isotherm: Toth isotherm is the Langmuir-based isotherm and considers a continuous distribution of site affinities. It is expressed as [29]

$$q_e = \frac{q_m \cdot C_e}{\left(K_{T_0} + C_e^n\right)^{1/n}}, \tag{15}$$

where K_{T_0} is the Toth model constant and n is the Toth model exponent ($0 < n \le 1$). It is obvious that, for $n = 1$, this isotherm reduces to the Langmuir equation.

2.6. Piecewise Linear Regression and Model Comparison.

The application of Weber's model often suffers from uncertainties caused by the multilinear nature of its plots. Malash and El-Khaiary [30] proposed the piecewise linear regression for the analysis of multilinearity in intraparticle diffusion and film diffusion. In this method, the experimental data could be fixed for one-, two-, three-, or four-linear-segment lines:

one-linear-segment line: $Y = B + A \cdot X$ (two parameters),

two-linear-segment line: $Y = B + A \cdot X + C \cdot (X - D) \cdot$ SIGN $(X - D)$ (four parameters),

three-linear-segment line: $Y = B + A \cdot X + C \cdot (X - D) \cdot$ SIGN $(X - D) + E \cdot (X - F) \cdot$ SIGN $(X - F)$ (six parameters),

four-linear-segment line: $Y = B + A \cdot X + C \cdot (X - D) \cdot$ SIGN $(X - D) + E \cdot (X - F) \cdot$ SIGN $(X - F) + G \cdot (X - H) \cdot$ SIGN $(X - H)$ (eight parameters),

where the values of A, B, C, D, E, F, G, and H are estimated by nonlinear regression. D, F, and H, called breakpoints, are the boundaries between the segments. The Microsoft Excel "SIGN" function is defined as follows:

$$\text{SIGN}(X - a) = \begin{pmatrix} 1 & \text{if } X > a \\ 0 & \text{if } X = a \\ -1 & \text{if } X < a \end{pmatrix}. \tag{16}$$

The example for the three-linear-segment equation is expressed as follows:

$$Y = \begin{pmatrix} A + C \cdot D + E \cdot F + X \cdot (B - C - E) & \text{if } X < D \\ A + E \cdot F + X \cdot (B - E) & \text{if } X = D \\ A - C \cdot D + E \cdot F + X \cdot (B + C - E) & \text{if } D < X < F \\ A - C \cdot D + X \cdot (B + C) & \text{if } X = F \\ A - C \cdot D - E \cdot F + X \cdot (B + C + E) & \text{if } X > F \end{pmatrix}. \tag{17}$$

Then, the linear equation of the first segment is $y = a_1 \cdot x + b_1$, where $b_1 = B - C - E$ and $a_1 = A + C \cdot D + E \cdot F$. The linear equation of the second segment is

$y = b_2 \cdot x + a_2$, where $b_2 = B + C - E$ and $a_2 = A + C \cdot D + E \cdot F$. The linear equation of the third segment is $y = b_3 \cdot x + a_3$, where $b_3 = B + C + E$ and $a_3 = A - C \cdot D - E \cdot F$.

The model's parameters are determined using the least squares method. This is calculated by minimizing the sum of squared errors, SSE_S, by numerical optimization techniques using the Solver function in Microsoft Excel. The function for minimization is

$$\text{SSE}_S = \sum_1^N \left(y_{\exp} - y_{\text{est}}\right)^2, \tag{18}$$

where $y_{\exp}$ is experimental datum and y_{est} is the value estimated from the model.

The determination coefficient, R^2, is obtained by the following expression:

$$R^2 = \frac{1 - \text{SSE}_S}{\text{SSE}_T}, \tag{19}$$

where SSE_T is the total sum of squares equal to $\sum_1^N \left(y_{\exp} - y_{\text{mean}}\right)^2$ (y_{mean} is the mean value of y).

We know that increasing the number of linear segments increases the number of regression parameters that almost universally led to the decrease of SSEs or the increase of R^2. Therefore, the model compatibility cannot be based only on SSE or R^2 but must also include the number of regression parameters as well as the experimental points. Akaike's information criterion (AIC) is one of the well-known statistical methods used in this case.

$$\text{AIC}_C = N \cdot \ln \frac{\text{SSE}}{N} + 2 \cdot N_p + \frac{2 \cdot N_p \cdot \left(N_p + 1\right)}{N - N_p - 1}, \tag{20}$$

where N_p is the parameter of the model. The other parameters are described above.

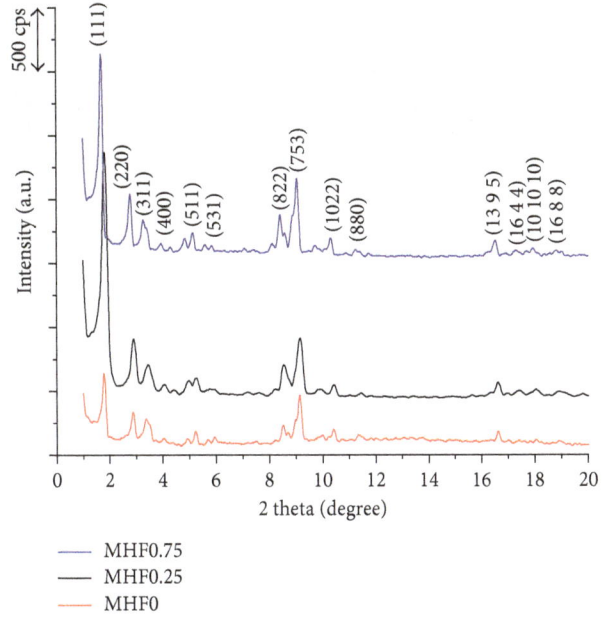

FIGURE 1: XRD patterns of MIL-101 synthesized with different molar ratios of HF/H_2BDC.

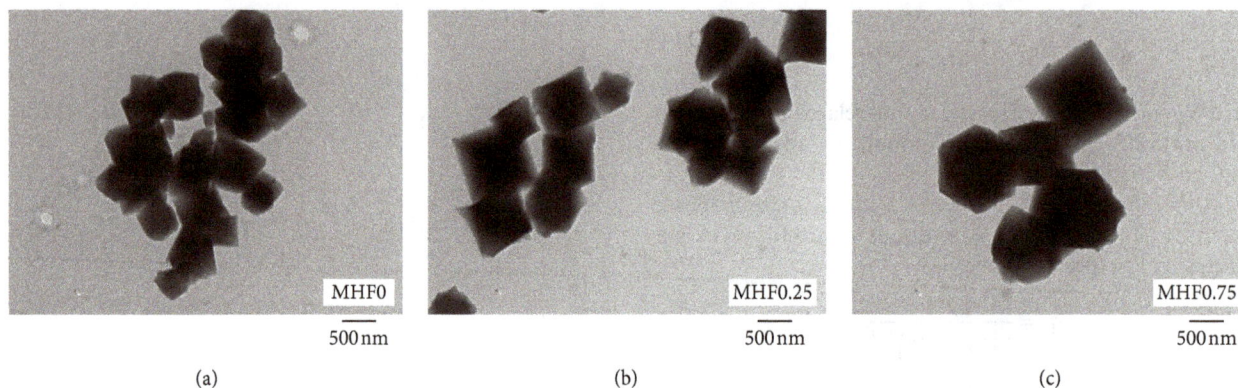

FIGURE 2: TEM images of MIL-101 synthesized with different molar ratios of HF/H$_2$BDC.

The AIC$_C$ is applied as N is small compared with N_p. AIC$_C$ is only computed as N is at least two units greater than N_p. The value of AIC$_C$ could be positive or negative, and the lower the AIC$_C$ value, the better it is. Another way of comparing AIC$_C$ is using the evidence ratio (ER) which is expressed as follows [17]:

$$ER = \frac{1}{e^{-0.5 \cdot \Delta}}, \qquad (21)$$

where Δ is the absolute value of the difference between AIC$_C$ and AIC$_S$ scores. ER means that the model with lower AIC is $1/e^{-0.5\Delta}$ times more likely to be correct than the alternative model.

In the present study, the comparison of the model will use R^2 or SSE if the models possess the same experimental points (N) and parameter numbers of models (N_p). Otherwise, AIC will be employed.

3. Results and Discussion

3.1. Characterization of MIL-101 Samples. Figure 1 shows the XRD patterns of MIL-101 synthesized with the HF/H$_2$BDC molar ratios of 0.00, 0.25, and 0.75. The characteristic diffractions of the samples matched well with the published XRD patterns of MIL-101 [18]. This means that the obtained materials are MIL-101. However, the peak intensity of the samples synthesized with HF is significantly higher than that of the sample synthesized without HF (MHF0). This could be due to fluorine that acts as a mineralizing agent in the hydrothermal synthesis for the formation of well crystalline microporous materials [31].

The morphology of the obtained material consists of octahedron-shaped crystals with smooth facets (Figure 2). The particle size of MIL-101 increases with the increase in the HF/H$_2$BDC ratio. The average particle size counted based on 50 particles is 234 nm for MHF0, 364 nm for MHF0.25, and 612 nm for MHF0.75 (Table 1).

The textural properties of the MIL-101 samples were investigated by using nitrogen adsorption/desorption isotherms. The isotherm curves belong to type IV according to IUPAC classification (Figure 3). The BET specific surface area for MHF0.25 is the highest (Table 1), and thus, MIL-101 synthesized with HF/H$_2$BDC = 0.25 was chosen for adsorption experiments.

TABLE 1: Textural properties of MIL-101 samples.

Notation	S_{BET} (m^2·g^{-1})	S_{Langmuir} (m^2·g^{-1})	V_{pore} (cm^3·g^{-1})	d_{TEM} (nm)
MHF0	2772	4652	1.45	234
MHF0.25	3586	5288	1.85	364
MHF0.75	2614	4381	1.43	612

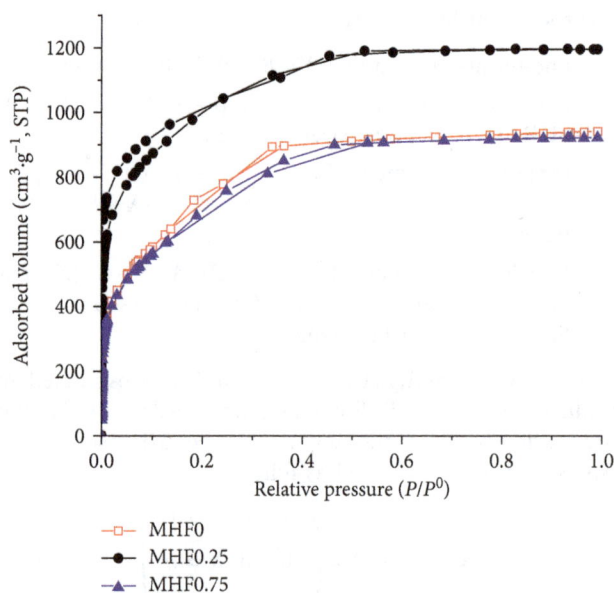

FIGURE 3: Nitrogen adsorption/desorption isotherms of MIL-101 synthesized with different molar ratios of HF/H$_2$BDC.

3.2. RDB Adsorption on MIL-101

3.2.1. Effect of Initial Concentrations. The RDB kinetics of adsorption on MIL-101 at different initial concentrations in the range of 25–600 ppm is illustrated in Figure 4(a). It is obvious that the adsorption capacity of RDB on MIL-101 increases when RDB initial concentration increases from 25 to 400 ppm. This might be due to the fact that, initially, the sites on the adsorbent surface are less occupied by the dye molecules, and increasing the concentration increases the interaction between the dye molecules and the adsorbent;

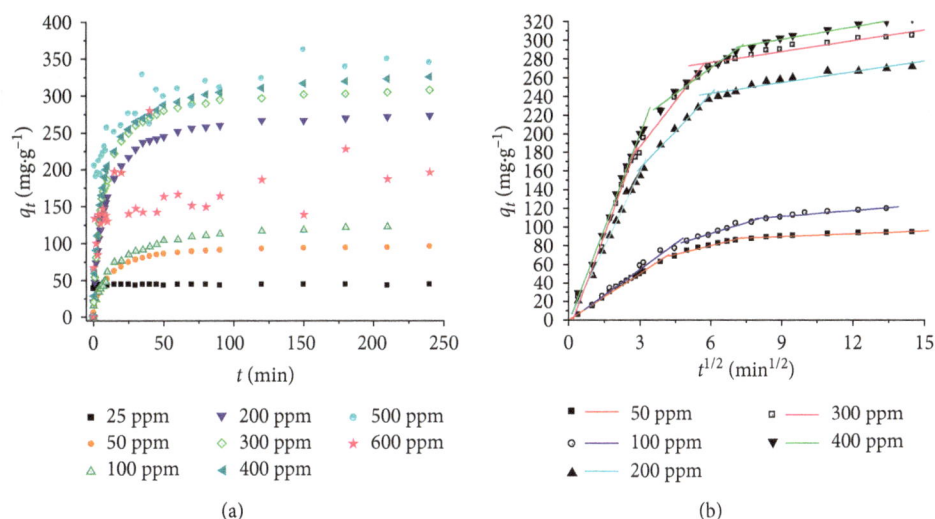

FIGURE 4: (a) RDB adsorption kinetics on MIL-101 at different concentrations; (b) stepwise lines of intraparticle diffusion plots for the adsorption of RDB on MIL-101 with different initial concentrations (adsorption condition: $V = 1000$ mL; $m_{MIL-101} = 0.5$ g; pH = 7.0; room temperature).

thus, more dye molecules adsorb on the surface [10, 32, 33]. In addition, the mass transfer driving force becomes larger as the initial concentration increases, and this results in higher adsorption capacity [21]. At higher concentrations (>400 ppm), the adsorption takes place very fast during the first 15 minutes, and then, it slows down and reaches equilibrium at about 150 minutes. Meanwhile, for concentration at 25 ppm, the adsorption reaches equilibrium practically immediately, just after about 10 minutes. This might be because at low dye concentration, the driving force is very small and the adsorption takes place only on the surface of MIL-101, whereas when the dye concentration is high with large driving force, the adsorption also occurs in the pore of the adsorbent, and due to high resistance in the pores, the adsorption becomes slower and reaches equilibrium after a longer period. Furthermore, adsorption might take place in several stages. At a very high dye concentration (600 ppm), the RDB adsorption on MIL-101 does not follow the same pattern as at lower initial dye concentration (Figure 4). This might be the result of forming a colloidal solution at a high concentration of RDB [20, 34].

Therefore, the RDB initial concentration from 50 to 400 ppm is suitable for the kinetic study of adsorption on MIL-101. In this kinetic study, Weber's intraparticle diffusion model [19] was applied to study the adsorption mechanism. The values of q_t at different times were analysed using piecewise linear regressions based on the assumption of one, two, three, and four linear segments. The AIC$_C$ value was the criterion for determining which one is the goodness of fit (Table 2).

The data indicate that the three-linear-segment model has the lowest AIC$_C$, and therefore, this Weber's model is the most accurate because for this criterion the lower the value, the more suitable the model. Three distinct steps can be seen on the kinetic curves: (i) instantaneous adsorption of RDB molecules within the first 7–21 minutes, (ii) a gradual attainment of the equilibrium due to the utilization of the all

TABLE 2: Comparison of piecewise models of one, two, three, and four linear segments using AIC for Weber's model.

C_0 (ppm)	Number of segments	N	N_p	SSE	AIC$_C$
50	1	27	2	4663116	330.103
	2	27	4	106	46.814
	3	27	6	26	15.678
	4	27	8	24	20.820
100	1	25	2	3435	127.600
	2	25	4	287	70.200
	3	25	6	149	61.350
	4	25	8	130	66.245
200	1	27	2	34824445	384.390
	2	27	4	1313	114.690
	3	27	6	270	78.333
	4	27	8	251	84.149
300	1	27	2	41762504	389.295
	2	27	4	1716	121.922
	3	27	6	642	101.746
	4	27	6	623	108.745
400	1	27	2	44322439	390.901
	2	27	4	1799	123.192
	3	27	6	245	75.756
	4	27	8	236	82.497

active sites on the adsorbent surface, and (iii) an equilibrium attainment of RDB molecules onto MIL-101 (Figure 4(b)). At the initial concentration of 50 mg·L^{-1}, for example, the intercept of the first linear segment is 0.05, and its 99% confidence interval is (−3.14; 3.23), indicating that the intercept is not significantly different from zero. This strongly suggests that intraparticle diffusion is the rate-controlling mechanism during the first 15 minutes of adsorption. The next two linear segments do not pass through the origin because the 99% confidence intervals of their intersects do not contain zero, indicating that the intraparticle diffusion is not the only rate-limiting step and film diffusion or chemical

TABLE 3: Parameters of Weber's three-linear-segment model (the values in parentheses are 99% confidence intervals).

C_0 (ppm)	50	100	200	300	400
Slope 1	16.70 (15.39; 18.02)	17.57 (14.71; 20.43)	53.65 (48.86; 58.44)	69.04 (57.27; 80.80)	67.13 (62.62; 71.64)
Intercept 1	0.05 (−3.14; 3.23)	1.16 (−7.37; 9.70)	−1.97 (−12.15; 8.21)	−12.18 (−34.26; 9.89)	−−2.04 (−12.14; 8.06)
Break point 1 (minutes)	16.46	21.15	9.77	7.53	11.08
Slope 2	6.39 (4.41; 8.38)	7.74 (3.66; 11.82)	26.60 (19.91; 33.28)	32.72 (20.63; 44.80)	18.29 (13.49; 23.08)
Intercept 2	41.67 (29.92; 53.42)	46.48 (19.63; 73.33)	82.60 (50.87; 114.33)	87.51 (38.08; 136.94)	160.50 (133.18; 187.82)
Break point 2 (minutes)	52.32	65.64	36.55	32.87	52.79
Slope 3	1.03 (0.58; 1.47)	2.11 (1.09; 3.13)	3.86 (2.50; 5.22)	3.90 (2.83; 4.98)	4.40 (3.20; 5.60)
Intercept 3	80.49 (75.64; 85.33)	92.23 (81.40; 103.07)	220.08 (206.55; 233.60)	252.69 (242.26; 263.12)	261.40 (248.27; 274.53)

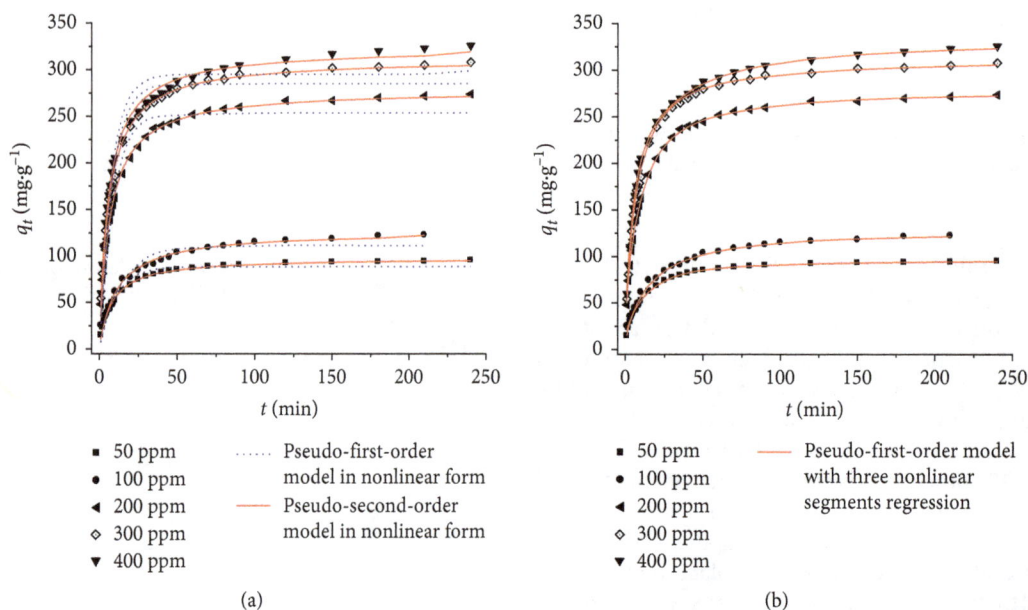

(a)

(b)

FIGURE 5: (a) Plots of pseudo-first- and pseudo-second-order adsorption kinetics; (b) plots of three pseudo-first-order adsorption kinetics of RDB on MIL-101.

reaction might take place during these periods of adsorption [20, 33]. This behavior is found for all concentrations. The intraparticle parameters are illustrated in Table 3.

The data indicate that the film thickness (intercept 2 and intercept 3) increases with the increase in initial RDB concentration. This suggests that film diffusion controls the adsorption process in the last two steps. In the first step, the intraparticle diffusion parameter, k_{p1}, increases as initial RDB concentration increases. These results suggest that intraparticle diffusion controls the rate in the initial step of the adsorption process.

In addition, the nature of the rate-limiting step is also confirmed by plotting the intraparticle diffusion constant, k_{p1}, versus the first power of the initial concentration. If k_{p1} is proportional to the initial dye concentration, the adsorption process is controlled by film diffusion, whereas, if intraparticle diffusion limits the adsorption process, the relationship between dye concentrations and k_{p1} will not be linear [20, 35]. For the adsorption of RDB onto MIL-101, the plot of k_{p1} versus the initial concentration (C_0) is not linear ($R^2 = 0.875$;

$p = 0.02 > 0.01$), confirming that the intraparticle diffusion mechanism controlled the adsorption in the initial step.

To study the formal kinetics of the RDB adsorption on MIL-101, the experimental data were subjected to pseudo-first-order and pseudo-second-order kinetics in the nonlinear form. The results are shown in Figure 5(a) and Table 4.

The experimental points in Figure 5(a) are very far from the first-order model curves, whereas they practically coincide with the second-order kinetic curves. From Table 4, we can see that the pseudo-second-order kinetic model has lower AIC values than the pseudo-first-order kinetic model. This means that the pseudo-second-order kinetic model explains the experimental data more appropriately. These results are consistent with those of other reports [30, 36, 37]. The rate constant k_2 calculated from the pseudo-second-order kinetic model decreases as the initial RDB concentration increases. This indicates that chemisorption is significant in the rate-limiting step, involving valence forces through sharing or exchange of electrons between RDB and MIL-101.

TABLE 4: Kinetic parameters for pseudo-first-order and pseudo-second-order kinetic models of RDB adsorption on MIL-101.

| C_0 (ppm) | $q_{e(exp)}$ (mg·g^{-1}) | Pseudo-first-order kinetics | | | Pseudo-second-order kinetics | | |
		$q_{e,cal}$ (mg·g^{-1})	$k_1 \cdot 10^2$ (min^{-1})	AIC$_C$	$q_{e,cal}$ (mg·g^{-1})	$k_2 \cdot 10^4$ (g·mg^{-1}·min^{-1})	AIC$_C$
50	95.51	88.89	9.52	88.654	98.42	12.76	35.537
100	122.78	111.65	6.92	112.215	126.91	6.78	90.873
200	273.80	253.99	11.12	149.668	279.09	5.36	108.087
300	308.00	286.07	11.99	146.452	313.32	5.33	86.556
400	325.99	295.34	12.81	156.611	323.38	5.33	108.258

TABLE 5: Kinetic parameters of three-nonlinear-segment regression for pseudo-first-order kinetic model of RDB adsorption on MIL-101.

C_0 (ppm)	q_1 (mg·g^{-1})	k_1 (min^{-1})	R^2	q_2 (mg·g^{-1})	k_2 (min^{-1})	R^2	q_3 (mg·g^{-1})	k_3 (min^{-1})	R^2	$q_{e,exp}$ (mg·g^{-1})
50	16.50	0.01	0.99	57.48	0.07	0.97	18.39	0.57	0.93	95.51
100	24.42	0.02	0.97	18.69	0.02	0.97	65.99	0.13	0.93	122.78
200	56.96	0.02	0.99	49.92	0.61	0.99	153.93	0.09	0.91	273.80
300	105.82	0.29	0.99	142.62	0.07	0.96	44.04	0.01	0.93	308.00
400	156.17	0.23	0.99	63.80	0.01	0.97	83.86	0.05	0.97	325.99

TABLE 6: AIC$_C$ values of the different kinetic models.

| Model | AIC$_C$ | | | | |
	50 ppm	100 ppm	200 ppm	300 ppm	400 ppm
Pseudo-first-order kinetics	88.65	112.22	149.67	146.45	156.61
Pseudo-second-order kinetics	35.54	90.87	108.09	86.56	108.26
Pseudo-first-order kinetics with three segments	−48.99	56.84	87.84	66.30	70.09

However, as mentioned earlier, in the present study, the adsorption process took place in three steps. Although the pseudo-second-order kinetic model accounts for the chemisorption nature well, it does not support the multi-segment adsorption process. Al-Ghouti et al. [20] mentioned this problem and tried to analyse the adsorption as a three-step process with three linear segments on the kinetic plots by using the graphical approach method. Therefore, in this study, we analyse the adsorption process in the same way; that is, the kinetic plot is also divided into three segments each of which follows the pseudo-first-order adsorption kinetics.

The three-step kinetic rate equation was expressed as follows:

$$q_t = q_0 + q_1 \cdot \left(1 - e^{-k_1 \cdot t}\right) + q_2 \cdot \left(1 - e^{-k_2 \cdot t}\right) + q_3 \cdot \left(1 - e^{-k_3 \cdot t}\right), \tag{22}$$

where q_1, q_2, and q_3 are the amount of dye adsorbed at time t after the subsequent kinetic steps (mg·g^{-1}); q_0 is the amount of dye adsorbed at time $t = 0$; and k_1, k_2, and k_3 are the kinetic rate constants associated with each kinetic step. At $t = \infty$, when the adsorption reaches equilibrium, $q_t = q_e$, and $q_e = q_0 + q_1 + q_2 + q_3$. Therefore, (22) can be written as

$$q_t = q_e - q_1 \cdot e^{-k_1 \cdot t} - q_2 \cdot e^{-k_2 \cdot t} - q_3 \cdot e^{-k_3 \cdot t}. \tag{23}$$

The values of q_1, q_2, and q_3 and k_1, k_2, and k_3 can be obtained using nonlinear regression by means of the Statistical Package for Scientific Social 20 (SPSS 20).

The three-nonlinear-segment regressions using the pseudo-first-order kinetic model for the RDB adsorption on MIL-101 are shown in Figure 5(b) and Table 5. High determination coefficients (0.91–0.99) indicate the appropriateness of the model. In addition, the AIC$_C$ values for the pseudo-first-order kinetics with three segments are also the lowest of the three kinetic models (Table 6). This further confirms the best fit of the pseudo-first-order kinetics with three segments with the experimental data. The finding is also consistent with the analysis of the three-step adsorption process using Weber's intraparticle diffusion model.

3.2.2. Effect of Particle Size and Agitation. To study diffusion kinetics of the RDB adsorption on MIL-101 in terms of particle size, the three-linear-segment regression for the intraparticle diffusion model was applied to analyse the experimental data (Figure 6(a)). The results were the same as those of the effect of initial concentration. Intraparticle diffusion limited the adsorption rate at the initial step, and film diffusion controlled the process in the next two steps. The values of k_{p1} are 14.04, 29.94, and 13.55 mg·g^{-1}·min$^{-0.5}$ for MHF0, MHF0.25, and MHF0.75, respectively. Theoretically, the intraparticle diffusion constant, k_{p1}, versus the inverse particle diameter, d^{-1}, did not give a straight line, and the conclusion for this is that the intraparticle diffusion was not the only operative mechanism [20]. In fact, in our study, this line was not linear ($R^2 = 0.008$, $p = 0.943$) although MIL-101 is a porous material; hence, we cannot rely on the particle size (external surface area) to confirm the adsorption mechanism. Therefore, the intraparticle diffusion

(a)

(b)

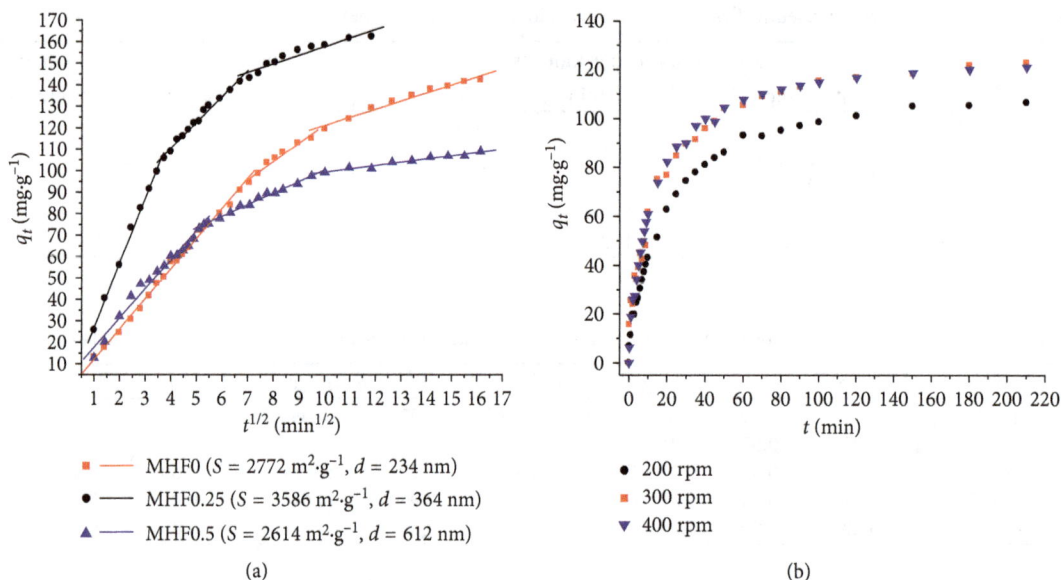

FIGURE 6: (a) Plots of the three-linear-segment regression of intraparticle diffusion model; (b) effect of stirring speed on the adsorption capacity of RDB on MIL-101.

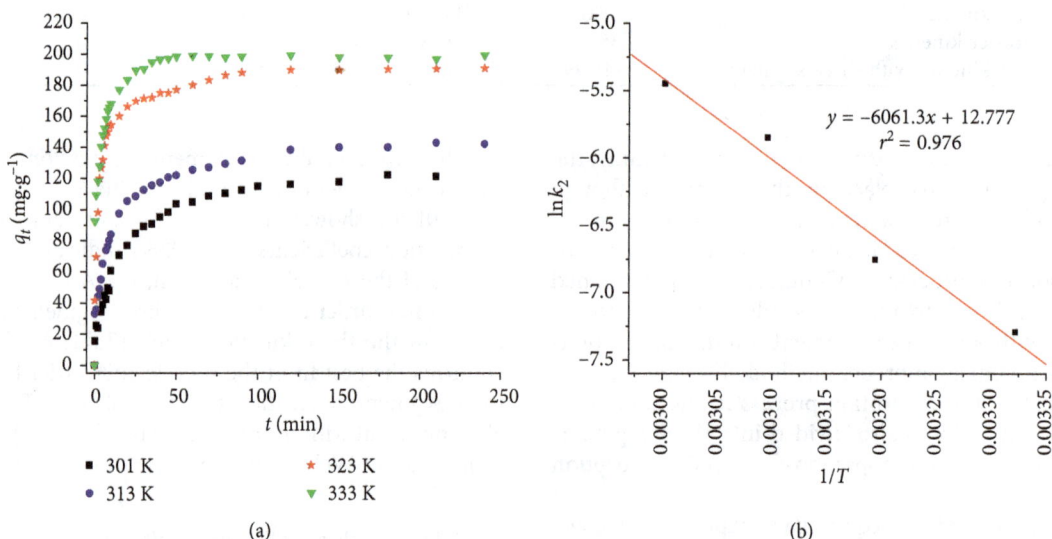

(a)

(b)

FIGURE 7: (a) RDB adsorption kinetics on MIL-101 at different temperatures; (b) Arrhenius plot.

mechanism controlled the initial step of the adsorption process. As can be seen from Figure 6(a), the adsorption capacity depends on the specific area rather than the particle size. This is because, for porous materials, the external surface contributes very little to the total surface area. In terms of particle size, only external surface area is concerned.

Stirring speed affects not only the distribution of the dye molecules in the bulk solution but also the formation of the external boundary film. Increasing stirring speed decreases the film thickness and thus the resistance to mass transfer around the adsorbent particle and increases the mobility of the whole system [38]. The effect of the stirring speed on RDB adsorption on MIL-101 was carried out with three values: 200, 300, and 400 rpm (Figure 6(b)). It is evident that

the adsorption capacity increased when the stirring speed increases from 200 rpm to 300 rpm and remained practically stable at 400 rpm. As the stirring speed increases the diffusion rate, the resistance of the solution becomes small. After a certain stirring rate, the external resistance no longer affects the sorption process.

3.2.3. *Thermodynamic Studies.* Temperature significantly affected the RDB adsorption over MIL-101. When the temperature increased from 301 K to 333 K, the adsorption capacity increased rapidly from $120\,mg\cdot g^{-1}$ to $190\,mg\cdot g^{-1}$ (Figure 7(a)). This indicates that the RDB adsorption on MIL-101 is an endothermic process. Similar results were

TABLE 7: Thermodynamic parameters of RDB adsorption on MIL-101.

Temperature (K)	ΔH^0 (kJ·mol^{-1})	ΔS^0 (J·mol^{-1}·T^{-1})	ΔG^0 (kJ·mol^{-1})
301			−1.014
313			−6.501
323	13.66	457.27	−11.074
333			−15.646

observed in the adsorption of uranine [9] and methyl orange [32] on MIL-101.

It is obvious that high temperature increased the diffusion rate of the dye molecules across the external boundary layer and in the internal pores of the adsorbent particle. This was the result of the decrease in the viscosity of the solution. In addition, the increase in adsorption capacity could also be ascribed to the increase in the number of active sites on the MIL-101 surface due to the decrease in the hydrogen bonding between adsorbed water and the adsorbent making more sites available for RDB molecules.

The pseudo-second-order kinetic model was more consistent with the kinetic data than the pseudo-first-order kinetic model in the temperature range of 301 K to 333 K. Hence, the rate constant k_2 was used to calculate the thermodynamic parameters. The E_a value obtained from the slope of the linear plot of ln k_2 versus T^{-1} ($F(3) = 59.15$; $R^2 = 0.98$, $p < 0.01$) (Figure 7(b)) was 50.39 kJ·mol^{-1}. This large activation energy (over 42 kJ·mol^{-1}) implies that chemisorption controlled the adsorption of RDB on MIL-101.

The thermodynamic parameters of the system, namely, ΔH^0, ΔS^0, and ΔG^0 were evaluated using the van't Hoff equation to determine the spontaneity of the adsorption process. The positive value of ΔH^0 (Table 7) suggested an endothermic adsorption process. The positive value of ΔS^0 indicated the increase in the randomness at the solid-liquid interface during the adsorption of RDB molecules on the adsorbent [39]. The large negative values of ΔG^0 strongly recommended the spontaneous RDB adsorption on MIL-101. The more negative value at higher temperatures suggested that the spontaneity increased with temperature. As the change of Gibbs free energy was negative and accompanied by the positive standard entropy change, the adsorption reaction was spontaneous with high affinity.

3.2.4. Effect of pH.

The pH of the solution affects the dye adsorption process because it can alternate both dye ionization and the ionic state of the surface of the adsorbent. Figure 8(a) presents the effect of pH on RDB adsorption from aqueous solutions. The RDB adsorption capacity of MIL-101 increased slightly with pH from 3 to 5, followed by a significant increase with pH from 5 to 9. The pH$_{pzc}$ (the point of zero charge) of MIL-101 determined by the pH drift method [40] is around 5 (Figure 8(b)). This pH$_{pzc}$ implies that the surface of the MIL-101 is positively charged when pH of the solution is below 5, whereas the surface of adsorbent becomes negatively charged at pH above 5.

Increasing pH led to an increase in adsorption capacity, suggesting that the adsorption could follow a mechanism other than electrostatic interaction. The π-π stacking interaction between the aromatic rings of the RDB and the aromatic rings of terephthalate in the MIL-101 framework was also thought to contribute to the RDB adsorption capacity. In addition, the coordination of the oxygen of the carboxyl group in the RDB molecules with the unsaturated Cr(III) ions in the MIL-101 framework is also responsible for more efficient adsorption. A possible mechanism of RDB adsorption on MIL-101 is illustrated in Scheme 2.

3.2.5. Adsorption Isotherms of RDB on MIL-101.

To describe the adsorption isotherms, Langmuir, Freundlich, Redlich–Peterson, Sips, and Toth equations were selected for use in this study. The determination of parameters of isotherm models is often based on the linear regression. However, linear regression requires the transformation of the original equation into a linear form that induces a problem related to abuse R^2. For example, the popular linear form of the Langmuir model is $C_e/q_e = C_e/q_m + 1/(K_L \cdot q_m)$, in which C_e is present in both independent and dependent variables [41]. Some papers [42, 43] reported that the linear form is less accurate than the nonlinear form in some cases of isotherm sorption as well as sorption kinetics. For these reasons, the isotherm equations in the nonlinear form are used in this study. Figure 9 shows the graphs that plot q_e versus C_e using the Langmuir, Freundlich, Redlich–Peterson, Sips, and Toth models. These models displayed lines around the experimental data, indicating that they all could describe the experimental data well.

The parameters of the isotherm models calculated using nonlinear regression are listed in Table 8. Except q_m derived from the Sips model (290.15 mg·g^{-1}), the values of q_m from the Redlich–Peterson and Toth models are fairly similar to that of q_m obtained from the Langmuir model due to the parameter of n being close to 1.

Based on the values of SSE as well as the coefficient of determination, we can see that the Langmuir, Redlich–Peterson, and Toth models provide a higher goodness of fit than the Sips and Freundlich models (Table 8).

To compare models with the same parameters and experimental points, SSE or R^2 are frequently utilized to estimate the goodness of fit. As a result, it is obvious that the experimental data fit the Langmuir model better than the Freundlich model. However, for models with a different degree of freedom, Akaike's information criterion [44] is used instead.

Table 9 shows the comparison of the Langmuir model with other models in the study. The value of ER (15.1, 15.8, 28.6, 1875.1) indicates that the Langmuir model is more appropriate than the Toth, Redlich–Peterson, and Freundlich models, implying that the monomolecular-layer nature is prevalent for the adsorption of RDB onto MIL-101.

It is obvious that the piecewise linear regression is a useful approach to analyse the multilinearity. In order to find out the parameters in regression equations, the initial variables should be provided. If the initial variable is as far

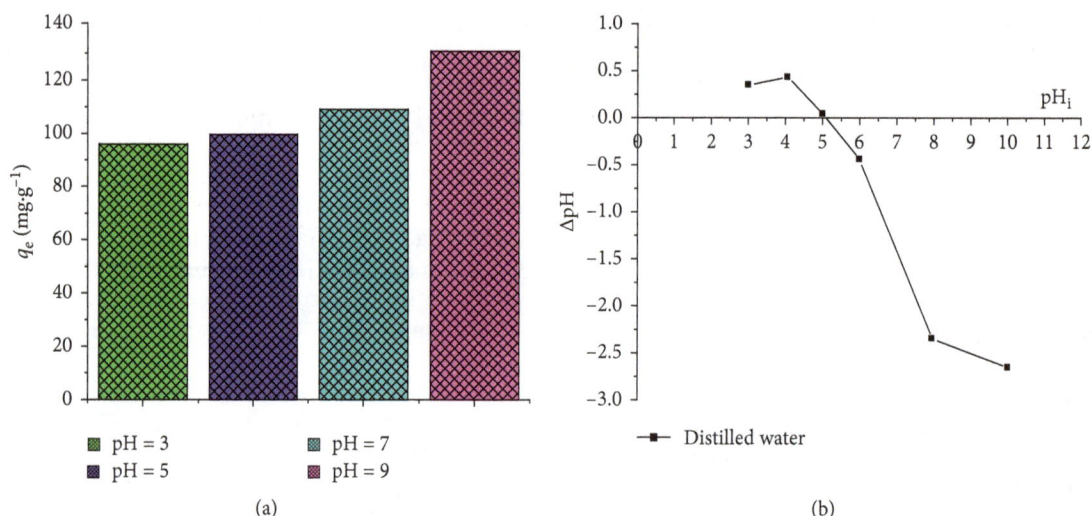

FIGURE 8: (a) Effect of pH on adsorption capacity; (b) the point of zero charge determined using the pH drift method.

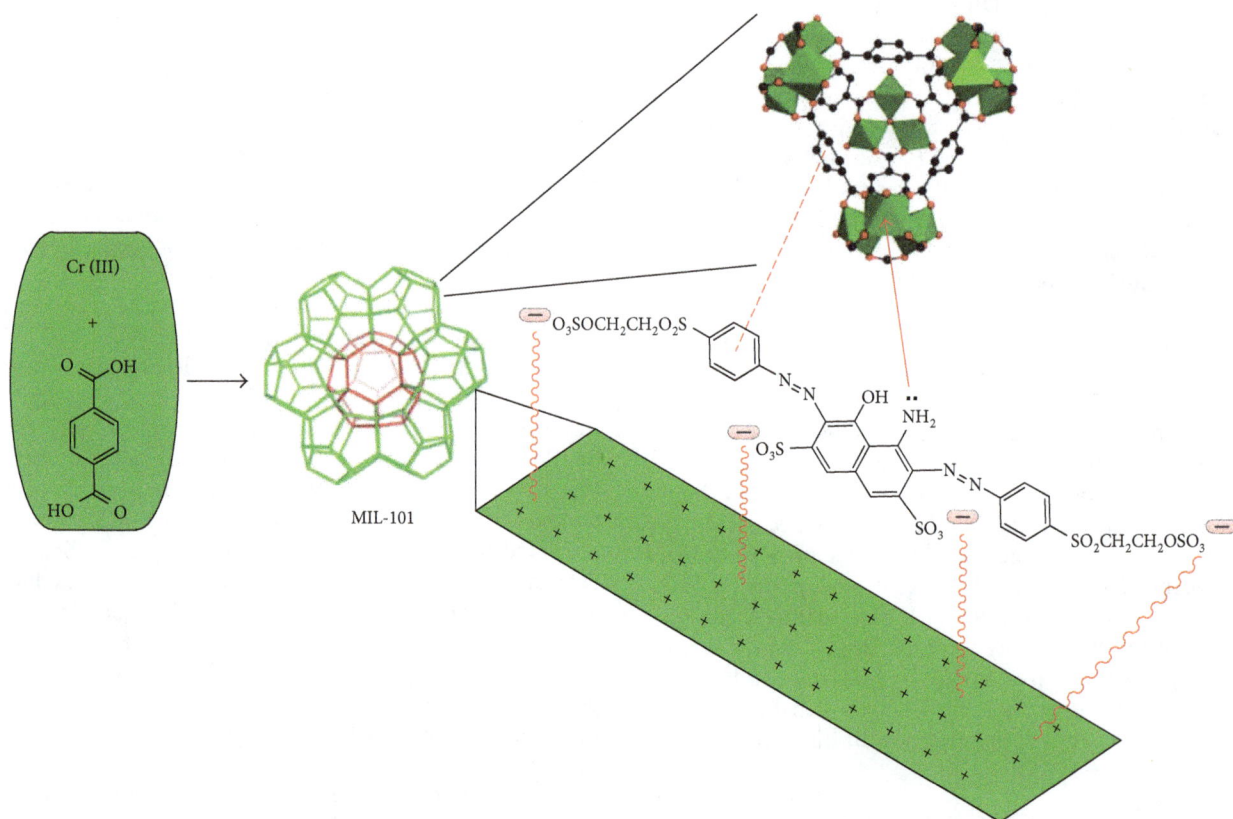

SCHEME 2: Proposed mechanisms of RDB adsorption on MIL-101.

away as the real parameters, the solving equation is easy to go wrong due to finding the local minimum. This problem needs to study further. In this study, using AIC is proved to be effective for evaluating the goodness of fit for models which have a different number of parameters and experimental points. However, the application of AIC in adsorption filed is limited. Clarification of the meaning behind the AIC should be clarified.

3.2.6. Reusability of MIL-101.
In order evaluate the reusability of MIL-101 in the removal of RDB, used MIL-101 samples were regenerated by washing with a 0.25 M NaOH solution for 5 h under sonication and drying for 10 hours at 100°C. Three generations were performed, and the RDB adsorption capacity decreases slightly (96.5%) and remained at around 120 mg·g^{-1} (Figure 10(a)). Furthermore, MIL-101 seems to be stable under adsorption conditions since the

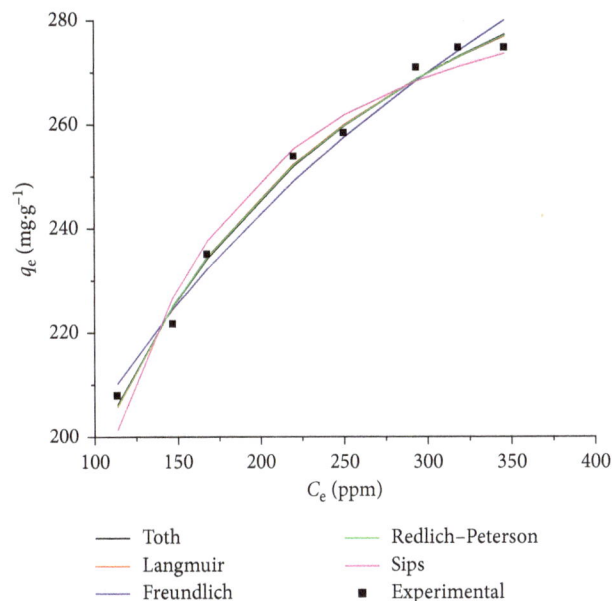

FIGURE 9: Plot of adsorption isotherms of RDB on MIL-101 using different models.

TABLE 8: Parameters of isotherm models.

Model	q_m (mg·g^{-1})	K_L, K_F, K_T, K_R, or K_S	a_R or a_S	n, b_R, or b_S	SSE	R^2
Langmuir	333.30	0.014	—	—	34.14	0.992
Freundlich	307.85	62.090	—	3.88	77.55	0.983
Toth	346.28	38.266	—	0.88	33.43	0.993
Redlich–Peterson	300.26	4.957	0.0165	0.98	33.81	0.993
Sips	290.15	0.144	0.0005	1.78	110.19	0.976

TABLE 9: Comparison of the Langmuir model with others using evidence ratio (ER).

Model comparison	AIC$_C$	N_p	ER
Langmuir/Toth	18.0 / 23.4	2 / 3	15.1
Langmuir/Redlich–Peterson	18.0 / 23.5	2 / 3	15.8
Langmuir/Sips	18.0 / 33.0	2 / 3	1785.1
Langmuir/Freundlich	18.0 / 24.6	2 / 2	26.6

XRD patterns of generated MIL-101 samples remained practically unchanged (Figure 10(b)).

4. Conclusions

MIL-101 was synthesized using the hydrothermal process. The particle size of MIL-101 could be controlled by adjusting the molar ratio of HF/H$_2$BDA. MIL-101 can serve as a useful adsorbent for RDB under batch conditions. The synthesized MIL-101 material displays high adsorption capacity and can be reused. Piecewise linear regression allows to objectively

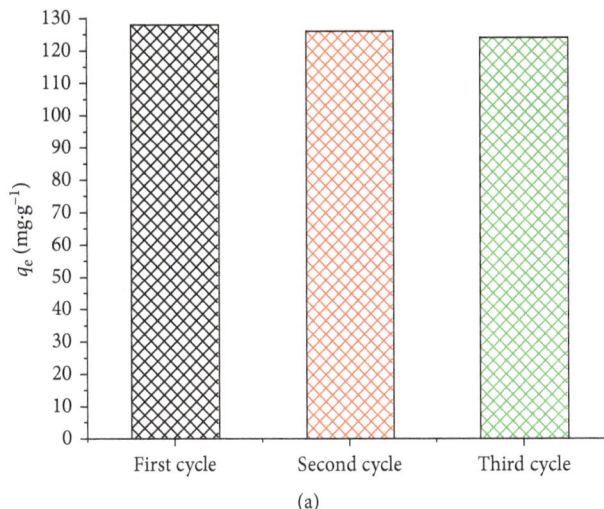

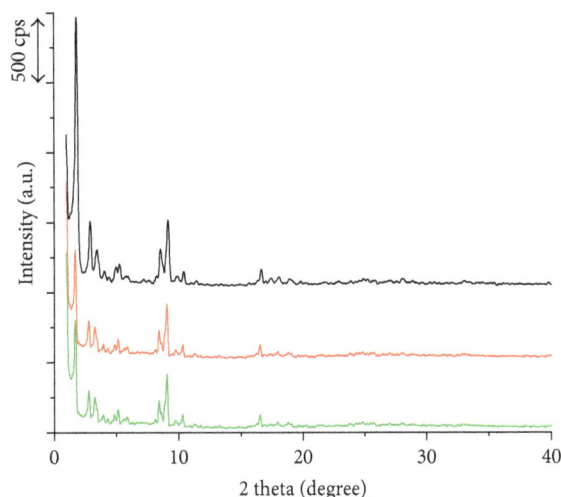

FIGURE 10: (a) Effect of regeneration cycles of ZIF-67 adsorbents on the adsorption of RDB; (b) the XRD patterns of ZIF-67 after three cycles ($V = 100$ mL; $C_{RDB} = 100$ mg·L^{-1}, $m_{MIL-101}/V_{RDB} = 0.1$ g/100 mL; shaking time = 12 hours).

analyse the experimental data using Weber's intraparticle diffusion and pseudo-kinetic adsorption models during the sorption process. The results of kinetic analysis suggested that the mechanism of the sorption of RDB on MIL-101 might take place throughout the three steps: (i) film diffusion that dominates at the beginning of the process, (ii) chemisorption that monitors the subsequent period of the process, and (iii) intraparticle diffusion, where the adsorption significantly slowed down. The best fit of the pseudo-first-order kinetics with three segments with the experimental data is consistent with the analysis of the three-step adsorption process using Weber's intraparticle diffusion model. Akaike's information criterion was employed to compare different isotherm models with a different degree of freedom. The equilibrium data of RDB onto MIL-101 fitted well to the Langmuir model rather than the Freundlich, Sips,

Toth, and Redlich–Peterson models. As the change of Gibbs free energy was negative and accompanied by the positive standard entropy change, the adsorption process was spontaneous with high affinity.

Conflicts of Interest

The authors declare that they have no conflicts of interest.

References

[1] D. Z. Van, P. Frank, and S. Villaverde, "Combined anaerobic–aerobic treatment of azo dyes—a short review of bioreactor studies," *Water Research*, vol. 39, no. 8, pp. 1425–1440, 2005.

[2] W. H. Cheung, Y. S. Szeto, and G. McKay, "Intraparticle diffusion processes during acid dye adsorption onto chitosan," *Bioresource Technology*, vol. 98, no. 15, pp. 2897–2904, 2007.

[3] J. Labanda, J. Sabaté, and J. Llorens, "Experimental and modeling study of the adsorption of single and binary dye solutions with an ion-exchange membrane adsorber," *Chemical Engineering Journal*, vol. 166, no. 2, pp. 536–543, 2011.

[4] S. Chen, J. Zhang, C. Zhang, Q. Yue, Y. Li, and C. Li, "Equilibrium and kinetic studies of methyl orange and methyl violet adsorption on activated carbon derived from Phragmites australis," *Desalination*, vol. 252, no. 1–3, pp. 149–156, 2010.

[5] Y. Li and R. T. Yang, "Hydrogen storage in metal-organic and covalent-organic frameworks by spillover," *AIChE Journal*, vol. 54, no. 1, pp. 269–279, 2008.

[6] P. L. Llewellyn, S. Bourrelly, C. Serre et al., "High uptakes of CO_2 and CH_4 in mesoporous metal-organic frameworks MIL-100 and MIL-101," *Langmuir*, vol. 24, no. 14, pp. 7245–7250, 2008.

[7] T. K. Trung, N. A. Ramsahye, P. Trens et al., "Adsorption of C_5–C_9 hydrocarbons in microporous MOFs MIL-100(Cr) and MIL-101(Cr): a manometric study," *Microporous and Mesoporous Materials*, vol. 134, no. 1–3, pp. 134–140, 2010.

[8] E. Haque, J. E. Lee, I. T. Jang et al., "Adsorptive removal of methyl orange from aqueous solution with metal-organic frameworks, porous chromium-benzenedicarboxylates," *Journal of Hazardous Materials*, vol. 181, no. 1–3, pp. 535–542, 2010.

[9] F. Leng, W. Wang, X. J. Zhao, X. L. Hu, and Y. F. Li, "Adsorption interaction between a metal–organic framework of chromium–benzenedicarboxylates and uranine in aqueous solution," *Colloids and Surfaces A: Physicochemical and Engineering Aspects*, vol. 441, pp. 164–169, 2014.

[10] C. Chen, M. Zhang, Q. Guan, and W. Li, "Kinetic and thermodynamic studies on the adsorption of xylenol orange onto MIL-101(Cr)," *Chemical Engineering Journal*, vol. 183, pp. 60–67, 2012.

[11] P. A. Soloman, C. A. Basha, M. Velan, V. Ramamurthi, K. Koteeswaran, and N. Balasubramanian, "Electrochemical degradation of Remazol black B dye effluent," *Clean–Soil, Air, Water*, vol. 37, no. 11, pp. 889–900, 2009.

[12] N. F. Cardoso, R. B. Pinto, E. C. Lima et al., "Removal of remazol black B textile dye from aqueous solution by adsorption," *Desalination*, vol. 269, no. 1–3, pp. 92–103, 2011.

[13] V. P. Ranjusha, R. Pundir, K. Kumar, M. G. Dastidar, and T. R. Sreekrishnan, "Biosorption of Remazol black B dye (Azo dye) by the growing *Aspergillus flavus*," *Journal of Environmental Science and Health. Part A*, vol. 45, no. 10, pp. 1256–1263, 2010.

[14] M. T. Thanh, T. V. Thien, V. T. T. Chau, P. D. Du, N. P. Hung, and D. Q. Khieu, "Synthesis of iron doped zeolite imidazolate framework-8 and its remazol deep black RGB dye adsorption ability," *Journal of Chemistry*, vol. 2017, Article ID 504597, 18 pages, 2017.

[15] M. T. Thanh, T. V. Thien, P. D. Du, and D. Q. Khieu, "Iron doped zeolitic imidazolate framework (Fe-ZIF-8): synthesis and photocatalytic degradation of RDB dye in Fe-ZIF-8," *Journal of Porous Materials*, vol. 25, no. 3, pp. 857–869, 2017.

[16] M. Roosta, M. Ghaedi, and A. Asfaram, "Simultaneous ultrasonic-assisted removal of malachite green and safranin O by copper nanowires loaded on activated carbon: central composite design optimization," *RSC Advances*, vol. 5, no. 70, pp. 57021–57029, 2015.

[17] H. Motulsky and A. Christopoulos, *Fitting Models to Biological Data Using Linear and Nonlinear Regression: A Practical Guide to Curve Fitting*, Oxford University Press, Oxford, UK, 2004.

[18] G. Férey, C. Mellot-Draznieks, C. Serre et al., "A chromium terephthalate-based solid with unusually large pore volumes and surface area," *Science*, vol. 309, no. 5743, pp. 2040–2042, 2005.

[19] W. J. Weber and J. C. Morris, "Kinetics of adsorption on carbon from solution," *Proceedings of the American Society of Civil Engineers/Journal of the Sanitary Engineering Division*, vol. 89, pp. 31–59, 1963.

[20] M. A. Al-Ghouti, M. A. M. Khraisheh, M. N. M. Ahmad, and S. Allen, "Adsorption behaviour of methylene blue onto Jordanian diatomite: a kinetic study," *Journal of Hazardous Materials*, vol. 165, no. 1–3, pp. 589–598, 2009.

[21] V. Vadivelan and K. V. Kumar, "Equilibrium, kinetics, mechanism, and process design for the sorption of methylene blue onto rice husk," *Journal of Colloid and Interface Science*, vol. 286, no. 1, pp. 90–100, 2005.

[22] M. F. Sawalha, J. R. Peralta-Videa, J. Romero-González, and J. L. Gardea-Torresdey, "Biosorption of Cd(II), Cr(III), and Cr (VI) by saltbush (*Atriplex canescens*) biomass: thermodynamic and isotherm studies," *Journal of Colloid and Interface Science*, vol. 300, no. 1, pp. 100–104, 2006.

[23] R. R. Sheha and A. A. El-Zahhar, "Synthesis of some ferromagnetic composite resins and their metal removal characteristics in aqueous solutions," *Journal of Hazardous Materials*, vol. 150, no. 3, pp. 795–803, 2008.

[24] I. Langmuir, "The constitution and fundamental properties of solids and liquids," *Journal of the American Chemical Society*, vol. 38, no. 11, pp. 2221–2295, 1916.

[25] H. M. F. Freundlich, "Over the adsorption in solution," *Journal of Physical Chemistry*, vol. 57, pp. 385–471, 1915.

[26] G. D. Halsey, "The role of surface heterogeneity in adsorption," *Advances in Catalysis*, vol. 4, pp. 259–269, 1952.

[27] O. Redlich and D. L. Peterson, "A useful adsorption isotherm," *Journal of Physical Chemistry*, vol. 63, no. 6, pp. 10–24, 1959.

[28] R. Sips, "On the structure of a catalyst surface," *Journal of Chemical Physics*, vol. 16, no. 5, pp. 490–495, 1948.

[29] J. Toth, "State equations of the solid gas interface layer," *Acta Chimica Hungarica*, vol. 69, pp. 311–317, 1971.

[30] G. F. Malash and M. I. El-Khaiary, "Piecewise linear regression: a statistical method for the analysis of experimental adsorption data by the intraparticle-diffusion models," *Chemical Engineering Journal*, vol. 163, no. 2010, pp. 256–263, 2010.

[31] T. Loiseau and G. Férey, "Crystalline oxyfluorinated open-framework compounds: silicates, metal phosphates, metal fluorides and metal-organic frameworks (MOF)," *Journal of Fluorine Chemistry*, vol. 128, no. 4, pp. 413–422, 2007.

[32] N. Mohammadi, H. Khani, V. K. Gupta, E. Amereh, and S. Agarwal, "Adsorption process of methyl orange dye onto mesoporous carbon material–kinetic and thermodynamic studies," *Journal of Colloid and Interface Science*, vol. 362, no. 2, pp. 457–462, 2011.

[33] X. Peng, D. Huang, T. Odoom-Wubah, D. Fu, J. Huang, and Q. Qin, "Adsorption of anionic and cationic dyes on ferromagnetic ordered mesoporous carbon from aqueous solution: Equilibrium, thermodynamic and kinetics," *Journal of Colloid and Interface Science*, vol. 430, pp. 272–282, 2014.

[34] S. J. Allen, L. J. Whitten, M. Murray, O. Duggan, and P. Brown, "The adsorption of pollutants by peat, lignite and activated chars," *Journal of Chemical Technology and Biotechnology*, vol. 68, no. 4, pp. 442–452, 1997.

[35] M. Korczak and J. Kurbiel, "New mineral-carbon sorbent: Mechanism and effectiveness of sorption," *Water Research*, vol. 23, no. 8, pp. 937–946, 1989.

[36] Z. Hasan, J. Jeon, and S. H. Jhung, "Adsorptive removal of naproxen and clofibric acid from water using metal-organic frameworks," *Journal of Hazardous Materials*, vol. 209–210, pp. 151–157, 2012.

[37] S. Lin, Z. Song, G. Che et al., "Adsorption behavior of metal–organic frameworks for methylene blue from aqueous solution," *Microporous and Mesoporous Materials*, vol. 193, pp. 27–34, 2014.

[38] C.-K. Lee, K.-S. Low, and L.-C. Chung, "Removal of some organic dyes by hexane-extracted spent bleaching earth," *Journal of Chemical Technology and Biotechnology*, vol. 69, no. 1, pp. 93–99, 1997.

[39] E. I. Unuabonah, K. O. Adebowale, and B. I. Olu-Owolabi, "Kinetic and thermodynamic studies of the adsorption of lead (II) ions onto phosphate-modified kaolinite clay," *Journal Hazardous Materials*, vol. 144, pp. 386–395, 2007.

[40] A. Kumar, B. Prasad, and I. M. Mishra, "Adsorptive removal of acrylonitrile by commercial grade activated carbon: Kinetics, equilibrium and thermodynamics," *Journal of Hazardous Materials*, vol. 152, no. 2, pp. 589–600, 2008.

[41] M. I. El-Khaiary and G. F. Malash, "Common data analysis errors in batch adsorption studies," *Hydrometallurgy*, vol. 105, no. 3–4, pp. 314–320, 2011.

[42] A. Günay, E. Arslankaya, and I. Tosun, "Lead removal from aqueous solution by natural and pretreated clinoptilolite: adsorption equilibrium and kinetics," *Journal of Hazardous Materials*, vol. 146, no. 1-2, pp. 362–371, 2007.

[43] K. V. Kumar and S. Sivanesan, "Pseudo-second order kinetic models for safranin onto rice husk: Comparison of linear and non-linear regression analysis," *Process Biochemistry*, vol. 41, no. 5, pp. 1198–1202, 2006.

[44] H. Akaike, "A new look at the statistical model dentification," *IEEE Transactions on Automatic Control*, vol. 19, no. 6, pp. 716–723, 1974.

Viscosities and Conductivities of [BMIM] Zn(Ac)$_x$Cl$_y$(x = 0, 1, 2, 3; y = 3, 2, 1, 0) Ionic Liquids at Different Temperatures

Hang Xu and Dandan Zhang

Chemical Engineering and Pharmaceutics School, Henan University of Science and Technology, Luoyang 471023, China

Correspondence should be addressed to Hang Xu; xhinbj@126.com

Academic Editor: Elena Gomez

Viscosity and conductivity data of [BMIM]Zn(Ac)$_x$Cl$_y$(x = 0, 1, 2, 3; y = 3, 2, 1, 0) ionic liquids were detected at temperature ranging from 323.15 to 353.15 K with an interval of 5 K. The conductivities of different ionic liquids at the same temperature followed the trend [BMIM][ZnAcCl$_2$] > [BMIM][ZnAc$_2$Cl] > [BMIM][ZnCl$_3$] > [BMIM][ZnAc$_3$]. The viscosities of different ionic liquid abided by the order [BMIM][ZnCl$_3$] > [BMIM][ZnAcCl$_2$] > [BMIM][ZnAc$_2$Cl] > [BMIM][ZnAc$_3$]. Acetate ion could reduce the viscosity of ionic liquids. The relationship between viscosity/conductivity and temperature obeyed the Arrhenius equation and Vogel-Fulcher-Tammann (VFT) equation very well with above 0.99 correlation coefficients.

1. Introduction

Ionic liquid (IL) is a kind of liquid which is composed of ions under room temperature. Physical properties of ionic liquids are being odorless, being noncombustible, strong electrical conductivity, and very low vapor pressure which can be used as solvent in high vacuum system and reduce the problems of environmental pollution due to volatility [1]. Compared with traditional organic solvents and electrolytes, ionic liquid has a great ability to dissolve a large number of inorganic and organic substances and, sometimes, exhibits a double functional role of solvent and catalyst [2]. Ionic liquid can be used in the fields of chemical reaction solvent and catalytic reaction carrier. For example, Mehnert et al. [3] used [BMIM][PF$_6$] as chemical solvent for homogeneous hydroformylation catalysis. Bagheri et al. [4] prepared heteropolytungstate-ionic liquid as carrier which was supported on the surface of silica coated magnetite nanoparticles. Wang et al. [5] prepared ZnCl$_2$-[BMIM]Cl which was used in cocatalyzed coupling reaction of CO$_2$ with epoxides. Sun and Zhao [6] used [BMIM]Cl/[FeCl$_3$] ionic liquid as catalyst for alkylation of benzene with 1-octadecene. Liu et al. [7] studied using ionic liquid [BMIM][Ac] as a catalyst for methanolysis of polycarbonate.

In most cases, the physical properties of ionic liquid, especially conductivity and viscosity, have a great influence for chemical reaction process. In this study, four ionic liquids of [BMIM][ZnCl$_3$], [BMIM][ZnAcCl$_2$], [BMIM][ZnAc$_2$Cl], and [BMIM][ZnAc$_3$] were prepared and those viscosities and conductivities were measured at different temperatures. Arrhenius equation and Vogel-Fulcher-Tammann (VFT) were used to describe the relationship between the conductivity/viscosity and temperature.

2. Experimental

2.1. Materials. 1-Butyl-3-methylimidazolium chloride ([BMIM]Cl) and 1-butyl-3-methylimidazolium acetic acid ([BMIM]Ac) ionic liquids were purchased from Linzhou Keneng Material Technology Co., Ltd. Purity grades of [BMIM]Cl and [BMIM]Ac were above 99% and they could be used after drying treatment. Zinc chloride and Zinc acetate were analytically pure and used after drying process.

2.2. Preparation of Ionic Liquids. The ionic liquids had been prepared based on [8]. [BMIM][ZnCl$_3$] ionic liquid was prepared by mixing 0.3 mol [BMIM]Cl and 0.3 mol ZnCl$_2$ at 90°C for 24 h. [BMIM][ZnAcCl$_2$] ionic liquid was prepared by mixing 0.3 mol [BMIM]Ac and 0.3 mol ZnCl$_2$ at 90°C for 24 h. [BMIM][ZnAc$_2$Cl] ionic liquid was prepared by mixing 0.3 mol [BMIM]Cl and 0.3 mol ZnAc$_2$ at 90°C for 24 h. [BMIM][ZnAc$_3$] ionic liquid was prepared by mixing 0.3 mol [BMIM]Ac and 0.3 mol ZnAc$_2$ at 90°C for 24 h. All the samples should be dried in vacuum drying oven at 90°C for 24 h.

2.3. Measurements of Conductivity and Viscosity. The conductivities of [BMIM][ZnCl$_3$], [BMIM][ZnAcCl$_2$], [BMIM][ZnAc$_2$Cl], and [BMIM][ZnAc$_3$] were measured by a Wayne-Kerr 6430B Autobalance Bridge fitted with a Shanghai DJS-1 electrode and the temperature was controlled within ±0.1 K using a HAAKEV26 temperature thermostat (Thermo Electron). In order to keep the ILs dry, the experiment should be carried out in a dry nitrogen atmosphere in the temperature range 323.15 to 353.15 K with an interval of 5 K. The viscosities of [BMIM][ZnCl$_3$], [BMIM][ZnAcCl$_2$], [BMIM][ZnAc$_2$Cl], and [BMIM][ZnAc$_3$] were measured using an Ostwald viscometer. We used a thermostatic bath to control the temperature to get a stability of ±0.1 K, which consumed 30 min to attain thermal equilibrium in the viscosity. Moreover, in order to prevent absorbing water from atmosphere, we should put the ILs in a dry nitrogen atmosphere when measuring the viscosity.

The water in the ionic liquid has a great influence on the data of conductivity and viscosity. So, moisture contents of ionic liquids were measured using a 851 Karl-Fischer Moisture Titrator (Metrohm). The results showed that the moisture contents of [BMIM][ZnCl$_3$], [BMIM][ZnAcCl$_2$], [BMIM][ZnAc$_2$Cl], and [BMIM][ZnAc$_3$] were 69.2 ppm, 83.5 ppm, 98.4 ppm, and 101.2 ppm, respectively. All the ionic liquids exhibited an extremely low moisture content in measuring process.

3. Results and Discussion

3.1. Conductivities of ILs at Different Temperature. The conductivities of [BMIM][ZnAc$_3$], [BMIM][ZnAc$_2$Cl], [BMIM][ZnAcCl$_2$], and [BMIM][ZnCl$_3$] ionic liquids are shown in Figure 1. From Figure 1, it is clearly seen that conductivities of ionic liquids increase with the rise of temperature. The conductivities of [BMIM][ZnAc$_3$] are 0.37 ms·cm^{-1} at 323.15 K and 2.05 ms·cm^{-1} at 353.15 K, respectively. The conductivities of [BMIM][ZnAc$_2$Cl] are 2.32 ms·cm^{-1} at 328.15 K and 4.88 ms·cm^{-1} at 348.15 K, respectively. The conductivities of [BMIM][ZnAcCl$_2$] are 3.36 ms·cm^{-1} at 333.15 K and 4.86 ms·cm^{-1} at 343.15 K, respectively. The conductivities of [BMIM][ZnCl$_3$] are 0.74 ms·cm^{-1} at 323.15 K and 1.86 ms·cm^{-1} at 338.15 K, respectively. With the rise of temperature, conductivities of ionic liquids have a great increase because the solvation degree and ion radius of ionic liquid are reduced. The ion motion is accelerated by the increase of temperature, which can also cause the improvement of the conductivity. Furthermore, the increase of temperature leads to the decrease of the solution viscosity and ions are easier to move. From Figure 1, the conductivities of different

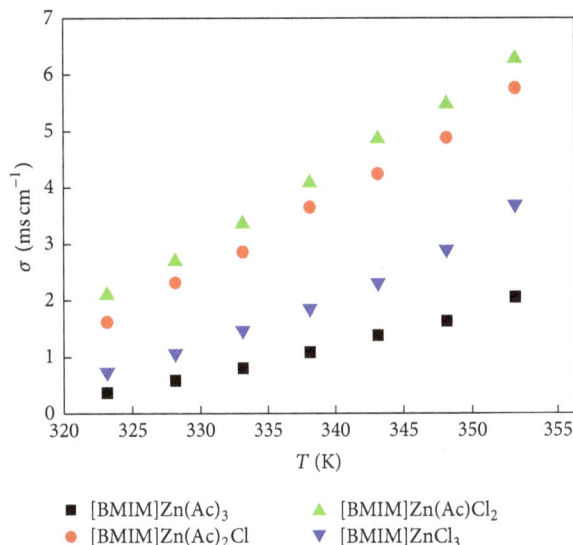

FIGURE 1: Conductivity of ILs at temperature from 323.15 K to 353.15 K.

ionic liquids at the same temperature followed the trend [BMIM][ZnAcCl$_2$] > [BMIM][ZnAc$_2$Cl] > [BMIM][ZnCl$_3$] > [BMIM][ZnAc$_3$]. For example, in 338.15 K, the conductivities of [BMIM][ZnAc$_3$], [BMIM][ZnAc$_2$Cl], [BMIM][ZnAcCl$_2$], and [BMIM][ZnCl$_3$] are 1.08 ms·cm^{-1}, 3.65 ms·cm^{-1}, 4.08 ms·cm^{-1}, and 1.86 ms·cm^{-1}, respectively. Compared with [ZnAc$_3$]$^-$, [ZnCl$_3$]$^-$ has the features of smaller ionic radius and lower mass which are advantage for ionic transportation. For another aspect, organic ion exhibits a weak ionization which is also bad for conductivity. So, conductivity of [ZnAcCl$_2$]$^-$ is better than [ZnAc$_2$Cl]$^-$. Contrary to [ZnCl$_3$]$^-$, the symmetry of [ZnAc$_2$Cl]$^-$ is low which is good for the ionic movement. The higher the symmetry in anion of ionic liquids, the lower the conductivity. The more the chlorine ion and the less the acetic acid, the greater the conductivity.

The relationship between conductivities of ionic liquid and temperature followed Arrhenius equation [9], which is described as

$$\ln \sigma = \ln \sigma_\infty - \frac{E_\sigma}{RT}. \tag{1}$$

Here, σ is the conductivity and σ_∞ is the empirical constant. E_σ notes for the activation energy. T is the temperature and R is 8.314 J·mol^{-1}·K^{-1}. According to the Arrhenius equation and experimental data, the relationship between conductivity and temperature is shown in Figure 2 and the calculated values of σ_∞, E_σ, and R^2 (correlation coefficient) are listed in Table 1. From Table 1, correlation coefficients (R^2) of four ionic liquids are above 0.99. The conductivities of ionic liquids at different temperatures follow Arrhenius relationship very well. The [BMIM]Zn(Ac)$_3$ ionic liquid has the maximal E_σ and σ_∞ compared to the other three ionic liquids. The E_σ values and σ_∞ values follow the order of [BMIM]Zn(Ac)$_3$ > [BMIM]ZnCl$_3$ > [BMIM]Zn(Ac)$_2$Cl > [BMIM]Zn(Ac)Cl$_2$.

TABLE 1: Fitted values of conductivity of σ_∞, E_σ, σ_0, B, T_0, and R_2 based on Arrhenius equation and VFT equation.

ILs	Arrhenius equation			VFT equation			
	$10^{-6} \cdot \sigma_\infty/\text{mS}\cdot\text{cm}^{-1}$	$E_\sigma/\text{kJ}\cdot\text{mol}^{-1}$	R^2	$\sigma_0/\text{mS}\cdot\text{cm}^{-1}$	B/K	T_0/K	R^2
[BMIM]Zn(Ac)$_3$	137.5	52.70	0.9917	22.70	177.3	280	0.9996
[BMIM]Zn(Ac)$_2$Cl	3.197	38.68	0.9904	29.64	116.3	283	0.9990
[BMIM]Zn(Ac)Cl$_2$	0.8267	34.47	0.9949	52.83	187.6	265	0.9998
[BMIM]ZnCl$_3$	76.78	49.40	0.9969	214.0	441.7	245	0.9992

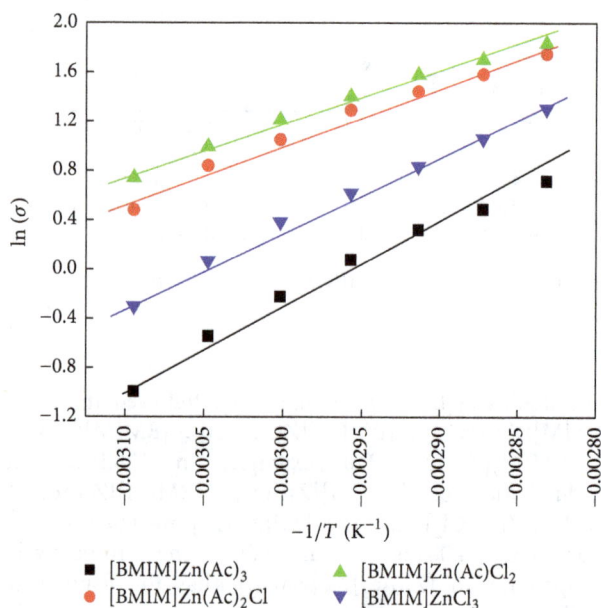

FIGURE 2: Relationship between conductivity and temperature based on Arrhenius equation.

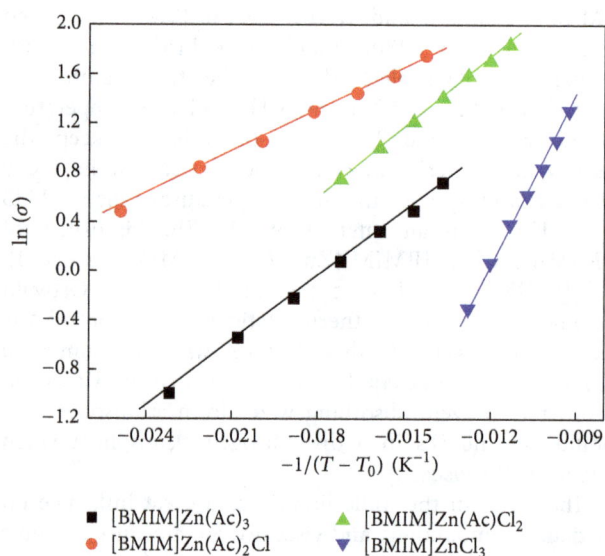

FIGURE 3: Relationship between conductivity and temperature based on VFT equation.

According to [10], the conductivity of ionic liquids also obeyed Vogel-Fulcher-Tammann (VFT) equation which is shown in

$$\sigma = \sigma_0 \exp \frac{-B}{T - T_0}. \qquad (2)$$

Here, σ_0, B, and T_0 are the empirical constants. According to the VFT equation, the relationship between conductivity and temperature is shown in Figure 3 and σ_0, B, T_0, and R^2 are calculated and listed in Table 1. From Table 1, it is clearly seen that all the R^2 values are above 0.99 which means VFT equation is better to describe the relationship between conductivity and temperature than Arrhenius equation. The biggest σ_0 and B values are [BMIM]ZnCl$_3$ ionic liquid with 214.0 mS·cm^{-1} and 441.7 K, respectively, compared to the other three ionic liquids. The smallest T_0 is [BMIM]ZnCl$_3$ ionic liquid with 245 K and the biggest is [BMIM]Zn(Ac)$_2$Cl with 283 K compared to others.

3.2. Viscosity of ILs at Different Temperature.

The viscosities of [BMIM][ZnAc$_3$], [BMIM][ZnAc$_2$Cl], [BMIM][ZnAcCl$_2$], and [BMIM][ZnCl$_3$] ionic liquids are shown in Figure 4. From Figure 4, it is clearly seen that viscosities of ionic liquids express a decrease in tendency with the rise of temperature.

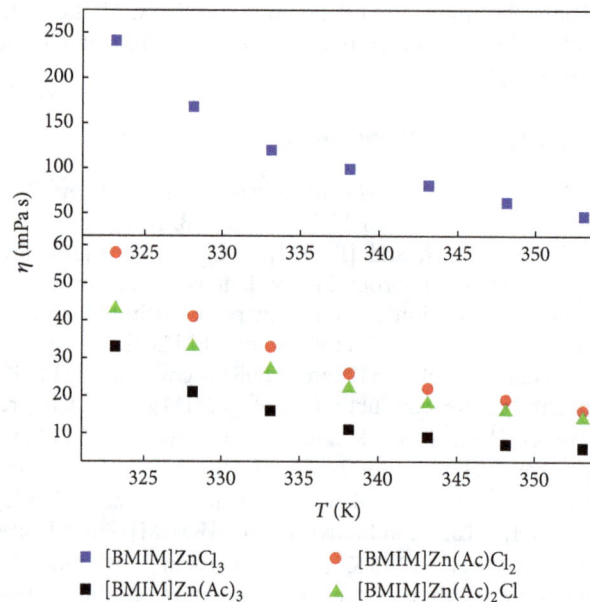

FIGURE 4: Viscosity of ILs at temperature from 323.15 K to 353.15 K.

The reason is that high temperature is a disadvantage for the intermolecular association. In low temperature, high intermolecular association can improve the viscosity of ionic liquid. For example, the viscosities of [BMIM][ZnAc$_3$] are

TABLE 2: Fitted values of viscosity of σ_∞, E_σ, σ_0, B, T_0, and R_2 based on Arrhenius equation and VFT equation.

ILs	Arrhenius equation			VFT equation			
	$10^6 \cdot \eta_\infty$/mPa·s	E_η/kJ·mol^{-1}	R^2	η_0/mPa·s	B/K	T_0/K	R^2
[BMIM]Zn(Ac)$_3$	0.06345	52.51	0.9922	0.3586	221.9	274	0.9989
[BMIM]Zn(Ac)$_2$Cl	75.29	34.73	0.9952	1.291	221.3	260	0.9994
[BMIM]Zn(Ac)Cl$_2$	22.01	38.66	0.9927	2.442	142.7	278	0.9993
[BMIM]ZnCl$_3$	2.282	49.48	0.9964	0.1022	914.0	205	0.9972

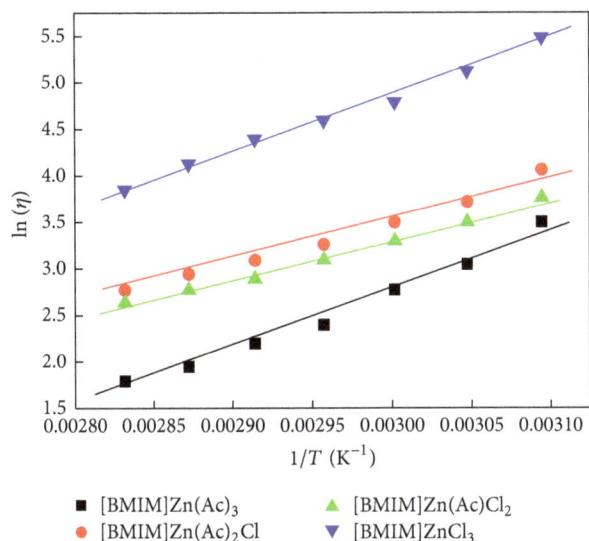

FIGURE 5: Relationship between viscosity and temperature based on Arrhenius equation.

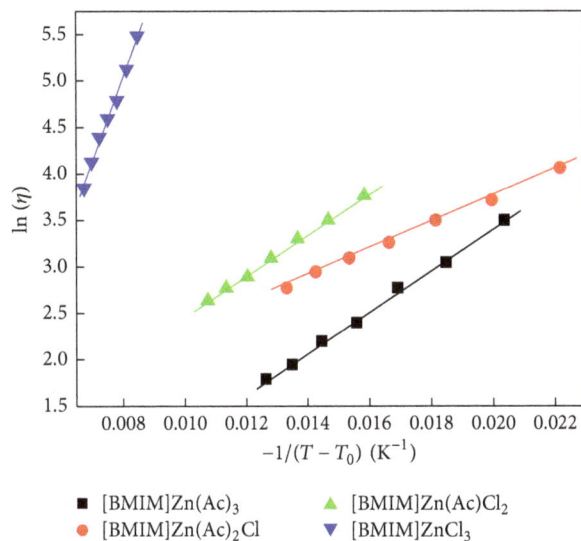

FIGURE 6: Relationship between viscosity and temperature based on VFT equation.

33 mPa·s at 323.15 K and 6 mPa·s at 353.15 K, respectively. The viscosities of [BMIM][ZnAc$_2$Cl] are 33 mPa·s at 328.15 K and 18 mPa·s at 348.15 K. The viscosities of [BMIM][ZnAcCl$_2$] are 58 mPa·s at 323.15 K and 22 mPa·s at 343.15 K. The viscosities of [BMIM][ZnCl$_3$] are 241 mPa·s at 323.15 K and 99 mPa·s at 338.15 K. The viscosities of [BMIM][ZnCl$_3$] are significantly higher than other ionic liquids. The viscosity order of four ionic liquids followed the tendency [BMIM][ZnCl$_3$] > [BMIM][ZnAcCl$_2$] > [BMIM][ZnAc$_2$Cl] > [BMIM][ZnAc$_3$]. Acetate ion can reduce the viscosity of ionic liquids. On the contrary, chloride ion can increase the viscosity of the ionic liquids.

According to [11], the Arrhenius equation exhibited in (3) can be used to describe the relationship between viscosity and temperature.

$$\ln \eta = \ln \eta_\infty + \frac{E_\eta}{RT}. \tag{3}$$

Here, η_∞ is the empirical constant and E_η denotes the activation energy for viscous flow. Figure 5 shows the relationship between viscosity of the logarithm and reciprocal of absolute temperature. From Figure 5, $\ln \eta$ and $1/T$ is linear correlation. The fitting result is shown in Table 2. From Table 2, the correlation coefficients are greater than

0.99 which indicated that Arrhenius formula is good enough to describe the relationship between viscosity and temperature of ionic liquids. The E_η values of [BMIM]Zn(Ac)$_3$ and [BMIM]ZnCl$_3$ are 52.51 kJ·mol^{-1} and 49.48 kJ·mol^{-1}, respectively. The E_η values of [BMIM] Zn(Ac)$_2$Cl and [BMIM]Zn(Ac)Cl$_2$ are 34.73 kJ·mol^{-1} and 38.66 kJ·mol^{-1}, respectively.

The viscosity values of four ionic liquids are also fitted by the VFT equation which is shown in

$$\eta = \eta_0 \exp\left(\frac{B}{T - T_0}\right). \tag{4}$$

Here, η_0, B, T_0 are the empirical constants. According to the VFT equation, relationship between dynamic viscosity and temperature is shown in Figure 6. From Figure 6, viscosity of ionic liquids at different temperature followed VFT equation very well. The fitting data was shown in Table 2. From Table 2, the correlation coefficients which are calculated by VFT equation are superior to Arrhenius model.

4. Conclusions

Four ionic liquids of [BMIM][ZnCl$_3$], [BMIM][ZnAcCl$_2$], [BMIM][ZnAc$_2$Cl], and [BMIM][ZnAc$_3$] were prepared and those viscosity and conductivity data of ILs were detected at temperature ranging from 323.15 to 353.15 K with an interval of 5 K. The conductivity order at the same temperature followed the trend: [BMIM][ZnAcCl$_2$] > [BMIM][ZnAc$_2$Cl] > [BMIM][ZnCl$_3$] > [BMIM][ZnAc$_3$]. The higher the symmetry in anion of ionic liquid, the lower the conductivity. The more the chlorine ion and the less the acetic acid, the greater the conductivity. The viscosity order of four ionic liquids is as follows: [BMIM][ZnCl$_3$] > [BMIM][ZnAcCl$_2$] > [BMIM][ZnAc$_2$Cl] > [BMIM][ZnAc$_3$]. Acetate ion could reduce the viscosity of ionic liquids. On the contrary, chloride ion could increase the viscosity of the ionic liquids. The relationship between viscosity/conductivity and temperature obeyed the Arrhenius equation and Vogel-Fulcher-Tammann (VFT) equation very well with above 0.99 correlation coefficients. According to the R^2 values, VFT equation was superior to Arrhenius model.

Conflicts of Interest

The authors declare that they have no conflicts of interest.

Acknowledgments

This work is supported by the National Natural Science Foundation of China (no. 21006057), Nature Science Foundation of Henan Province (no. 162300410083), and Youth Teacher Support Program of Henan Provincial University (no. 2014GGJS-055).

References

[1] A. R. Ferreira, L. A. Neves, J. C. Ribeiro et al., "Removal of thiols from model jet-fuel streams assisted by ionic liquid membrane extraction," *Chemical Engineering Journal*, vol. 256, pp. 144–154, 2014.

[2] W. Ochędzan-Siodłak and K. Dziubek, "Metallocenes and post-metallocenes immobilized on ionic liquid-modified silica as catalysts for polymerization of ethylene," *Applied Catalysis A: General*, vol. 484, pp. 134–141, 2014.

[3] C. P. Mehnert, R. A. Cook, N. C. Dispenziere, and M. Afeworki, "Supported ionic liquid catalysis—a new concept for homogeneous hydroformylation catalysis," *The Journal of the American Chemical Society*, vol. 124, no. 44, pp. 12932-12933, 2002.

[4] M. Bagheri, M. Masteri-Farahani, and M. Ghorbani, "Synthesis and characterization of heteropolytungstate-ionic liquid supported on the surface of silica coated magnetite nanoparticles," *Journal of Magnetism and Magnetic Materials*, vol. 327, pp. 58–63, 2013.

[5] F. Wang, C. Xu, Z. Li, C. Xia, and J. Chen, "Mechanism and origins of enantioselectivity for [BMIM]Cl ionic liquids and ZnCl2 co-catalyzed coupling reaction of CO$_2$ with epoxides," *Journal of Molecular Catalysis A: Chemical*, vol. 385, pp. 133–140, 2014.

[6] X. Sun and S. Zhao, "[bmim]Cl/[FeCl3] ionic liquid as catalyst for alkylation of benzene with 1-octadecene," *Chinese Journal of Chemical Engineering*, vol. 14, no. 3, pp. 289–293, 2006.

[7] F. Liu, L. Li, S. Yu, Z. Lv, and X. Ge, "Methanolysis of polycarbonate catalysed by ionic liquid [Bmim][Ac]," *Journal of Hazardous Materials*, vol. 189, no. 1-2, pp. 249–254, 2011.

[8] H. Xu, T. Yu, M. Li, and Z. Liang, "Viscosities and conductivities of [BMIM][ZnxCly] (x = 1, 1.5, 2; y = 3, 4, 5) ionic liquids at different temperatures," *Asian Journal of Chemistry*, vol. 27, no. 1, pp. 349–352, 2015.

[9] J. Restolho, A. P. Serro, J. L. Mata, and B. Saramago, "Viscosity and surface tension of 1-ethanol-3-methylimidazolium tetrafluoroborate and 1-methyl-3-octylimidazolium tetrafluoroborate over a wide temperature range," *Journal of Chemical and Engineering Data*, vol. 54, no. 3, pp. 950–955, 2009.

[10] H. Tokuda, K. Hayamizu, K. Ishii, M. A. B. H. Susan, and M. Watanabe, "Physicochemical properties and structures of room temperature ionic liquids. 2. Variation of alkyl chain length in imidazolium cation," *Journal of Physical Chemistry B*, vol. 109, no. 13, pp. 6103–6110, 2005.

[11] C. Schreiner, S. Zugmann, R. Hartl, and H. J. Gores, "Temperature dependence of viscosity and specific conductivity of fluoroborate-based ionic liquids in light of the fractional walden rule and angell's fragility concept," *Journal of Chemical and Engineering Data*, vol. 55, no. 10, pp. 4372–4377, 2010.

22

Enthalpies of Combustion and Formation of Histidine Stereoisomers

A. Neacsu [ID],[1] D. Gheorghe [ID],[1] I. Contineanu,[1] A. M. Sofronia,[1] F. Teodorescu,[2] and S. Perişanu [ID][3]

[1]Institute of Physical Chemistry "Ilie Murgulescu" of the Romanian Academy, 202 Splaiul Independentei St., 060021 Bucharest, Romania
[2]Centre of Organic Chemistry "C.D. Nenitescu" of the Romanian Academy, 202 B Splaiul Independentei St., 060023 Bucharest, Romania
[3]Department of General Chemistry, Polytechnic University of Bucharest, 1 St. Polizu, Bucharest, Romania

Correspondence should be addressed to D. Gheorghe; gheorghedanny2@gmail.com and S. Perişanu; stefan.perisanu@upb.ro

Academic Editor: João Paulo Leal

The combustion energy of histidine enantiomers (L and D) and of their racemic mixture was measured experimentally. The following values for the enthalpies of formation corresponding to the crystalline state were derived (L = −451.7, D = −448.7, DL = −451.5 kJ·mol^{-1}), and information concerning their stability was obtained by correlating the values of the above thermochemical quantity with the structure of the molecules by using the group additivity scheme. The samples were characterized using a simultaneous thermogravimetry (TG) coupled with differential scanning calorimetry (DSC) techniques in the temperature range between ambient and beyond melting-decomposition, and the corresponding parameters were calculated. The high values of the decomposition temperatures highlight the stability of the compounds. The decomposition reactions are discussed in terms of DSC and TG data, obtained by us and other researchers.

1. Introduction

The aim of this study was a calorimetric characterization of L-, D-, and DL-histidine isomers.

Histidine is a heterocyclic amino acid containing an imidazole ring. The imidazole ring is aromatic, as it contains six π electrons. Histidine is a proteinogenic amino acid, and it is able to form π stacking interactions. Consequently, the side chain can have a role in stabilizing the folded structures of proteins. Histidine is an essential amino acid that is not synthesized *de novo* in humans and is needed for growth and tissue repair [1].

The imidazole ring allows histidine to act as a coordinating ligand in metal proteins, including certain enzymes. It is important in hemoglobin, as well. Histidine is involved in stabilizing oxyhemoglobin and destabilizing CO-bound hemoglobin. Histidine is important for maintenance of myelin membranes that protect nerve cells and is metabolized to the neurotransmitter histamine [2].

Two divergent values of the enthalpy of formation of L-histidine are found in the literature: −441.8 ± 2.6 kJ·mol^{-1} [3] and −466.7 ± 2.8 kJ·mol^{-1} [4]. Vasilev et al. [3] were aware that their data differ from previously reported values [4] and considered that the difference is due to the inadequate purity of the sample of the later authors. No data about the enthalpy of formation of D-histidine and for the racemic are found in the literature. The main purpose of this work is to determine these parameters and to compare them with those calculated by the group additivity method. The paper brings more information about the thermal behavior of the investigated compounds in function of the isomer type.

2. Experimental

2.1. Materials. D-histidine and DL-histidine were obtained commercially from Sigma-Aldrich, mass fraction purities ≥98% and 99%, respectively, and L-histidine, assay >99%, from Fluka with $M = 155.15\,\text{g·mol}^{-1}$. Samples were dried in a vacuum oven for 3 hours at 90°C and preserved in a desiccator before use, in order to eliminate adsorbed water.

The purity of samples was tested by DSC and polarimetry.

Our DSC data confirm the purity of over 99% for the DL- and D-isomers (99.06 and 99.25%, respectively) while for the D-histidine, the purity was 98.86%. The data are shown in Table 1 of Supplementary materials.

Specific rotations $[\alpha]_\lambda^{25}$ of the investigated compounds were determined on solutions in deionized water for checking the amino acid optical purity. A 341 PerkinElmer polarimeter was used in the D line of sodium, with glass cells (1 cm path length), at 25°C. Table 1 contains our values compared to literature data ([5], p. C768).

Like in the case of other amino acids, the only impurities amounting at least 0.1% (other than water) certified by the manufacturer consist of other amino acids, with similar values of the massic heat of combustion.

2.2. Methods

2.2.1. Combustion Calorimetry.
The combustion experiments were performed using a Parr Instruments model 6200 microprocessor controlled isoperibol oxygen bomb calorimeter. Temperatures are measured with a high-precision electronic thermometer using a specially designed thermistor sensor sealed in a stainless steel probe which is fixed in the calorimeter cover. Measurements were taken with 0.0001 K resolution. The jacket temperature is held constant for isoperibol operation. The semimicro kit handling samples from 25 to 200 mg was used because of the small amounts of the studied compounds. High-purity oxygen 99.998% was used for combustion. Calorific grade benzoic acid supplied by Parr, with heat of combustion $26,454\,\text{J·g}^{-1}$, was used for the standardization of the combustion calorimeter. The determined calorimeter constant was $\varepsilon_{\text{calor}} = 2326.9 \pm 1.9\,\text{J·K}^{-1}$.

The samples were pressed into pellets of 3 mm diameter. The pellets were weighed with a Mettler–Toledo analytical balance, model XP6 with an accuracy of $\pm 2 \cdot 10^{-6}\,\text{g}$.

The final solution from the bomb was analyzed for the presence of nitric acid (about 20% from the total nitrogen) by titration with solution of Na_2CO_3 $0.1\,\text{mol·L}^{-1}$. The heat due to nitric acid formation was obtained using the value of the enthalpy of formation of nitric acid solution, $\Delta_f H_{HNO_3}$, aq = $-58.8\,\text{kJ·mol}^{-1}$ [6].

2.2.2. Thermal Analysis.
For the thermal characterization of histidines, a simultaneous thermogravimetry (TG) and differential scanning calorimetry (DSC) TGA/DSC Setaram Setsys Evolution 17 analyzer was employed. Thermal properties (temperatures, enthalpies, and mass losses) associated with melting and/or decomposition processes of the histidine stereoisomers were measured in the temperature ranging from 20 to 600°C with a scanning rate of $10°\text{C min}^{-1}$ in alumina crucibles, using Ar flow. Standard metallic substances of 99.999% purity (In, Sn, Pb, Zn, and Al) were used for the

calibration in temperature. The melting onset temperatures and heats of fusion of standard materials were used for temperature correction and energy calibration. The sample mass for simultaneous TG-DSC measurements was about 1-2 mg. The error of TG measurement is $\pm 0.154\%$. All thermal analysis (TG-DSC) data were processed using Calisto software.

3. Results

3.1. Combustion Energy.
At least 6 runs were retained for each isomer. Some runs were rejected because of doubt about combustion completeness. In runs used in data calculation, there was no evidence of soot formation in the bomb. The data regarding the combustion measurements for the three isomers are given in Tables 2–4 in Supplementary materials. The assigned uncertainties are twice the standard error of the mean. ΔU (fuse) and ΔU (ign) were calculated from the mass of cotton and $\Delta_c u$ (cotton) = $16240 \pm 20\,\text{J·g}^{-1}$ [7] and from the mass of the fire and $\Delta_c u$ (Ni–Cr) = $5.86\,\text{kJ·g}^{-1}$ (certified by the fabricant), respectively. The values obtained experimentally for the combustion energy were reported to the standard state ($T = 298.15$ K and $p = 101.325\,\text{kPa}$). Corrections were performed using Washburn methodology [8].

In order to calculate the enthalpies of formation, the following values were used: $\Delta_f H_{CO_2\,(g)}^\circ = -393.51 \pm 0.13\,\text{kJ·mol}^{-1}$ and $\Delta_f H_{H_2O\,(l)}^\circ = -285.83 \pm 0.042\,\text{kJ·mol}^{-1}$ [9]. In Table 2 are presented our data for the solid-state enthalpies of formation, together with literature values [3, 4].

The values of the enthalpies of formation of the L-enantiomer are quasi-identical with that of racemic, while that of D-histidine is more negative (within the cumulated experimental errors).

3.2. DSC.
Figure 1 presents the DSC curves of L-, D-, and DL-histidine isomers. A single-sharp peak is recorded for all three stereoisomers.

The temperature ranges in which the decomposition processes (decomposition prevails due to the high temperature) of the three stereoisomers take place are similar (274–290°C).

Our peak temperatures of the enantiomers (Table 3) are in reasonable agreement with the values reported by Olafsson and Bryan [10] (288°C), Weiss et al. [11] (272°C), and Anandan et al. [12] (275°C) as well as with that included in the Handbook of Chemistry and Physics ([5], p. C445) (287°C), but not with the value of Wesolowski and Erecinska [13] (250°C). The "melting" points usually found for amino acids are irrelevant since they decompose, so that the temperatures may vary according to the morphology of the sample and to the experimental conditions used by the researchers [13].

3.3. Thermogravimetry.
Figure 2 shows the temperatures and weight losses in the TG and DTG curves. Thermogravimetric records for samples show a first-step fast weight loss of 17-18% starting above 240°C, followed by a continuous mass decrease of the sample. At 600°C, a mass reduction of over 53% is observed for both L- and

TABLE 1: Polarimetric data of histidine stereoisomers.

Isomeric histidines	Concentration (g 10^{-2} mL^{-1})	α_λ^{25}	$[\alpha]_\lambda^{25}$	$[\alpha]_\lambda$ literature [5]
L-histidine	1.128	−0.044	39.0	−39.01 (25°C)
D-histidine	2.66	+0.103	38.72	+39.80 (23°C)
DL-histidine	1.128	0	0	—

TABLE 2: Enthalpies of combustion and formation in solid state of isomeric histidines.

Isomeric histidines	$-\Delta_c U°$ (kJ·mol^{-1})	$-\Delta_c H°$ (kJ·mol^{-1})	$-\Delta_f H_{cr}°$ (kJ·mol^{-1}) (this work)[a]	$-\Delta_f H_{cr}°$ (kJ·mol^{-1}) (literature)	$-\Delta_f H_{cr}°$ (kJ·mol^{-1}) (calculated)[b]
L-histidine	3196.2 ± 2.4	3195.6 ± 2.4	451.7 ± 3.4	441.8 ± 2.6 [3] 466.7 ± 2.8 [4]	
D-histidine	3199.2 ± 2.3	3198.6 ± 2.3	448.7 ± 3.3		450.68
DL-histidine	3196.4 ± 2.3	3195.8 ± 2.3	451.5 ± 3.3		

[a]Uncertainty included the uncertainties of the enthalpies of formation of the reaction products H_2O and CO_2. [b]Estimated value by means of the group additivity method, with parameters recommended by Domalski and Hearing [14].

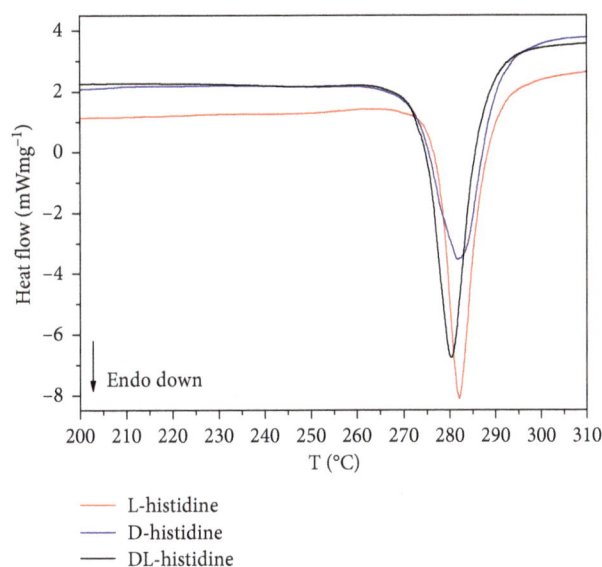

FIGURE 1: DSC curves for L-, D-, and DL-histidines (10°C min^{-1}).

TABLE 3: DSC data of isomeric histidines (10°C min^{-1}).

Isomeric histidines	$T_{on\ set}$ (°C)	T_{max} (°C)	$T_{end\ set}$ (°C)	ΔH (kJ·mol^{-1})
L-histidine	277.1	282.1	287.6	72.6 (melting-decomposition)
D-histidine	272.8	282.4	290.3	70.2 (melting-decomposition)
DL-histidine	274.2	280.4	286.6	75.4 (melting-decomposition)

D-enantiomers and the racemic (Table 4). The difference between the initial mass of samples and the mass loss due to decomposition was consistent with the mass of solid residue.

Table 5 shows comparative data of thermal analysis on histidine reported by different authors.

4. Discussion

The formation enthalpies in solid state experimentally found were compared with those obtained by means of the group additivity method, with parameters recommended by Domalski and Hearing [14]. The value of the solid-state group parameter corresponding to an imine nitrogen atom bound to a carbon atom (N_I-(C)) is missing. A value of 67 kJ·mol^{-1} was assigned to this parameter, taking into account the values of the same parameter for the liquid and gaseous states. Generally, because of the presence of an α-amino acid moiety, a zwitterion contribution is considered. A comparison between experimental and calculated values in solid state is shown in Table 2.

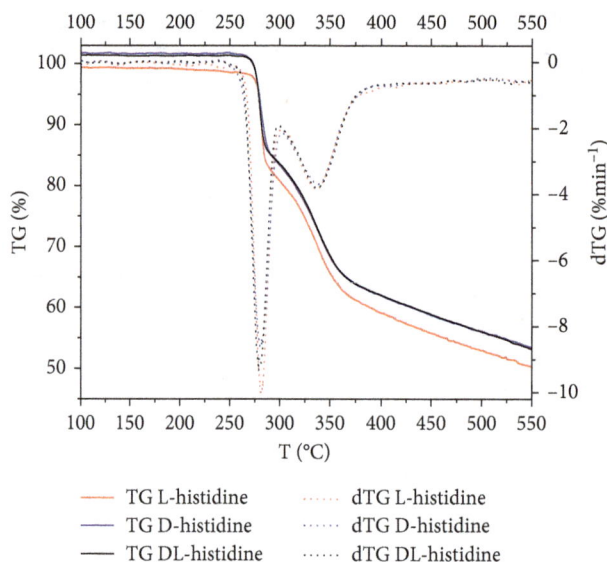

FIGURE 2: TG-DTG curves for L-, D-, and DL-histidines.

TABLE 4: Thermogravimetric data of isomeric histidines ($10°C\,min^{-1}$).

Isomers histidines	Δm (%)/Δt (°C)	Δm (%)/Δt (°C)	Δm (%)/Δt (°C)	Δm (%)/(600°C)
L-histidine	−17.98/244.1–300.1	−21.6/301.3–403.5	−11.36/402.5–594.1	−52.37
D-histidine	−18.82/246.6–301.6	−21.17/301.6–402.0	−10.63/402.0–594.2	−48.76
DL-histidine	−17.53/241.0–298.6	−21.79/298.6–400.0	−10.97/400.0–595.0	−48.96

TABLE 5: Comparative values of melting-decomposition parameters of L-histidine reported by different researchers.

$T_{on\,set}$ (°C)	T_{max} (°C)	$T_{end\,set}$ (°C)	Weight loss (%)	ΔH ($kJ\cdot mol^{-1}$)	Method	Literature reference
277.1	282.1	287.6	17.98	72.6	TG/DSC	This work
288					DSC	[10]
272	280				DSC/TGA	[11]
275	296		13	82	TG/DTA	[12]
287						[5]
250					TG/DTA	[13]

The calculated enthalpy of formation [14] agrees fairly with our experimental values. Only four hydrogen bonds per molecule were reported, less than for other amino acids. One of them is intramolecular in the case of enantiomers [15, 16] while all four are intermolecular for DL [17]. Three of them are taken into account in the group additivity calculations.

The enthalpy of formation in the ideal gas state of the histidine stereoisomers was calculated by means of the same quantity in the crystalline state and of the standard enthalpy of sublimation (Table 6). Gaffney et al. [18] have derived a value of $142 \pm 8\,kJ\cdot mol^{-1}$ from vapor pressure measurements in the temperature range 392–492 K. A positive correction to the standard state of $6 \pm 2\,kJ\cdot mol^{-1}$ is obtained by means of the estimation methods, recommended by Chickos et al., for phase-change enthalpies and heat capacities [19, 20]. A much larger value of $182\,kJ\cdot mol^{-1}$ is predicted by Badelin et al. [21] by quantum chemical computations.

As it may be easily seen, the discrepancies with respect to the calculated value are mainly due to the large uncertainties in the evaluation of the standard enthalpy of sublimation.

A small molecule (possibly CO) is evolved during the first step and other gaseous molecules (H_2O and NH_3) at higher temperatures.

The processes revealed by DSC and TG are essentially, if not exclusively due to decomposition. Bryan and Olafsson [22] state that the main decomposition reaction is decarboxylation, possibly preceded by deamination. Weiss et al. [11] consider that a single main reaction occurs, taking into account that a sharp peak is obtained during the DSC run. The weight loss in the case of decarboxylation reaction was the only one that would be about 28.4% (higher than our experimental value, about 18%). The calculated thermal effect of the decarboxylation reaction, by means of the group additivity method [14], is about $79\,kJ\cdot mol^{-1}$ (in standard conditions) comparable with our experimental values (Table 6).

TABLE 6: Ideal gas state enthalpies of formation of isomeric histidines.

Isomeric histidines	$\Delta_{sub}H°$ ($kJ\cdot mol^{-1}$) [18–20]	$\Delta_{sub}H°$ ($kJ\cdot mol^{-1}$) [21]	$-\Delta_f H_g°$ ($kJ\cdot mol^{-1}$)[a]	$-\Delta_f H_g°$ ($kJ\cdot mol^{-1}$)[b]	$-\Delta_f H_g°$ ($kJ\cdot mol^{-1}$) (calculated)[c]
L-histidine	148 ± 10	182	304 ± 14	270	
D-histidine	148 ± 10	182	301 ± 14	267	284.60
DL-histidine	148 ± 10	182	304 ± 14	270	

[a]Calculated with enthalpy of sublimation from column 2. [b]Calculated with enthalpy of sublimation from column 3. [c]Estimated value by means of the group additivity method, with parameters recommended by Domalski and Hearing [14].

Weiss et al. [11] obtained totally different results by means of mass spectroscopy, i.e., histidine ejects 1 mol H_2O in the following reaction:

$$His = C_6H_9N_3O_2 \longrightarrow H_2O + C_6H_7N_3O \qquad (1)$$

This observation was confirmed in their opinion by a weight loss of 13–15%, in contradiction with our value of about 18% and incompatible with decarboxylation. The above authors consider that inner cyclization seems likely. However, the proposed structure of the residue would be in this case $C_6H_9N_3O$ (139 Da), so that a difference of 2 Da is arising.

Conflicts of Interest

The authors declare that they have no conflicts of interest.

Acknowledgments

This contribution was carried out within the research program "Chemical Thermodynamics" of the "Ilie Murgulescu" Institute of Physical Chemistry of the Romanian Academy. Support of the EU (ERDF) and Romanian Government for the acquisition of the research infrastructure under Project INFRANANOCHEM- No. 19/ 01.03.2009 is gratefully acknowledged. F.T. thanks the financial support of Executive Agency for Higher Education, Research, Development and Innovation (UEFISCDI) under Bridge Grant Contract 27BG/2016, AromaVer.

References

[1] L. Wang, N. Sun, S. Terzyan, X. Zhang, and D. R. Benson, "A histidine/tryptophan π-stacking interaction stabilizes the heme-independent folding core of microsomal apocytochrome b_5 relative to that of mitochondrial apocytochrome b_5," Biochemistry, vol. 45, no. 46, pp. 13750–13759, 2006.

[2] R. A. Ingle, "Histidine biosynthesis," The Arabidopsis Book: American Society of Plant Biologists, vol. 9, article e0141, 2011.

[3] V. P. Vasilev, V. A. Borodin, and S. B. Kopnyshev, "Standard enthalpies of formation of L-histidine and L-proline," Russian Journal of Physical Chemistry A, vol. 63, pp. 891-892, 1989.

[4] S. R. Wilson, I. D. Watson, and G. N. Malcolm, "Enthalpies of formation of solid cytosine, L-histidine, and uracil," Journal of Chemical Thermodynamics, vol. 11, no. 9, pp. 911-912, 1979.

[5] R. C. Weast, Handbook of Chemistry and Physics, CRC Press, Cranwood Parkway, Cleveland, OH, USA, 58th edition, 1977.

[6] J. D. Cox and G. Pilcher, Thermochemistry of Organic and Organometallic Compounds, Academic Press, London, UK, 1970.

[7] J. Coops, R. S. Jessup, and K. van Nes, "Calibration of calorimeters for reactions in a bomb at constant volume," in Experimental Thermochemistry (Chapter 3), F. D. Rossini, Ed., Vol. 1, Interscience, New York, NY, USA, 1956.

[8] W. N. Hubbard, D. W. Scott, and G. Waddington, "Standard states and corrections for combustions in a bomb at constant volume," in Experimental Thermochemistry, F. D. Rossini, Ed., vol. 1, pp. 75–128, Interscience, New York, NY, USA, 1956.

[9] CODATA Bulletin, Recommended Key Values for Thermodynamics, CODATA Bulletin, Paris, France, vol. 28, 1977.

[10] P. G. Olafsson and A. M. Bryan, "Evaluation of thermal decomposition temperatures of amino acids by differential enthalpic analysis," Mikrochimica Acta, vol. 58, no. 5, pp. 871–878, 1970.

[11] I. M. Weiss, C. Muth, R. Drumm, and H. O. K. Kirchner, "Thermal decomposition of the amino acids glycine, cysteine, aspartic acid, asparagine, glutamic acid, glutamine, arginine and histidine," BMC Biophysics, vol. 11, no. 1, pp. 1–15, 2018.

[12] P. Anandan, M. Arivanandhan, Y. Hayakawa et al., "Investigations on the growth aspects and characterization of semiorganic nonlinear optical single crystals of L-histidine and its hydrochloride derivative," Spectrochimica Acta Part A: Molecular and Biomolecular Spectroscopy, vol. 121, pp. 508–513, 2014.

[13] M. Wesolowski and J. Erecinska, "Relation between chemical structure of amino acids and their thermal decomposition," Journal of Thermal Analysis and Calorimetry, vol. 82, no. 2, pp. 307–313, 2005.

[14] E. S. Domalski and E. D. Hearing, "Estimation of the thermodynamic properties of C-H-N-O-S-Halogen compounds at 298.15 K," Journal of Physical and Chemical Reference Data, vol. 22, no. 4, pp. 805–1160, 1993.

[15] J. J. Madden, E. L. McGandy, and N. C. Seeman, "The crystal structure of the orthorombic form of L-(+)-histidine," Acta Crystallographica Section B Structural Crystallography and Crystal Chemistry, vol. 28, no. 8, pp. 2377–2382, 1972.

[16] J. J. Madden, E. L. McGandy, N. C. Seeman, M. M. Harding, and A. Hoy, "The crystal structure of the monoclinic form of L-histidine," Acta Crystallographica Section B Structural Crystallography and Crystal Chemistry, vol. 28, pp. 2382–2389, 1972.

[17] P. Edington and M. M. Harding, "The crystal structure of DL-histidine," Acta Crystallographica Section B Structural Crys-

tallography and Crystal Chemistry, vol. 30, no. 1, pp. 204–206, 1974.

[18] J. S. Gaffney, R. C. Pierce, and L. Friedman, "Mass spectrometer study of evaporation of alpha-amino acids," *Journal of the American Chemical Society*, vol. 99, no. 13, pp. 4293–4298, 1977.

[19] J. S. Chickos, W. Acree Jr., and J. F. Liebman, "Estimating phase change entropies and enthalpies," in *ACS Symposium Series 677, Computational Thermochemistry, Prediction and Estimation of Molecular Thermodynamics*, D. Frurip and K. Irikura, Eds., pp. 63–93, ACS, Washington, DC, USA, 1998.

[20] J. S. Chickos, D. G. Hesse, and J. F. Liebman, "A group Additivity approach for the estimation of heat capacities of organic liquids and solids," *Structural Chemistry*, vol. 4, no. 4, pp. 261–268, 1993.

[21] V. G. Badelin, E. Yu. Tyunina, G. V. Girichev, N. I. Giricheva, and O. V. Pelipets, "Relationship between the molecular structure of amino acids and dipeptides and thermal sublimation effects," *Journal of Structural Chemistry*, vol. 48, no. 4, pp. 647–653, 2007.

[22] A. M. Bryan and P. G. Olafsson, "Analysis of thermal, decomposition patterns of aromatic and heteroaromatic amino acids," *Analytical Letters*, vol. 2, no. 10, pp. 505–513, 1969.

23

Removal of Methylene Blue from Aqueous Solution Using Agricultural Residue Walnut Shell: Equilibrium, Kinetic, and Thermodynamic Studies

Ranxiao Tang,[1] Chong Dai,[2] Chao Li,[1] Weihua Liu,[1] Shutao Gao,[1] and Chun Wang[1]

[1]*College of Science, Agricultural University of Hebei, Baoding 071001, China*
[2]*Department of Civil & Environmental Engineering, University of Houston, Houston, TX 77004, USA*

Correspondence should be addressed to Chun Wang; chunwang69@126.com

Academic Editor: Khalid Z. Elwakeel

Walnut shell (WS), as an economic and environmental-friendly adsorbent, was utilized to remove methylene blue (MB) from aqueous solutions. The effects of WS particle size, solution pH, adsorbent dosage and contact time, and concentration of NaCl on MB removal were systematically investigated. Under the optimized conditions (i.e., contact time ~ 2 h, pH ~ 6, particle size ~ 80 mesh, dye concentration 20 mg/L, and 1.25 g/L adsorbent), the removal percentages can achieve ~97.1%, indicating WS was a promising absorbent to remove MB. Other supplementary experiments, such as Fourier transform infrared spectroscopy (FTIR), dynamic light scattering (DLS), and Brunauer-Emmett-Teller (BET) method, were also employed to understand the adsorption mechanisms. FTIR confirmed that the successful adsorption of MB on WS particles was through functional groups of WS. Using DLS method, the interactions between WS particles and dyes under various pH were investigated, which can be ascribed to the electrostatic forces. Kinetic data can be well fitted by the pseudo-second-order model, indicating a chemical adsorption. The adsorption isotherms were well described by both Langmuir and Freundlich models. Dubinin-Radushkevich model also showed that the adsorption process was a chemical adsorption. Thermodynamic data indicated that the adsorption was spontaneous, exothermic, and favorable at room temperature.

1. Introduction

Dyes have been widely used in various fields, such as textile, paper, rubber, plastic, leather, cosmetic, food, and drug industries. However, the extensive use of dyes produces a large amount of dye wastewater, which threatens our environment. What is more, most of the dyes or their metabolites are toxic and some of them are considered carcinogenic for human health [1, 2]. Therefore, it is imperative to treat the dye wastewater before releasing it into the groundwater [3, 4].

The removal of dyes from wastewater has been extensively studied for decades, and many technologies have been developed, including oxidative degradation [5], biochemical degradation [6, 7], photodegradation [8, 9], electrocoagulation [10], electrochemical degradation [11], and adsorption [12, 13]. Among these methods, adsorption has been found to be one of the most well-known and economic techniques for dye removal due to its easy operation, high efficiency, low cost, and recyclability [14]. Most recently, much attention has been paid to the development of crude biomass materials, such as peanut husk [15], coconut husk [16], potato peel [17], rice husk [18], and pomegranate peel [19], for the removal and separation of dyes from the wastewater.

Walnut shell (WS), as an abundant agricultural by-product, has good chemical stability and mechanical strength. Meanwhile, WS can be easily grinded into particles with desired particle sizes, and grinded WS has been demonstrated to be an effective absorbent for the removal of organic pollutants [20], heavy metals [21, 22], malachite green [23], and reactive brilliant red (K-2BP) [24]. However, to the best of our knowledge, few studies have been conducted to utilize walnut shells (WS) to remove methylene blue (MB) from aqueous solution. In this study, WS was used to investigate its capabilities to remove methylene blue (MB) from the aqueous solution for the first time, aiming to develop the efficient and low-cost treatment of methylene blue as well as promote the resource utilization of walnut shell.

To fill these information gaps, the objectives of this study were 3-fold: (1) batch adsorption experiments were conducted to measure the removal percentages of MB by WS particles with different sizes, in order to find the most effective WS particles for removing MB from aqueous solution; (2) batch adsorption experiments were conducted to systematically study the effects of pH, contact time, adsorbent dosage, and initial dye concentration on MB removal; (3) batch adsorption experiments were carried out for the kinetic under various dye concentrations and thermodynamics studies under various temperatures. Thermodynamic parameters (e.g., ΔG, ΔH, and ΔS) were calculated, and the results were analyzed to gain the mechanistic understanding of MB removal by WS under various experimental conditions. Some other supplementary techniques, including dynamic light scattering (DLS), Fourier transform infrared spectrometry (FTIR), and Brunauer-Emmett-Teller method (BET), were also utilized.

2. Materials and Methods

2.1. Preparation of Solutions.
All the chemicals are of analytical grade. The stock solution of methylene blue (MB) (0.5 g/L) was prepared in distilled water (DI water). The experimental solutions were prepared by diluting the stock solution with the distilled water to the desired concentrations (i.e., 20, 40, 60, 80, and 100 mg/L).

2.2. Preparation of the Adsorbent.
Walnut shells (WS) were washed with tap water multiple times, to remove dust and soluble impurities. Then, WS were rinsed with distilled water and dried in an oven at 105°C to constant weight. The dried and cleaned WS were grounded, sieved (20, 60, 80, and 120 mesh), and stored in a desiccator for further use.

2.3. Adsorbent Characterization.
The chemical bonding states of WS, before and after MB adsorption, were measured by Fourier transform infrared spectrometry (FTIR) (WQF-520A). The specific surface areas of WS with different sizes (20, 60, 80, and 120 mesh) were measured by nitrogen adsorption using the Brunauer-Emmett-Teller method (V-Sorb2800). Dynamic light scattering (DLS, Zetasizer Nano-series, Malvern Instruments) was used to measure the zeta potential values (ζ) of WS particles in aqueous solutions under various pH conditions [30, 31]. According to previous work, the WS particles of 0.1 g were suspended in DI water (pH was adjusted from 2 to 12 using 0.1 mol/L HCl or NaOH), and the zeta potential measurements were conducted every 1 min for 30 min. The average values and standard deviations of zeta potential values were calculated after the readings became stable [32, 33].

2.4. Batch Adsorption Experiments.
Batch adsorption experiments were carried out by mixing a certain amount of MB particles with 20.0 mL MB solution of desired concentrations

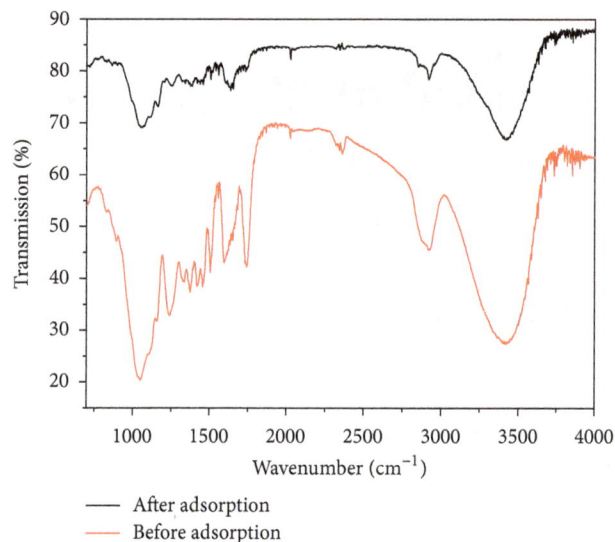

FIGURE 1: FTIR spectrum of walnut shell particles, before and after adsorption.

and then shaking on a platform shaker until the equilibrium was achieved. The effects of the WS particle sizes (20–120 mesh), adsorbent dosages (0.10–2.50 g/L), contact time (0–120 min), initial pH (2–11), initial MB concentration (20–100 mg/L), concentration of NaCl (1, 5, and 10%), and temperature (298, 308, 318, and 328 K) on the adsorption of MB on WS particles were also investigated. After adsorption experiments, the solutions were filtered and the MB concentrations in the supernatant were analyzed using a UV/Vis-DR 3900. The wavelength of 568 nm was chosen, where the maximum absorbance can be achieved.

The adsorption capacity (q_t) and MB removal percentage (R) were calculated as (1) and (2), respectively.

$$q_t = \frac{c_0 - c_t}{m} \times V \tag{1}$$

$$R = \frac{c_0 - c_t}{c_0} \times 100\%, \tag{2}$$

where q_t (mg/g) represents the adsorption capacity at time t; c_0 and c_t (mg/L) are the MB concentration at initial solution and at time t (min), respectively; V (L) is the volume of the solution; and m (g) is the weight of the WS particles.

3. Results and Discussion

3.1. Characterization of Walnut Shell.
Figure 1 shows the FTIR spectrum of walnut shell before and after MB adsorption. Before MB adsorption, the spectrum contained several peaks, corresponding to –OH stretching (at 3422.5 cm^{-1}), C–H stretching (at 2923.7 cm^{-1}), C=O stretching (at 1739.0 cm^{-1}), N–H bending (at 1631.4 cm^{-1}), CH$_2$ deformation (at 1162.3 cm^{-1}), and –COO symmetric stretching (at 1053.9 cm^{-1}). After MB adsorption, the shifts of bands were observed: –OH stretching (at 3428.7 cm^{-1}), C–H stretching (at 2936.1 cm^{-1}), C=O stretching (at 1745.3 cm^{-1}), N–H bending

TABLE 1: BET-N_2 specific surface area of walnut shell (WS) particles with different sizes.

Particle size/mesh	20	60	80	120
Surface area/(m^2/g)	1.01	1.24	2.82	3.12

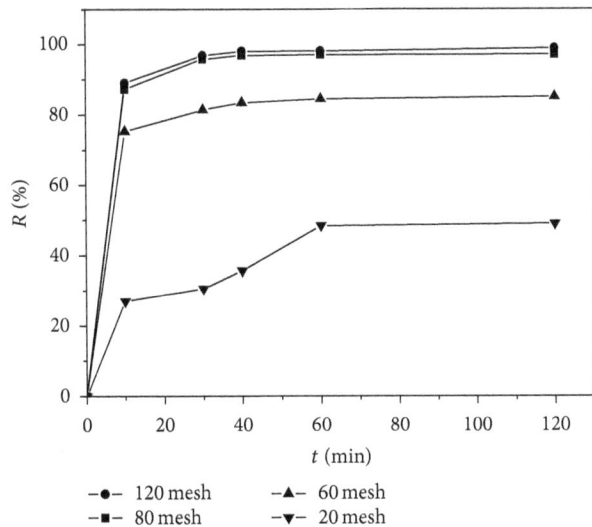

FIGURE 2: The effect of particle sizes on the adsorption of MB on WS. Conditions: temperature (298 K), MB concentration (20 mg/L), and WS adsorbent dosage (1.25 g/L).

(at $1601.6\,cm^{-1}$), CH_2 deformation (at $1240.2\,cm^{-1}$), and –COO symmetric stretching (at $1047.7\,cm^{-1}$). The shifts of these bands indicated MB adsorption onto WS was via functional groups of WS [34].

Table 1 shows the BET-N_2 specific surface area of WS particles with different sizes, which was obtained by N_2 adsorption and desorption isotherm. Results showed that the surface area of WS increased with the decreasing of WS particle sizes.

3.2. Effect of Particle Size and Contact Time. At the contact time of 2 h, the MB removal percentages (%) of 120, 80, 60, and 20 mesh particles were 98.9, 97.1, 85.1, and 49.0%, respectively (Figure 2), indicating that MB removal percentage (%) increased with the decreasing of WS particle sizes. As discussed earlier, the small particles had higher surface areas than the big particles, which contributed to more available adsorption active sites, thus resulting in higher removal percentages [35]. When the particle size of WS particles reached ~80 mesh, the removal percentage of MB on WS can achieve 97.1%, which met the requirement of dye removal from wastewater [16, 19]. When the particle sizes of WS particles were smaller than 80 mesh (i.e., 120 mesh), the dye removal percentages were similar as 80 mesh; therefore, the particles of 80 mesh were used for the rest of the experiments.

As shown in Figure 2, within the first 30 min, for all particles, the adsorption capacities of MB on WS particles with different sizes significantly increased with the increased contact time. After 30 min, the adsorption processes slowed down until the equilibrium was achieved after 2 h for all particles. Therefore, a contact time of 2 h was chosen as for

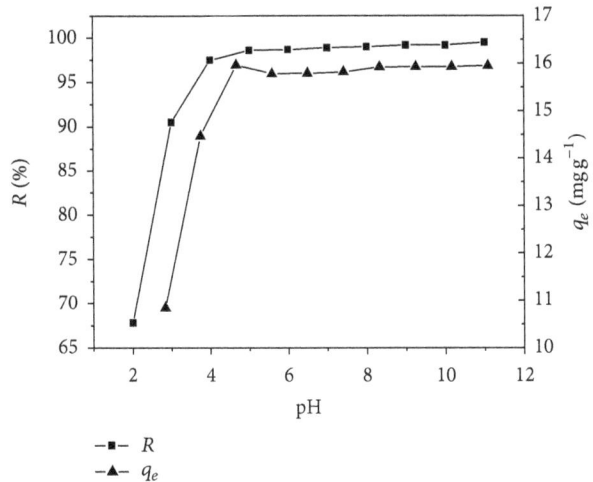

FIGURE 3: Effect of initial pH on MB adsorption (conditions: temperature (298 K), dye concentration (20 mg/L), particle size (80 mesh), adsorbent dosage (1.25 g/L), and contact time (2 h)).

the following experiments. The adsorption of MB on WS particles occurred very rapidly during the first 30 min due to the sufficient available number of active sites at walnut shell surface; then the adsorption slowed down when the remaining active sites were less available until the equilibrium phase was achieved [36].

3.3. Effect of Initial pH. The aqueous solution pH was found to significantly affect the adsorption capacities of dyes onto adsorbents [12]. Accordingly, the adsorption capacities of MB on WS were investigated under various aqueous solution pH. Figure 3 shows that, under acidic conditions (pH = 2~5), MB dye removal percentage (from 67.8 to 98.6%) and q_e (from 10.84 to 15.78 mg/g) on WB increased significantly with pH increasing, while under base conditions (pH = 6~11), MB dye removal percentage (from 98.7 to 99.5%) and q_e (from 15.79 to 15.95 mg/g) on WB were similar.

The differences in adsorption behavior under various pH values can be explained well from the electrostatic forces between the surface charge of WS particles and MB dyes. The zeta potentials (ζ) of WS under varying pH values are shown in Table 2. Results showed that the WS particles were all negatively charged under our pH conditions (pH = 2~11). However, the WS surface was significantly less negatively charged at lower pH than at higher pH. Therefore, there was less electrostatic attractive force between the dye cations and the adsorbent surface, which can result in lower adsorption capacity at lower pH. Also, under acidic conditions, large amounts of H_3O^+ ions existed, which can compete with dye cations adsorbing on the active sites of WS, resulting in the significant decreasing in the amount of adsorbed dye.

To sum up, the adsorption capacities of WS particles for MB were highly dependent on solution pH. In order to get high removal rates of MB dyes, the pH of above 6 of aqueous solutions was suggested. The pH value of MB initial solution was measured to be ~6.15, and MB solutions without adjusting pH were used for studying the effect of adsorbent dosage as well as the kinetic and thermodynamic studies.

TABLE 2: The zeta potentials of WS at different pH.

pH	2.02	2.98	3.98	4.89	5.80	7.04	7.88	8.97	10.00	11.93
ζ/mV	−4.217	−12.19	−18.51	−22.99	−23.15	−24.96	−25.01	−25.18	−26.68	−33.68

TABLE 3: Kinetic parameters for MB adsorption on WS.

c_0/mg/L	q_e, exp/mg/g	Pseudo-first order equation			Pseudo-second order equation			Intraparticle diffusion model		
		R^2	q_e, cal/mg/g	k_1/g/mg·min	R^2	q_e, cal/mg/g	k_2/g/mg·min	R^2	C/mg/g	k_3/mg/g·min$^{1/2}$
20	15.70	0.697	0.310	0.0290	1.000	15.72	0.097	0.778	14.91	0.074
40	28.88	0.748	2.494	0.0223	0.998	28.99	0.025	0.843	25.67	0.307
60	40.52	0.011	0.865	0.0039	1.000	40.49	0.102	0.336	39.46	0.097
80	48.22	0.567	5.733	0.0316	0.998	48.31	0.013	0.191	45.22	0.196
100	48.77	0.676	3.645	0.0129	1.000	48.08	0.020	0.710	43.32	0.471

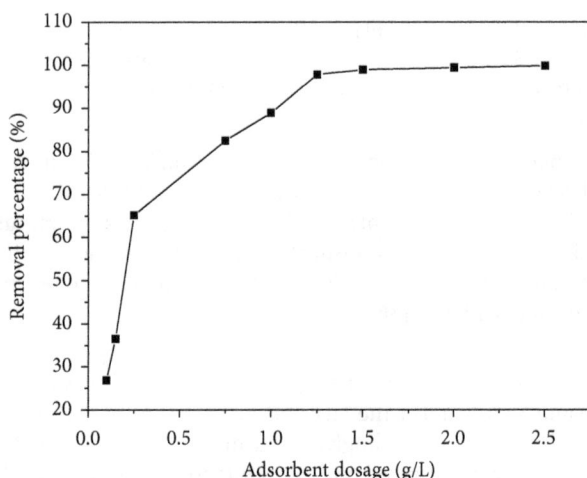

FIGURE 4: Effect of adsorbent dosage on the adsorption of MB by WS particles. Conditions: temperature (298 K), dye concentration (20 mg/L) and particle size (80 mesh), and contact time (2 h).

3.4. Effect of Adsorbent Dosage.

The effect of walnut shell dosage on the MB removal percentage (%) was examined for a contact time of 2 h. As shown in Figure 4, MB dye removal percentage (%) at equilibrium increased from 26.8 to 99.8% with the increasing of the adsorbent dosage from 0.10 to 1.25 g/L. Above 1.25 g/L of adsorbent dose, the dye removal percentage did not significantly improve, which indicated that the removal percentage of MB by WS particles reached an optimal value at 1.25 g/L. With the increasing of adsorbent dosage, more adsorption sites are available. When the adsorption equilibrium was reached, all the available adsorption sites of WS particles were almost saturated with MB; therefore, any further increase of adsorbent dose only slightly affected the removal percentage of MB by WS particles. Accordingly, considering both the high removal percentage and low cost, the optimal adsorbent dose value of 1.25 g/L was selected to carry out the following kinetic and thermodynamic adsorption experiments.

3.5. Kinetic Analysis.

The kinetic experiments were conducted at the optimal conditions (pH ~ 6, 1.25 g/L WS particles with 80 mesh, and contact time of 2 h) under various dye concentrations (i.e., 20, 40, 60, 80, and 100 mg/L). The kinetics of MB adsorption onto WS particles were investigated using three common models, being pseudo-first-order model (3), pseudo-second-order model (4), and intraparticle diffusion model (5), respectively.

$$\log\left(q_e - q_t\right) = \log q_e - \frac{k_1 t}{2.303} \qquad (3)$$

$$\frac{t}{q_t} = \frac{1}{k_2 q_e^2} + \frac{t}{q_e} \qquad (4)$$

$$q_t = k_3 t^{1/2} + C, \qquad (5)$$

where q_t (mg/g) and q_e (mg/g) represent the adsorption capacities at time t and at equilibrium, respectively; k_1 (g/(mg·min)), k_2 (g/(mg·min)), and k_3 (g/(mg·min$^{1/2}$)) are the first-order, second-order, and intraparticle diffusion rate constants, respectively; C is the intercept of intraparticle diffusion model.

The fitting results are summarized in Figure 5 and Table 3. For all dye concentrations (20, 40, 60, 80, and 100 mg/L), the correlation for pseudo-second-order model (>0.998) was much larger than that for the pseudo-first-order model (<0.697), indicating that MB adsorption onto WS can be described as pseudo-second-order model. Furthermore, the adsorption capacities calculated by pseudo-second-order model were close to those determined by the experiments. Contrarily, the adsorption capacities calculated by pseudo-first-order model were quite different with the experimental data. Therefore, we concluded that MB adsorption on WS can be described as pseudo-second-order model, which indicated that the adsorption of MB on WS can be described as chemical adsorption [18, 37].

For the intraparticle diffusion model, none of the regions has C values equal to zero, indicating that these lines did not pass through the origins. This suggested that intraparticle diffusion was present but may not be the rate limiting step [38].

3.6. Adsorption Isotherms.

Three commonly used models (i.e., Langmuir, Freundlich, and Dubinin-Radushkevich (D-R)) were utilized to analyze the adsorption isotherms of MB

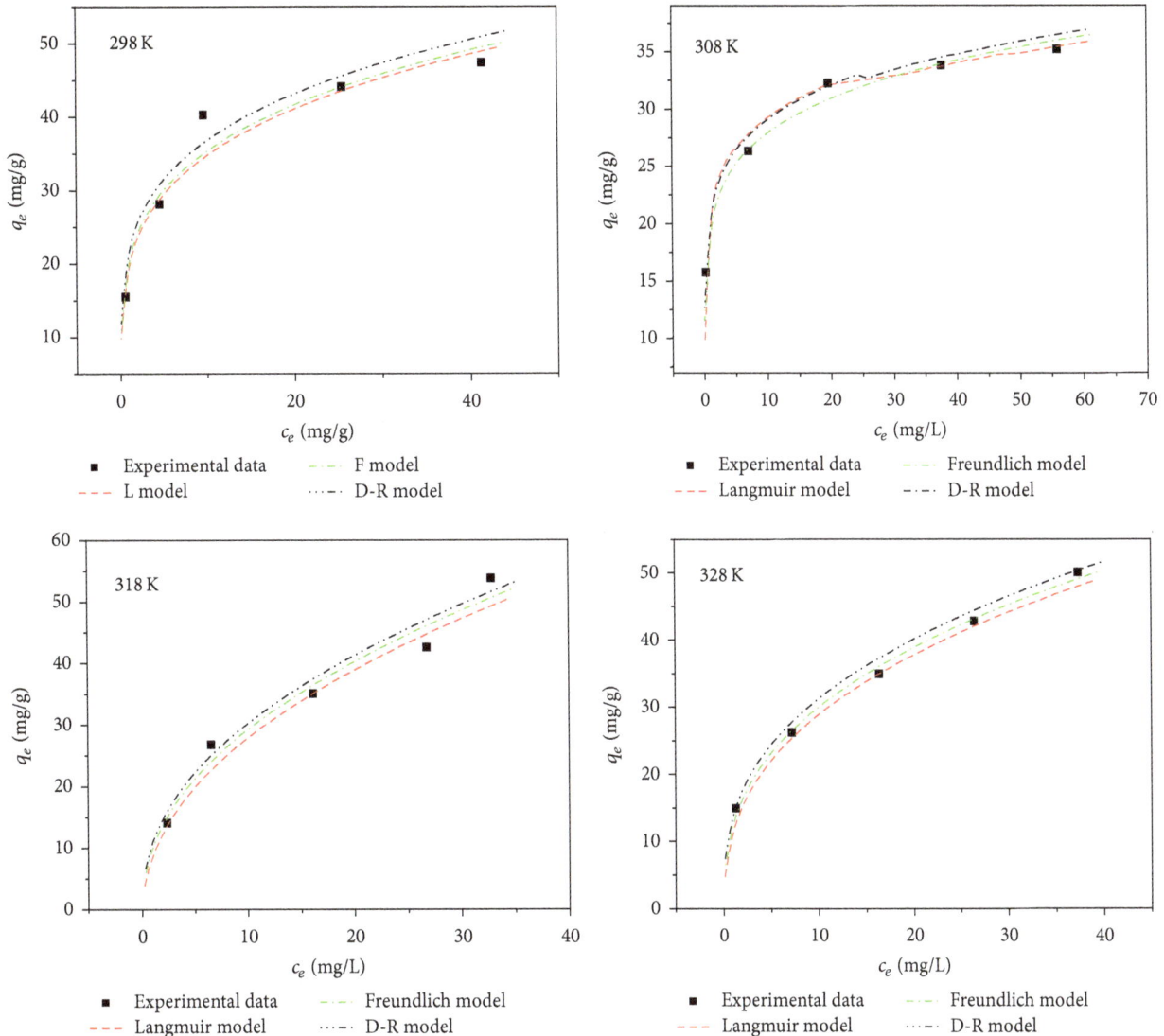

FIGURE 6: The adsorption isotherms at different temperature (i.e., 298, 308, 318, and 328 K). The dash lines are the fitting curves of Langmuir, Freundlich, and D-R models.

percentages. Also, the presence of Na^+ ions can compete with the adsorption of dye cations on the active sites of WS, which resulted in the significant decreasing in MB removal percentages.

3.9. Comparison of WS Adsorption Capacity with Other Adsorbents. The maximum adsorption capacity (q_{max}) of MB dye on WS can reach 51.55 mg/g under the optimal conditions (i.e., contact time ~ 2 h, pH ~ 6, particle size ~ 80 mesh, and 1.25 g/L adsorbent). Table 6 listed the maximum adsorption capacity (q_{max}) of some environmental-friendly and low-cost adsorbents for MB dye [25–29]. We found that the maximum adsorption capacity (q_{max} = 51.55 mg/g) of MB dye on walnut shell (WS) was much higher than that of most of the low-cost adsorbents (e.g., orange peel, wheat shells, and rice husk), indicating that WS was a very promising and environmental-friendly adsorbent.

4. Conclusions

Our study proposed to use an economic and environmental-friendly walnut shell as the adsorbent to remove methylene blue for the first time. The effects of particle size, contact time, pH, adsorbent dosage, and the concentration of salt (NaCl) were investigated. By utilizing FTIR, the successful adsorption of MB on WS particles was confirmed to occur through functional groups. BET analysis showed that smaller particles have larger surface area than the bulk particles, resulting in more active adsorption sites, thus significantly improving the removal percentages. The solution pH was found to be a very important factor controlling the adsorption processes: the adsorption was favored under base conditions (pH > 6). This can be explained by the electrostatic forces between WS particles and WB dyes. DLS results showed that WS surfaces were all negatively charged, while the surface of WS under base conditions was more negatively charged than under acidic

TABLE 6: Comparison of MB adsorption capacity with some environmental-friendly adsorbents.

Adsorbent	Dye name	Adsorption capacity (mg/g)	Reference
Orange peel	MB	18.6	[25]
Peanut hull	MB	68.03	[26]
Rice husk	MB	40.59	[27]
Cherry Sawdust	MB	39.84	[28]
Wheat shells	MB	16.56	[29]
Walnut shell (WS)	MB	51.55	This work

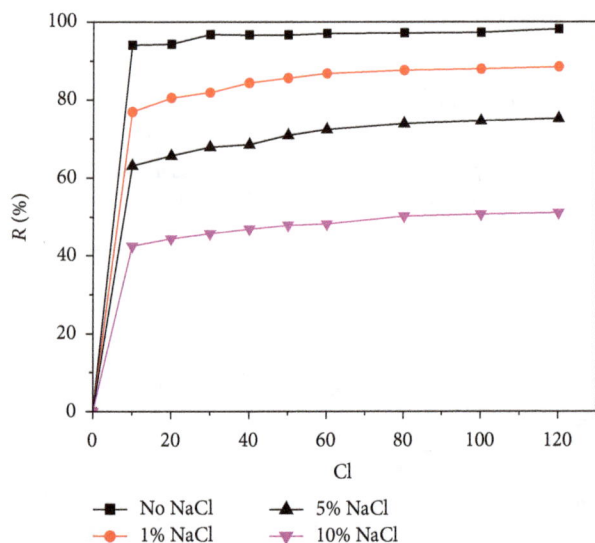

FIGURE 7: The effect of common salt (NaCl) on the MB removal percentages ($R\%$).

conditions. Therefore, under base conditions (pH > 6), more attractive forces existed between cationic MB dyes and WS particles, resulting in more adsorption of MB on WS particles. The presence of NaCl was found to decrease the adsorption capacity of WS for MB due to the adsorption of Na^+ on WS particle surfaces. Under the optimized experimental conditions, the removal efficiency of 97.1% for MB by WS can be achieved, indicating that walnut shell was a promising and environmental-friendly adsorbent to remove cationic dyes.

To further understand the adsorption mechanisms, this study also systematically investigated the adsorption of MB on walnut shell from both kinetic and thermodynamic analysis. From the kinetic view, the rapid adsorption process fitted well with the pseudo-second-order kinetic model within the dye concentrations range investigated, indicating the adsorption process was a chemical adsorption. Intraparticle diffusion model showed that intraparticle was present but may not be the rate limiting factor. The positive correlation coefficient suggested that MB adsorption on WS could be best described by both the Langmuir isotherm (R^2 > 0.97) and the Freundlich isotherm (R^2 > 0.96). D-R model also showed that the main mechanism for MB sorption was a chemical reaction. From the thermodynamic view, the adsorption process was spontaneous and exothermic. Meanwhile, the adsorption was favored at

room temperature, which can be used in various industrial applications.

Conflicts of Interest

The funding did not lead to any conflict of interests regarding the publication of this manuscript.

Acknowledgments

This work was supported by the National Natural Science Foundation of China (no. 31671930), the Natural Science Foundation of Hebei Province (no. B2016204136), and the Program of Study Abroad for Young Teachers by Agricultural University of Hebei and Science & Engineering Program by Agricultural University of Hebei (no. LG20150402).

References

[1] J. Duan, R. Liu, T. Chen, B. Zhang, and J. Liu, "Halloysite nanotube-Fe_3O_4 composite for removal of methyl violet from aqueous solutions," *Desalination*, vol. 293, pp. 46–52, 2012.

[2] J. Pierce, "Colour in textile effluents—the origins of the problem," *Journal of the Society of Dyers & Colourists*, vol. 110, no. 4, pp. 131–133, 1994.

[3] V. Rocher, J.-M. Siaugue, V. Cabuil, and A. Bee, "Removal of organic dyes by magnetic alginate beads," *Water Research*, vol. 42, no. 4-5, pp. 1290–1298, 2008.

[4] G. Crini, "Non-conventional low-cost adsorbents for dye removal: a review," *Bioresource Technology*, vol. 97, no. 9, pp. 1061–1085, 2006.

[5] B. Gözmen, B. Kayan, A. M. Gizir, and A. Hesenov, "Oxidative degradations of reactive blue 4 dye by different advanced oxidation methods," *Journal of Hazardous Materials*, vol. 168, no. 1, pp. 129–136, 2009.

[6] A. N. Kagalkar, U. B. Jagtap, J. P. Jadhav, V. A. Bapat, and S. P. Govindwar, "Biotechnological strategies for phytoremediation of the sulfonated azo dye Direct Red 5B using *Blumea malcolmii* Hook," *Bioresource Technology*, vol. 100, no. 18, pp. 4104–4110, 2009.

[7] S. B. Jadhav, S. S. Phugare, P. S. Patil, and J. P. Jadhav, "Biochemical degradation pathway of textile dye Remazol red and subsequent toxicological evaluation by cytotoxicity, genotoxicity and oxidative stress studies," *International Biodeterioration and Biodegradation*, vol. 65, no. 6, pp. 733–743, 2011.

[8] D. K. Gardiner and B. J. Borne, "Textile waste waters: treatment and environmental effects," *Journal of the Society of Dyers and Colourists*, vol. 94, no. 8, pp. 339–348, 1978.

[9] B. Li, Y. Hao, X. Shao et al., "Synthesis of hierarchically porous metal oxides and Au/TiO$_2$ nanohybrids for photodegradation of organic dye and catalytic reduction of 4-nitrophenol," *Journal of Catalysis*, vol. 329, pp. 368–378, 2015.

[10] K. S. P. Kalyani, N. Balasubramanian, and C. Srinivasakannan, "Decolorization and COD reduction of paper industrial effluent using electro-coagulation," *Chemical Engineering Journal*, vol. 151, no. 1–3, pp. 97–104, 2009.

[11] A. Thiam, E. Brillas, J. A. Garrido, R. M. Rodríguez, and I. Sirés, "Routes for the electrochemical degradation of the artificial food azo-colour Ponceau 4R by advanced oxidation processes," *Applied Catalysis B: Environmental*, vol. 180, pp. 227–236, 2016.

[12] P. Janoš, H. Buchtová, and M. Rýznarová, "Sorption of dyes from aqueous solutions onto fly ash," *Water Research*, vol. 37, no. 20, pp. 4938–4944, 2003.

[13] A. G. Espantaleón, J. A. Nieto, M. Fernández, and A. Marsal, "Use of activated clays in the removal of dyes and surfactants from tannery waste waters," *Applied Clay Science*, vol. 24, no. 1-2, pp. 105–110, 2003.

[14] A. Mittal, J. Mittal, A. Malviya, D. Kaur, and V. K. Gupta, "Adsorption of hazardous dye crystal violet from wastewater by waste materials," *Journal of Colloid and Interface Science*, vol. 343, no. 2, pp. 463–473, 2010.

[15] R. P. Han, P. Han, Z. H. Cai, Z. H. Zhao, and M. S. Tang, "Kinetics and isotherms of neutral red adsorption on peanut husk," *Journal of Environmental Sciences*, vol. 20, no. 9, pp. 1035–1041, 2008.

[16] K. Low and C. Lee, "The removal of cationic dyes using coconut husk as an adsorbent," *Pertanika*, vol. 13, no. 2, pp. 221–228, 1990.

[17] E.-K. Guechi and O. Hamdaoui, "Sorption of malachite green from aqueous solution by potato peel: kinetics and equilibrium modeling using non-linear analysis method," *Arabian Journal of Chemistry*, vol. 9, supplement 1, pp. S416–S424, 2016.

[18] V. Vadivelan and K. V. Kumar, "Equilibrium, kinetics, mechanism, and process design for the sorption of methylene blue onto rice husk," *Journal of Colloid and Interface Science*, vol. 286, no. 1, pp. 90–100, 2005.

[19] N. K. Amin, "Removal of direct blue-106 dye from aqueous solution using new activated carbons developed from pomegranate peel: adsorption equilibrium and kinetics," *Journal of Hazardous Materials*, vol. 165, no. 1-3, pp. 52–62, 2009.

[20] E. Lorenc-Grabowska, G. Gryglewicz, and M. A. Diez, "Kinetics and equilibrium study of phenol adsorption on nitrogen-enriched activated carbons," *Fuel*, vol. 114, pp. 235–243, 2013.

[21] E. Pehlivan and T. Altun, "Biosorption of chromium(VI) ion from aqueous solutions using walnut, hazelnut and almond shell," *Journal of Hazardous Materials*, vol. 155, no. 1-2, pp. 378–384, 2008.

[22] A. Almasi, M. Omidi, M. Khodadadian, R. Khamutian, and M. B. Gholivand, "Lead(II) and cadmium(II) removal from aqueous solution using processed walnut shell: kinetic and equilibrium study," *Toxicological and Environmental Chemistry*, vol. 94, no. 4, pp. 660–671, 2012.

[23] M. K. Dahri, M. R. R. Kooh, and L. B. L. Lim, "Water remediation using low cost adsorbent walnut shell for removal of malachite green: equilibrium, kinetics, thermodynamic and regeneration studies," *Journal of Environmental Chemical Engineering*, vol. 2, no. 3, pp. 1434–1444, 2014.

[24] J.-S. Cao, J.-X. Lin, F. Fang, M.-T. Zhang, and Z.-R. Hu, "A new absorbent by modifying walnut shell for the removal of anionic dye: kinetic and thermodynamic studies," *Bioresource Technology*, vol. 163, pp. 199–205, 2014.

[25] G. Annadurai, R.-S. Juang, and D.-J. Lee, "Use of cellulose-based wastes for adsorption of dyes from aqueous solutions," *Journal of Hazardous Materials*, vol. 92, no. 3, pp. 263–274, 2002.

[26] R. Gong, M. Li, C. Yang, Y. Sun, and J. Chen, "Removal of cationic dyes from aqueous solution by adsorption on peanut hull," *Journal of Hazardous Materials*, vol. 121, no. 1-3, pp. 247–250, 2005.

[27] V. Vadivelan and K. Vasanth Kumar, "Equilibrium, kinetics, mechanism, and process design for the sorption of methylene blue onto rice husk," *Journal of Colloid and Interface Science*, vol. 286, no. 1, pp. 90–100, 2005.

[28] F. Ferrero, "Dye removal by low cost adsorbents: hazelnut shells in comparison with wood sawdust," *Journal of Hazardous Materials*, vol. 142, no. 1-2, pp. 144–152, 2007.

[29] Y. Bulut and H. Aydin, "A kinetics and thermodynamics study of methylene blue adsorption on wheat shells," *Desalination*, vol. 194, no. 1–3, pp. 259–267, 2006.

[30] C. Dai, A. G. Stack, A. Koishi, A. Fernandez-Martinez, S. S. Lee, and Y. Hu, "Heterogeneous nucleation and growth of barium sulfate at organic-water interfaces: interplay between surface hydrophobicity and Ba^{2+} adsorption," *Langmuir*, vol. 32, no. 21, pp. 5277–5284, 2016.

[31] Y. Hu, C. Neil, B. Lee, and Y.-S. Jun, "Control of heterogeneous Fe(III) (Hydr)oxide nucleation and growth by interfacial energies and local saturations," *Environmental Science and Technology*, vol. 47, no. 16, pp. 9198–9206, 2013.

[32] C. Dai and Y. Hu, "Fe(III) hydroxide nucleation and growth on quartz in the presence of Cu(II), Pb(II), and Cr(III): metal hydrolysis and adsorption," *Environmental Science and Technology*, vol. 49, no. 1, pp. 292–300, 2015.

[33] C. Dai, X. Zuo, B. Cao, and Y. Hu, "Homogeneous and heterogeneous (Fe$_x$, Cr$_{1-x}$)(OH)$_3$ precipitation: implications for Cr sequestration," *Environmental Science & Technology*, vol. 50, no. 4, pp. 1741–1749, 2016.

[34] Z. Bekçi, Y. Seki, and L. Cavas, "Removal of malachite green by using an invasive marine alga Caulerpa racemosa var. cylindracea," *Journal of Hazardous Materials*, vol. 161, no. 2-3, pp. 1454–1460, 2009.

[35] M. Arami, N. Yousefi Limaee, and N. M. Mahmoodi, "Investigation on the adsorption capability of egg shell membrane towards model textile dyes," *Chemosphere*, vol. 65, no. 11, pp. 1999–2008, 2006.

[36] A. B. Pérez Marín, M. I. Aguilar, V. F. Meseguer, J. F. Ortuño, J. Sáez, and M. Lloréns, "Biosorption of chromium (III) by orange (Citrus cinensis) waste: batch and continuous studies," *Chemical Engineering Journal*, vol. 155, no. 1-2, pp. 199–206, 2009.

[37] J. Liu, X. Wu, Y. Hu, C. Dai, Q. Peng, and D. Liang, "Effects of Cu(II) on the adsorption behaviors of Cr(III) and Cr(VI) onto kaolin," *Journal of Chemistry*, vol. 2016, Article ID 3069754, 11 pages, 2016.

[38] Y. Yao, F. Xu, M. Chen, Z. Xu, and Z. Zhu, "Adsorption behavior of methylene blue on carbon nanotubes," *Bioresource Technology*, vol. 101, no. 9, pp. 3040–3046, 2010.

[39] I. Langmuir, "The constitution and fundamental properties of solids and liquids. Part I. Solids," *The Journal of the American Chemical Society*, vol. 38, no. 2, pp. 2221–2295, 1916.

[40] H. Freundlich, "Over the adsorption in solution," *The Journal of Physical Chemistry A*, vol. 57, no. 385471, pp. 1100–1107, 1906.

[41] F. Helfferich, *Ion Exchange*, McGraw-Hill Book, New York, NY, USA, 1962.

[42] C. Wang, C. Feng, Y. Gao, X. Ma, Q. Wu, and Z. Wang, "Preparation of a graphene-based magnetic nanocomposite for the removal of an organic dye from aqueous solution," *Chemical Engineering Journal*, vol. 173, no. 1, pp. 92–97, 2011.

[43] J. Romero-Gonzalez, J. Peralta-Videa, E. Rodrıguez, S. Ramirez, and J. Gardea-Torresdey, "Determination of thermodynamic parameters of Cr (VI) adsorption from aqueous solution onto *Agave lechuguilla* biomass," *The Journal of Chemical Thermodynamics*, vol. 37, no. 4, pp. 343–347, 2005.

[44] E. Oguz, "Adsorption characteristics and the kinetics of the Cr(VI) on the Thuja oriantalis," *Colloids and Surfaces A: Physicochemical and Engineering Aspects*, vol. 252, no. 2-3, pp. 121–128, 2005.

Permissions

List of Contributors

Qiqi Guo and Jianzhong Pei
School of Highway, Chang'an University, Xi'an 710064, China

Rui Li
School of Highway, Chang'an University, Xi'an 710064, China
School of Materials Science and Engineering, Nanyang Technological University, Singapore 639798

Hui Du
School of Transportation Engineering, Southeast University, Nanjing 210096, China

Rui F. J. Pereira
Chemistry Research Unit (CIQUP), Departamento de Química e Bioquímica, Faculdade de Ciências da Universidade do Porto, R. Campo Alegre 687, 4169-007 Porto, Portugal

Luís Pinto da Silva
Chemistry Research Unit (CIQUP), Departamento de Química e Bioquímica, Faculdade de Ciências da Universidade do Porto, R. Campo Alegre 687, 4169-007 Porto, Portugal
LACOMEPHI, Departamento de Geociências, Ambiente e Ordenamento do Território, Faculdade de Ciências da Universidade do Porto, R. Campo Alegre 687, 4169-007 Porto, Portugal

Joaquim C. G. Esteves da Silva
Chemistry Research Unit (CIQUP), Departamento de Química e Bioquímica, Faculdade de Ciências da Universidade do Porto, R. Campo Alegre 687, 4169-007 Porto, Portugal
LACOMEPHI, Departamento de Geociências, Ambiente e Ordenamento do Território, Faculdade de Ciências da Universidade do Porto, R. Campo Alegre 687, 4169-007 Porto, Portugal
Chemistry Research Unit (CIQUP), Departamento de Geociências, Ambiente e Ordenamento do Território, Faculdade de Ciências da Universidade do Porto, R. Campo Alegre 687, 4169-007 Porto, Portugal

María Camila Hoyos-Sánchez and Angie Carolina Córdoba-Pacheco
Department of Biology, University of Tolima, Altos de Santa Helena, C. P. 730006 Ibaguè, Colombia

Luis Fernando Rodríguez-Herrera
Department of Chemistry, University of Tolima, Altos de Santa Helena, C. P. 730006 Ibaguè, Colombia

Ramiro Uribe-Kaffure
Department of Physics, University of Tolima, Altos de Santa Helena, C. P. 730006 Ibaguè, Colombia

Abhinay Man Shrestha and Sanjila Neupane
Department of Environment Science and Engineering, School of Science, Kathmandu University, Dhulikhel, Nepal

Gunjan Bisht
Department of Chemical Science and Engineering, School of Engineering, Kathmandu University, Dhulikhel, Nepal

Lihong Cheng, Wenkui Li, Zhiqin Chen, Jianping Ai and Zehua Zhou
Key Laboratory of Surface Engineering of Jiangxi Province, Jiangxi Science and Technology Normal University, Nanchang, Jiangxi 330013, China

Tianliang Xu
Zhengzhou Institute of Finance and Economics, Zhengzhou, China

Jianwen Liu
National Supercomputing Center in Shenzhen, Shenzhen 518055, China

Abdelhalim I. A. Mohamed
Petroleum Engineering Department, University of Wyoming, Laramie, WY 82071, USA

Abdullah S. Sultan
Petroleum Engineering Department and Center for Petroleum & Minerals, King Fahd University of Petroleum & Minerals, Dhahran 31261, Saudi Arabia

Ibnelwaleed A. Hussein
Gas Processing Center, College of Engineering, Qatar University, Doha, Qatar

Ghaithan A. Al-Muntasheri
EXPEC Advanced Research Center, Saudi Aramco, Dhahran 31311, Saudi Arabia

Qing Chen, Jiping She and Yang Xiao
College of Energy, Chengdu University of Technology, Chengdu, Sichuan 610059, China

Jianlong Wang
Laboratory of Environmental Technology, Institute of Nuclear and New Energy Technology, Tsinghua University, Beijing 100084, China

Chen Lv
Laboratory of Environmental Technology, Institute of Nuclear and New Energy Technology, Tsinghua University, Beijing 100084, China
Key Laboratory of Songliao Aquatic Environment, Ministry of Education, Jilin Jianzhu University, Changchun 130118, China

Ming Li, Shuang Zhong and Lei Wu
Key Laboratory of Songliao Aquatic Environment, Ministry of Education, Jilin Jianzhu University, Changchun 130118, China

Huan Wang, Xiaohong Gao, Sa Lv, Xuefeng Chu, Chao Wang, Lu Zhou and Xiaotian Yang
Jilin Provincial Key Laboratory of Architectural Electricity & Comprehensive Energy Saving, Department of Materials Science, Jilin Jianzhu University, Changchun 130118, China

Yaodan Chi
Jilin Provincial Key Laboratory of Architectural Electricity & Comprehensive Energy Saving, Department of Materials Science, Jilin Jianzhu University, Changchun 130118, China
College of Instrumentation & Electrical Engineering, Jilin University, Changchun 130012, China

Dora Alicia Solis-Casados, Lizbeth Serrato-Garcia and Alejandro Dorazco-Gonzalez
Centro Conjunto de Investigaciòn en Química Sustentable UAEM-UNAM, Km 14.5 Carretera Toluca-Atlacomulco, Unidad San Cayetano, 50200 Toluca, MEX, Mexico

Luis Escobar-Alarcon
Departamento de Física, Instituto Nacional de Investigaciones Nucleares, 11801 Mexico City, Mexico

Antonia Infantes-Molina and Enrique Rodriguez-Castellon
Departamento de Quimica Inorganica, Facultad de Ciencias, Universidad de Malaga, 29071 Malaga, Spain

Tatyana Klimova
Departamento de Ingenieria Quimica, UNAM, Mexico City, Mexico

Susana Hernandez-Lopez
Facultad de Química, Universidad Autònoma del Estado de México, Paseo Colon esq Paseo Tollocan Col Nueva la Moderna, 50000 Toluca, MEX, Mexico

Leandro Marques Correia, Célio Loureiro Cavalcante Jr. and Rodrigo Silveira Vieira
Grupo de Pesquisa em Separações por Adsorção (GPSA), Departamento de Engenharia Química, Universidade Federal do Cearà (UFC), Campus do Pici, Bl. 709, 60455-760 Fortaleza, CE, Brazil

Juan Antonio Cecilia and Enrique Rodríguez-Castellón
Departamento de Química Inorgànica, Cristalografía y Mineralogía, Facultad de Ciencias, Universidad de Màlaga, Campus de Teatinos, 29071 Màlaga, Spain

Siong Fong Sim and Szewei Elaine Tai
Faculty of Resource Science & Technology, Universiti Malaysia Sarawak, 94300 Kota Samarahan, Sarawak, Malaysia

Karolina Kafarska, Michał Gacki and Wojciech M. Wolf
Institute of General and Ecological Chemistry, Faculty of Chemistry, Lodz University of Technology, 116 Zeromskiego Street, 90-924 Lodz, Poland

In-Hwan Yang, Hee-Chul Yang and Hyung-Ju Kim
Decontamination & Decommissioning Research Division, Korea Atomic Energy Research Institute, Daejeon 34057, Republic of Korea

Wen Zhou and Xianghao Meng
State Key Laboratory of Oil and Gas Reservoir Geology and Exploration, Chengdu University of Technology, Chengdu 610059, China

Qing Chen, Yuanyuan Tian and Changhui Yan
State Key Laboratory of Oil and Gas Reservoir Geology and Exploration, Chengdu University of Technology, Chengdu 610059, China
College of Energy Resource, Chengdu University of Technology, Chengdu 610059, China

Li Zheng, Hucheng Deng and Peng Li
College of Energy Resource, Chengdu University of Technology, Chengdu 610059, China

Yu Pang
Petroleum Engineering, Texas Tech University, Lubbock, TX, USA

Eric M. Garcia, Hosane A. Taroco, Ana Paula C. Madeira, Amauri G. Souza, Rafael R. A. Silva, Júlio O. F. Melo, Cristiane G. Taroco and Quele C. P. Teixeira
DECEB, Federal University of São João del-Rei, Campus Sete Lagoas, MG-424, Km 45, 35701-970 Sete Lagoas, MG, Brazil

Shobha Regmi, Balmukunda Regmi, Shiva Pathak, Bishnu Prasad Bhattarai and Saroj Kumar Sah
Department of Pharmacy, Institute of Medicine, Tribhuvan University, Maharajgunj Medical Campus, Kathmandu, Nepal

Sajan Lal Shyaula
Nepal Academy of Science and Technology (NAST), Khumaltar, Lalitpur, Nepal

Bing Xie
College of Materials Science and Engineering, Chongqing University, Chongqing 400044, China

Hao Liu
College of Materials Science and Engineering, Chongqing University, Chongqing 400044, China
School of Metallurgy and Materials Engineering, Chongqing University of Science and Technology, Chongqing 401331, China

Yue-lin Qin
School of Metallurgy and Materials Engineering, Chongqing University of Science and Technology, Chongqing 401331, China

Xiangwei Sun, Feiyue Wu, Yan Luo, Mengjun Huang, Yuntao Li, Xiang Liu and Kunpeng Liu
College of Materials and Chemical Engineering, Chongqing University of Arts and Sciences, Honghe Road 319, Chongqing 402160, China

Vo Thi Thanh Chau
Tran Quoc Tuan High School, Quảng Ngãi 570000, Vietnam

Huynh Thi MinhThanh
Department of Chemistry, Qui Nhon University, B`ınh Định 590000, Vietnam

Pham Dinh Du
Faculty of Natural Sciences, Qui Dau Mot University, Thủ Dầu Một 820000, Vietnam

Tran Thanh Tam Toan, Tran Ngoc Tuyen, Tran Xuan Mau and Dinh Quang Khieu
University of Sciences, Hue University, Hue 530000, Vietnam

Hang Xu and Dandan Zhang
Chemical Engineering and Pharmaceutics School, Henan University of Science and Technology, Luoyang 471023, China

A. Neacsu, D. Gheorghe, I. Contineanu and A. M. Sofronia
Institute of Physical Chemistry "Ilie Murgulescu" of the Romanian Academy, 202 Splaiul Independentei St., 060021 Bucharest, Romania

F. Teodorescu
Centre of Organic Chemistry "C.D. Nenitescu" of the Romanian Academy, 202 B Splaiul Independentei St., 060023 Bucharest, Romania

S. Perişanu
Department of General Chemistry, Polytechnic University of Bucharest, 1 St. Polizu, Bucharest, Romania

Ranxiao Tang, Chao Li, Weihua Liu, Shutao Gao and Chun Wang
College of Science, Agricultural University of Hebei, Baoding 071001, China

Chong Dai
Department of Civil & Environmental Engineering, University of Houston, Houston, TX 77004, USA

Index

A

Activated Carbon, 18, 25, 133, 148, 174-175
Activated Charcoal, 133, 142-143, 146-148
Activated Sludge, 59-60, 63
Adsorption, 18-20, 23-25, 35-37, 39-42, 75-77, 83-85, 88, 124-134, 136-139, 141-148, 162-175, 187-196
Adsorption Isotherms, 124-125, 128-130, 133, 142, 144, 146, 171, 173, 187, 190, 192-193
Adsorption-desorption Isotherms, 75, 77, 85
Ag2o, 73-81
Ag-modified V2o5, 73-74, 77
Al2o3, 35-41, 82, 84, 93
Alkali Metal Salts, 114, 117, 119-122
Amperometric Formaldehyde Gas Sensor, 64-65, 72
Amperometric Gas Sensor, 64, 69
Anionic Dye, 195
Aqueous Media, 18-19, 24
Atmospheric Evaporation, 114
Atomic Absorption Spectrometry, 19, 108
Azo Dye, 174, 194

B

B2o3, 39-41
Batch Reactor, 59, 63
Benzene, 176, 180
Biochemical Oxygen Demand, 95, 97
Biodiesel Production, 83-84, 86, 88-89, 92-94
Biosorption, 24-25, 133, 141, 163, 174, 195

C

Calcined Quail Eggshell, 83-90, 92
Capillary Condensation, 124-125, 127-128, 130
Carbon Nanotubes, 41, 195
Carcinogenic, 162, 187
Cd (II) Adsorption, 18-20, 23-24
Chemical Oxygen Demand, 95
Chemiexcitation, 10-17
Chitosan, 84, 93, 174
Co(II), 107, 109-110, 113
Condensation Coefficient, 114-115, 119-123
Corrosion Inhibitors, 47
Cr(III), 24, 171, 174, 195
Crystal Structure, 9, 82, 113, 161, 185-186
Cu(II), 107, 109, 111, 113, 195
Cyclohexane, 82

D

Density Functional Theory, 12, 17, 35-36, 40-41
Dextromethorphan, 142-145, 147-148
Differential Scanning Calorimetry (DSC), 181-182
Dissolved Oxygen, 60, 95, 106
Dubinin-radushkevich Model, 187
Dy(SSA)3phen Complex, 155-160

E

Effluents, 24, 26-31, 33-34, 84, 100, 106, 194
Electrocoagulation, 187
Electron Microscopy, 74-76, 84, 88, 137-138
Electron-hole Pair, 73-74, 80
Emulsification, 42-43, 49, 51
Emulsion Thermal Stability, 46, 50
Environmental Degradation, 26
Equilibrium Solid Phase, 53, 55-56
Equilibrium Solid Phases, 53, 55, 57
Ethylenediamine-tetrol, 44-46

F

Fecal Coliform Count (FCC), 95, 97
Flame Ionization Detector, 86
Formaldehyde, 64-66, 68-72
Formic Acid Oxidation, 72
Fourier Transform Infrared Spectroscopy, 187

G

Gas Chromatography, 64, 71
Groundwater Pollution, 26, 33

H

Heavy Metals, 18, 24-25, 27, 29-30, 73, 96, 187
High-performance Liquid Chromatography, 64
Histidine Stereoisomers, 181-184
Hydraulic Retention Time, 59, 61
Hydrogen Bonding, 156, 171
Hydrophilic-lipophilic Balance, 42, 44, 50

I

Industrial Effluents, 26, 28-31, 33-34, 106

L

Langmuir Model, 162, 171, 173, 191, 193
Li-ion Batteries (LIBS), 135
Lib Spent Cathode, 135-140

M

Magnetron Sputtering, 74

Malachite Green, 73, 75, 80, 174, 187, 195

Malachite Green Dye, 73, 80

Mefenamic Acid, 107-109, 112-113

Mefenamic Ligand, 107, 109

Membrane Filtration, 97

Metal Complexes, 107, 112-113

Methyl Orange, 162, 171, 174-175

Methylene Blue, 135-141, 149-153, 174-175, 187-188, 192-193, 195

Methylene Blue (MB), 135-136, 187

Methylene Blue Discoloration, 135, 137-138, 140

Mn(II), 107, 109

Model Dioxetanone, 10-11, 13-15

Montmorillonite, 149-153

N

Ni(II), 107, 109-110, 112

O

Organic Pollution, 96

P

Phosphine, 161

Phospholipids, 87

Photocatalysis, 10-11, 15-16, 73, 81-82

Photocatalytic Degradation, 73, 80-81, 163, 174

Photoluminescence, 75, 80, 82, 161

Polyethylene Glycol, 44-49

Principal Component Analysis, 100

R

Raman Spectroscopy, 77

Ruthenium Complex, 16

S

Salicylic Acid, 155, 160-161

Scanning Electron Microscopy, 74-76, 84, 88, 137-138

Supernatant Liquid, 164

Superoxide Dismutase, 113

Surface Hydrophobicity, 195

Suspended Solids, 29, 31, 33, 59-60, 95, 97

T

Thermogravimetric Analysis, 19, 83-84, 87, 155

Total Dissolved Solids, 26, 31, 43-44

Transesterification, 83-86, 88-89, 92-94

W

Water Quality Index, 95-98, 100, 106

X

X-ray Diffraction, 66, 73-74, 83-84, 86-87, 109, 136, 138, 140

Z

Zn(II), 107, 109, 111